# *NEW* Math Workbook for
# SAT * I

### LAWRENCE S. LEFF

Assistant Principal, Mathematics Supervision
Franklin D. Roosevelt High School
Brooklyn, New York

BARRON'S

*SAT is a registered trademark of the College Entrance Examination Board,
which does not endorse this book.

SAT I directions and grids for student-produced response questions selected from *Introducing the New SAT,* College Entrance Examination Board, 1993. Reprinted by permission of Educational Testing Service, the copyright owner of the test questions.

Permission to reprint the above material does not constitute review or endorsement by Educational Testing Service or the College Board of this publication as a whole or of any other questions or testing information it may contain.

*All inquiries should be addressed to:*
Barron's Educational Series, Inc.
250 Wireless Boulevard
Hauppauge, New York 11788

Library of Congress Catalog Card No. 96-11451

International Standard Book No. 0-8120-9285-6

**Library of Congress Cataloging-in-Publication Data**

Leff, Lawrence S.
  New math workbook for SAT I / Lawrence S. Leff.
    p.    cm.
  Summary : A study guide covering basic arithmetic skills, algebra, word problems, and geometry. Includes practice tests.
  ISBN 0-8120-9285-6
  1. Mathematics—Examinations, questions, etc.    2. Scholastic Aptitude Test—Study guides.    [1. Mathematics—Examinations, questions, etc.    2. Scholastic Aptitude Test—Study guides.]    I. Title.
  QA43.L467    1996
  513'.1'076—dc20

                                              96-11451
                                                  CIP
                                                   AC

PRINTED IN THE UNITED STATES OF AMERICA
98765

# Contents

## PART III: TAKING PRACTICE TESTS

# Dedication

*To Rhona:*

*For the understanding,*
*for the sacrifices,*
*and*
*for the love*

# Preface

This workbook is organized around a simple and easy-to-follow five-step study program:

1. Know what to expect on test day.
2. Become testwise.
3. Learn special math strategies.
4. Review SAT I math topics.
5. Take practice exams under test conditions.

As you work your way through this book you will find:

- Hundreds of practice SAT I type math questions with solutions.
- Special math strategies that will help you to get the right answers to different types of problems that you may not know how to solve.
- All the math you need to know for the SAT I, conveniently organized into sets of easy-to-read lessons. These lessons summarize and illustrate key concepts and formulas. Each lesson is followed by a special set of SAT I Tune-Up Exercises that give practice in answering the three different types of SAT I questions while reinforcing the math concepts covered in the lesson.
- "Tips for Scoring High," which will help you develop speed and accuracy.
- Two complete SAT I math practice tests with detailed solutions.

Lawrence S. Leff
January 1996

# PART I

## LEARNING ABOUT SAT I MATH

Getting Acquainted
with the SAT I

# Knowing What You're Up Against

The abbreviation SAT stands for *Scholastic Assessment Test*. The newest version of this college admissions test is called the SAT I: Reasoning Test. After reading this chapter, you will know how the math sections of the SAT I are organized. You will also become familiar with the three basic types of SAT I math questions and the instructions that accompany them. Learning these instructions *now* will save you valuable time on test day. Knowing the simple test-taking strategies that are featured in this chapter will earn you additional SAT math points.

---

## LESSONS IN THIS CHAPTER

---

---

# Getting Acquainted with the SAT I

## OVERVIEW

*The SAT I is a 3 hour exam that many colleges require you to take as part of the admissions process. Since the format and mathematical content of the SAT I are predictable, you can prepare for this important exam. The chapters in this book comprise a complete and easy-to-follow study program for the math portion of the test. As you move from chapter to chapter, you will acquire the knowledge, skill, and confidence that can lead to large gains in your SAT I math score.*

## WHAT DOES THE SAT I MEASURE?

The SAT I measures your verbal and mathematical reasoning abilities. It does not test intelligence, creativity, or motivation. The math questions test your reasoning abilities by asking you to solve unfamiliar-looking problems using the basic math facts and skills that every high school student is expected to know and to be able to apply.

## WHY DO COLLEGES REQUIRE THE SAT I?

College admissions officers know that the students who apply to their colleges come from a wide variety of high schools that may have different grading systems and academic standards. SAT I scores make it possible for colleges to compare the course preparation and the performances of applicants by using one academic yardstick. Your SAT I score, together with your high school grades and other information you or your high school may be asked to provide, helps college admission officers to predict your chances of success in the college courses you will take.

## HOW IS THE SAT I ORGANIZED?

The SAT I is a 3-hour exam that is divided into seven timed sections:

- Three verbal sections consisting of two 30-minute and one 15-minute sections.
- Three math sections consisting of two 30-minute and one 15-minute sections.
- One "experimental" verbal or math section that takes 30 minutes but does not count toward your score. You will not know which of the seven test sections is the experimental section.

These seven sections can appear in any order.

## HOW ARE SCORES REPORTED?

After you take the test, you will receive a report that contains a score for the verbal questions and a separate score for the math questions. Raw scores are converted to scores that range from 200 to 800, with 500 representing the average SAT I math score.

## WHAT MATH DO YOU NEED TO KNOW?

The math portion of the SAT I consists of 60 questions on arithmetic, algebra, and geometry. Some questions may involve interpreting tables and graphs, probability and counting, and logical analysis. Key reference formulas that you may need to answer some of the questions are provided at the beginning of each math section. Although you are allowed to use a calculator during the test, no question will require its use.

The SAT I will *not* have any questions that involve:

- the quadratic formula,
- complicated operations with radicals,
- complicated factoring,
- operations with fractional exponents,
- formal geometric proofs, or
- trigonometric ratios.

## CAN YOU STUDY FOR THE SAT I?

Since the format, the types of questions, and the math background needed to answer SAT I math questions do not change from exam to exam, effective preparation before you take the test will boost your math score. To make the best use of your study time, however, you need a plan. Here is a five-step study plan for the SAT I math sections that you can follow simply by reading this book from beginning to end.

---

### A FIVE-STEP STUDY PLAN FOR SAT I MATH

Step 1.   Know what to expect on test day.
Step 2.   Become testwise.
Step 3.   Learn special math strategies.
Step 4.   Review SAT I math topics.
Step 5.   Take practice exams under test conditions.

---

### Step 1.  Know What to Expect on Test Day

Know the format, the types of math questions, and the special directions that appear on every SAT I exam. This information will save you valuable testing time when you take the SAT I. It will also build your confidence and prevent errors that may arise from not understanding the directions. Chapter 1 gives you this important edge.

## Step 2.  Become Testwise

Learn the test-taking tips that will help you gain points by increasing your speed and accuracy. Chapter 1 contains many SAT I test-taking tips highlighted in a boxed display.

## Step 3.  Learn Special Math Strategies

Many SAT math questions are not like the standard types of textbook problems you have solved in high school math classes. To assist you with these SAT questions, Chapter 2 reveals special math strategies that you may not have learned. These strategies can help you arrive at the correct answer when you get stuck or when traditional methods take too long.

## Step 4.  Review SAT I Math Topics

Although SAT I math questions stress problem solving and reasoning skills, you are expected to remember important math concepts taught in arithmetic, elementary algebra, and beginning geometry courses. Chapters 3–7 serve as brief math refresher courses. In these chapters you will also find a large number of SAT-type math questions organized by lesson topic.

## Step 5.  Take Practice Exams Under Test Conditions

Practice makes perfect! Chapter 8 features two full-length SAT I math practice tests with Answer Keys and detailed explanations of answers. Each test consists of three math sections that closely resemble the three math sections of an actual SAT I exam. These practice tests should be taken under test-like conditions, using the sample SAT I answer forms provided at the back of this book. Be sure to use the same calculator that you will bring to the testing room. This activity will provide an opportunity to try out, under test conditions, the tips and strategies that you have learned in this book. It will also give you an idea of how well you can manage your time on the various math sections. Finally, studying the solutions to each of the exam questions you missed or skipped over will help to eliminate any remaining weak spots in your preparation.

# Regular Multiple-Choice Questions

## OVERVIEW

*The SAT I groups math problems by question type. More than half of the math questions that appear on the SAT I are standard multiple-choice questions, each of which offers five answer choices from which you must pick the correct one.*

## THE MOST COMMON TYPE OF SAT I MATH QUESTION

Thirty-five of the 60 math questions that appear on the SAT I are multiple-choice questions that come with five answer choices, labeled from (A) to (E). After you figure out which one of the five answer choices is correct, you must locate the corresponding question number for that section on a separate answer sheet. Using a pencil, you then fill in the oval on the answer form that contains the same letter as the correct choice. Since answer forms are scored by a machine, make sure to fill in completely each oval you choose as a correct answer, as shown below.

Ⓐ ● Ⓒ Ⓓ Ⓔ          Ⓐ Ⓑ Ⓒ Ⓓ Ⓔ          Ⓐ Ⓑ Ⓒ Ⓓ Ⓔ          Ⓐ Ⓑ Ⓒ Ⓓ Ⓔ

right              wrong              wrong              wrong

Here is an example of a regular multiple-choice question that involves simple algebra.

**EXAMPLE:**  If $3x - 12 = 15$, then $x - 4 =$

(A) 5
(B) 6
(C) 7
(D) 8
(E) 10

**SOLUTION:**  Don't waste time by isolating $x$ on one side of the equation. Instead, solve the equation for $x - 4$ by dividing each term of the equation by 3. Thus,

$$\frac{\overset{1}{\cancel{3}}x}{\cancel{3}} - \frac{12}{3} = \frac{15}{3}$$

$$x - 4 = 5$$

The correct choice is (A).

On your answer sheet you must darken oval A, as shown.

● Ⓑ Ⓒ Ⓓ Ⓔ

## ROMAN NUMERAL MULTIPLE-CHOICE QUESTIONS

In a special type of multiple-choice question, three possible statements are numbered as I, II, and III. Each Roman numeral statement is either true or false for the particular question. Different combinations of these statements are presented as the standard set of five answer choices. You need to figure out which answer choice includes all of the Roman numeral statements that are true.

**EXAMPLE:**    For the ordered set of numbers 3, 5, $x$, 7, 10, which of the following statements are true when the arithmetic mean (average) is 6?

   I. mean > median
  II. mode > mean
 III. mode = median

(A) I only
(B) II only
(C) I and III only
(D) II and III only
(E) I, II, and III

**SOLUTION:**    The arithmetic mean or average of a set of numbers is the sum of the numbers in the set divided by the number of values in the set. Since the mean or average is 6:

$$\frac{3 + 5 + x + 7 + 10}{5} = 6$$

$$\frac{25 + x}{5} = 6$$

Multiplying each side of this equation by 5 gives $25 + x = 30$, so $x = 5$ and the original list of numbers is 3, 5, 5, 7, 10.

The question states that the *mean* is 6. You need to determine the mode and the median of the list 3, 5, 5, 7, and 10.

Since the *mode* is the value in the list that occurs most often, 5 is the mode.

Since the *median* is the middle value in a list of numbers arranged in size order, the median is 5.

Now determine whether each of the three Roman numeral statements is true or false:

- I. Is the mean > median? Since the mean is 6 and the median is 5, statement I is true.
- II. Is the mode > mean? Since the mode is 5 and the mean is 6, statement II is false. Hence, you can eliminate answer choices (B), (D), and (E).
- III. Is the mode = median? Since both the mode and the median are 5, statement III is true.

Only Roman numeral statements I and III are true. Hence, the correct choice is (C).

On your answer sheet you must darken oval C, as shown.

## TIPS FOR SCORING HIGH

1. Don't keep moving back and forth from the question page to the answer sheet. Record the answer next to each question. After you accumulate a few answers, transfer them to the answer sheet at the same time. This strategy will save you time.

2. After you record a group of answers, check to make sure that you didn't accidentally skip a line and enter the answer to question 3, for example, in the space for question 4. This strategy will save you from a disaster!

3. On the answer sheet, be sure to fill one oval for each question you answer. If you need to change an answer, erase it completely. If the machine that scans your answer sheet "reads" what looks like two marks for the same question, the question will not be scored.

4. Do your calculations wherever you find room on the question pages of the test booklet.

5. In the test booklet, put a question mark (?) to the left of any question that you skip over. This will allow you to quickly identify the questions in a section that you need to come back to, if time permits.

6. Don't panic or become discouraged if you have to omit some questions. Very few students will be able to answer all of the questions in a section correctly.

7. Analyze Roman numeral multiple-choice questions by:
   - determining whether each of the three Roman numeral statements is true or false,
   - eliminating any answer choice that includes a false statement, and
   - picking the answer choice that gives the correct combination of true statements.

# Quantitative Comparison Questions

## OVERVIEW

*Quantitative Comparison questions (QCs for short) represent the second type of SAT math question. QCs don't look like the math questions that you are used to solving in your high school math classes. To answer this type of question, you must compare a boxed quantity in Column A with another boxed quantity in Column B. Your job is to figure out which quantity, if either, is bigger. QCs tend to emphasize mathematical reasoning rather than routine computations.*

## ALL QC QUESTIONS HAVE THE SAME FOUR ANSWER CHOICES

You must answer each of the 15 QC questions that appear on the SAT I by picking:

- choice (A) if the quantity in Column A is *always* bigger; *or*
- choice (B) if the quantity in Column B is *always* bigger, *or*
- choice (C) if the two quantities are *always* equal; *or*
- choice (D) if one of the other three possible answers—(A), (B), or (C)—is only *sometimes* correct.

After you decide which is the correct choice, you must fill in the corresponding oval on the answer sheet. Never fill in oval E as the answer for a QC question. If you do, that question will not be scored.

Since the four answer choices are the same for each QC question, they do not appear after each question. Instead, the four answer choices are summarized in the directions at the top of each page of the test booklet that contains QC questions.

## COMMON VARIABLES MEAN THE SAME THING IN BOTH COLUMNS

If the same variable or symbol appears in Column A and in Column B, you can assume that it represents the same thing in both columns.

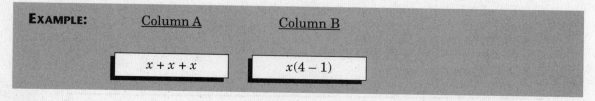

**EXAMPLE:**

| Column A | Column B |
|----------|----------|
| $x + x + x$ | $x(4 - 1)$ |

**SOLUTION:** In Column A, $x + x + x = 3x$. In Column B, $x(4 - 1) = x(3) = 3x$. Since $x$ represents the same number in both columns, the two quantities are equal. The correct choice is (C).

## YOU NEED TO KNOW CERTAIN COMPARISON SYMBOLS

If $a$ and $b$ stand for numbers, then

- $a > b$ is read as "$a$ is greater than $b$." For example, $5 > 3$.
- $a < b$ is read as "$a$ is less than $b$." For example, $4 < 7$.

The notation $a \geq b$ means that a is greater than $b$ _or_ is equal to $b$.

Similarly, $a \leq b$ states that $a$ is less than $b$ _or_ is equal to $b$.

## ADDITIONAL FACTS ARE CENTERED BUT NOT BOXED

If additional information or a diagram that applies to one or both of the boxed quantities is given, it appears centered above the two columns but is not boxed.

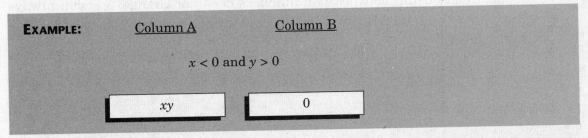

**EXAMPLE:**     Column A          Column B

$x < 0$ and $y > 0$

$xy$      $0$

**SOLUTION:** The quantity in Column A is the product of $x$ and $y$. Since $x$ is negative ($x < 0$) and $y$ is positive ($y > 0$), their product is negative. Thus, $0 > xy$. Since the quantity in Column B is greater than the quantity in Column A, the correct choice is (B).

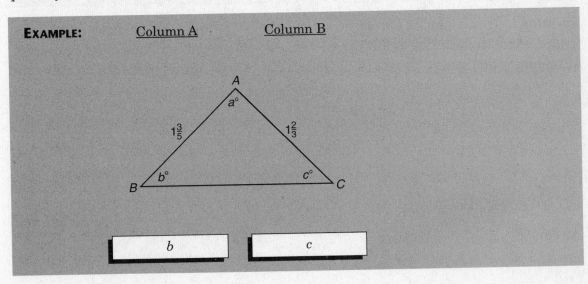

**EXAMPLE:**     Column A          Column B

$b$      $c$

**SOLUTION:** If two sides of a triangle are unequal in length, the measures of the angles opposite them are also unequal. The angle with the larger degree measure lies opposite the longer side. In decimal form, $1\frac{3}{5} = 1.6$ and $1\frac{2}{3} = 1.66+$. Thus, side $AC$ is longer than side $AB$. Since $\angle B$ lies opposite $AC$ and $\angle C$ lies opposite $AB$, $b > c$. Hence, the correct choice is (A).

## KNOW WHEN TO PICK CHOICE (D) AS THE ANSWER

Sometimes the size relationship between the quantities being compared cannot be determined.

- Choice (D) *may* be the correct answer if at least one of the quantities to be compared contains a variable.
- Choice (D) is *always* the correct answer when you can show that the size relationship between the quantity in Column A and the quantity in Column B can change for different numerical replacements of a variable.
- Choice (D) is *never* the correct answer when two numbers are being compared. The size relationship between two numbers does not change. You can always tell whether two numbers are equal or whether one number is larger than the other number.

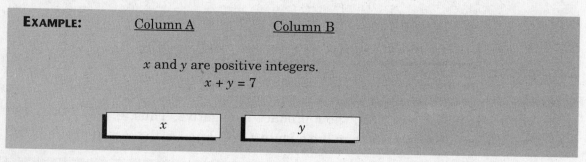

**EXAMPLE:**      Column A                Column B

x and y are positive integers.
$x + y = 7$

|     x     |     y     |

**SOLUTION:** If the sum of two positive integers is 7, then one of these numbers is larger than the other. For instance, when $x = 4$, $y = 3$, so choice (A) is correct. However, when $x = 3$, $y = 4$, so choice (B) is correct. Since the size relationship between $x$ and $y$ can change for particular values of $x$ and $y$, you can't be sure which column contains the bigger value. Therefore, the correct choice is (D).

## TIPS FOR SCORING HIGH

1. Memorize the four answer choices for QC questions.
2. Forget choice (D) when both columns of a QC question contain only numbers. Choose (D), however, when replacing variables in Columns A and B with different numbers makes different answer choices true.
3. Keep in mind that, if you find yourself doing a lot of calculations for a QC question, you are probably on the wrong track and need to change your approach.
4. Be aware that you can sometimes make a QC question simpler by plugging in numbers for variables. Since some variable quantities behave differently for certain types of numbers, don't forget to use negative numbers, fractions, 0, 1, and so forth.

# Grid-In Questions

## OVERVIEW

*The third and last type of* SAT *math question is the "grid-in" or "student-produced response" question. Grid-in questions are similar to standard multiple-choice questions except that no answer choices are provided. For each grid-in question you must come up with your own answer and then record that answer on a special four-column grid. Ten grid-in questions are included in the test section that begins with 15 quantitative comparison questions.*

## BE FAMILIAR WITH THE ANSWER GRID

When you figure out the answer to a grid-in question, you must record it on a four-column answer grid like the one in Figure 1.1. The answer grid can accommodate whole numbers from 0 to 9999, as well as fractions and decimals. No grid-in question has a negative answer or an answer that contains a variable.

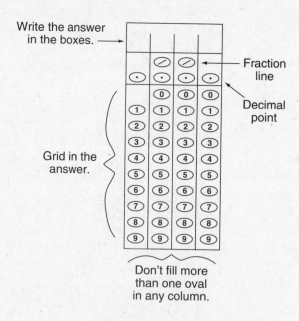

**Figure 1.1**  *An Answer Grid*

To grid in an answer:

- Write the answer in the top row of column boxes of the grid. A decimal point or fraction bar (slash) requires a separate column. Although writing the answer in the column boxes is not required, it will help guide you when you grid the answer in the matching ovals below the column boxes.
- Fill the ovals that match the answer you wrote in the top row of column boxes. Make sure that no more than one oval is filled in any column. Columns that are not needed should be left blank. If you forget to fill in the ovals, the answer that appears in the column boxes will *NOT* be scored.

Here are some examples:

Answer: 0.237          Answer: 23.7          Answer: $\frac{23}{7}$

Answers that need fewer than four columns, except 0, may start in any column, provided that the whole answer fits.

**EXAMPLE:**          If $a * b = a^3 - b^2 + 1$, what is the value of $3 * 2$?

**SOLUTION:**   Plug in 3 for $a$ and 2 for $b$:

$$3 * 2 = 3^3 - 2^2 + 1$$

Since $3^3 = 3 \times 3 \times 3 = 27$ and $2^2 = 2 \times 2 = 4$,

$$3 * 2 = 27 - 4 + 1 = 24$$

The answer is 24. Here are one recommended and two acceptable ways of gridding in 24.

## START ALL ANSWERS, EXCEPT 0, IN THE FIRST COLUMN

If you get into the habit of always starting answers in the first column of the answer grid, you won't waste time thinking about where a particular answer should begin. Note, however, that the first column of the answer grid does not contain 0. Therefore a zero answer can be entered in any column after the first. If your answer is a decimal number less than 1, don't bother writing the answer with a 0 in front of the decimal point. For example, if your answer is 0.126, write .126 in the four column boxes on the answer grid.

## ENTER MIXED NUMBERS AS FRACTIONS OR AS DECIMALS

The answer grid cannot accept mixed numbers. You *must* change a mixed number into an improper fraction or a decimal before you grid it in. For example, to enter a value of $1\frac{3}{5}$, grid in either 8/5 or 1.6.

Answer: $\frac{8}{5}$          Answer: 1.6          Answer: $1\frac{3}{5}$

Acceptable          Acceptable          *WRONG* since the computer interprets the answer as $\frac{13}{5}$

## IF YOUR ANSWER FITS THE GRID, DON'T CHANGE ITS FORM

The machine that scores your paper is concerned only with the *value* of an answer, not its form. If your answer is a fraction that fits the grid, enter it in that form. Don't waste time, and risk making a careless error, by trying to reduce the fraction or by rewriting it as a decimal number.

## KNOW WHEN TO CHANGE THE FORM OF AN ANSWER

If the answer is a decimal or fraction that does not fit the grid, then you must change the form of the answer.

- If the answer is a long decimal number, enter only the first three digits and the decimal point in its correct position. For example, if you obtain the repeating decimal 1.7777 . . . as your answer, keep the first three digits and enter 1.77 on the grid.

Answer: 1.7777 . . . .

- If the answer is a fraction that needs more than four columns, reduce the fraction, if possible, or enter the decimal form of the fraction. For example, since the fractional answer $\frac{17}{25}$ does not fit the grid and cannot be reduced, use a calculator to divide the denominator into the numerator. Then enter the decimal value that results. Since $17 \div 25 = 0.68$, grid in .68 for $\frac{17}{25}$.

Answer: $\dfrac{17}{25}$

## KNOW HOW TO HANDLE GRID-IN QUESTIONS WITH MORE THAN ONE CORRECT ANSWER

When a grid-in question has more than one correct answer, enter any correct answer.

**EXAMPLE:**     What is one possible value of $x$ for which $\frac{3}{4} < x < \frac{4}{5}$?

**SOLUTION:**   Since

$$\frac{3}{4} = 0.75 \text{ and } \frac{4}{5} = 0.8,$$

then $0.75 < x < 0.80$. Any value of $x$ that is between 0.75 and 0.80 satisfies this inequality. Thus, 0.76 is a correct answer. When a grid-in question has more than one correct answer, enter only one.

## TIPS FOR SCORING HIGH

1. Keep in mind that an answer to a grid-in question may not include a negative sign, variable, comma, or other special symbol. If your answer contains a negative sign or a variable, you know that you solved the problem incorrectly.
2. Be aware that an answer grid is scored by a scanning machine, not by a person.
   - If you enter the correct answer in the first row of column boxes but forget to fill in the ovals below, your answer will *not* be scored.
   - If you know the right answer but do not fill in the ovals correctly, your answer will be scored as incorrect.
   - If you do not erase thoroughly, your new answer may not be scored.
3. Check your gridding to make sure you don't mark more than one oval in any column. A slash or a decimal point should never be marked in the same column as a digit.

4. Enter an answer in its original form, fraction or decimal, if it fits the grid. A mixed number must be changed to a fraction or decimal, however, before gridding it in.

5. Begin answers in the leftmost columns of the answer grids. Do NOT grid zeros before decimal points. If an answer is a repeating decimal number, enter the most accurate value by filling the answer grid completely with the first three digits of the number and the decimal point.

# How the SAT I Math Sections Are Organized, Timed, and Scored

## OVERVIEW

*The 60 SAT I math questions that are scored are divided into three timed sections. Two of these sections contain only regular multiple-choice questions. The other math section contains a set of quantitative comparison questions followed by a set of grid-in questions.*

## THE THREE TIMED MATH SECTIONS

Table 1.1 shows how the three types of math questions are distributed in the three math sections that are scored. Remember that the math and verbal SAT sections can appear in any order in your test booklet. If you complete a section before the time allowed for it runs out, you may review questions only in that section. You may not go back to a preceding section or start a new one.

### TABLE 1.1  BREAKDOWN OF THE THREE SCORED MATH SECTIONS

| Math Section Type | Number of Questions | Time Allowed |
|---|---|---|
| 1. Regular multiple choice | 25 | 30 minutes |
| 2. Quantitative comparison and grid-ins | 15 QCs and 10 grid-ins | 30 minutes |
| 3. Regular multiple choice | 10 | 15 minutes |
| Total | 60 questions | 1 hour 15 minutes |

## THE ARRANGEMENT OF QUESTIONS

In each math section, the questions gradually become more difficult. Expect easier questions at the beginning of each section and harder questions near the end. In the math section that contains both QCs and grid-ins, the QC questions are arranged in their approximate order of difficulty, as are the grid-in questions that follow.

Only a very small percentage of test-takers answer correctly the last few questions in each math section. If you are not aiming for a math score over 700, you can make the best use of your time by paying little attention to the final questions in each section.

## FIGURING OUT SAT I RAW SCORES

Your raw SAT I math score is the number of correct answers minus a penalty deduction for any incorrect answers to questions that have answer choices. No deduction is made for unanswered questions.

- One-fourth $\left(\frac{1}{4}\right)$ of a point is deducted for each incorrect answer to a standard multiple-choice question.
- One-third $\left(\frac{1}{3}\right)$ of a point is deducted for each incorrect answer to a quantitative comparison question.
- No penalty deduction is made for an incorrect answer to a grid-in question.

If, after a penalty deduction has been made, the new score includes a fraction, then this score is rounded off to the nearest whole number. The number that results is converted into a score that can range from 200 to 800. The conversion process has been recently adjusted to make certain that a score of 500 represents the average SAT I math score.

## WHEN YOU SHOULD GUESS

The more choices you can rule out as definitely wrong, the greater the chances of guessing the correct answer. Also, since there is no penalty for an incorrect answer to a grid-in question, try guessing when you don't know the answer to a grid-in question. There is a catch, however: since grid-in questions offer no answer choices, it may be very difficult to make a reasonable guess.

## TIPS FOR SCORING HIGH

1. Don't be concerned about the order or number of SAT math sections.
2. Don't get hung up on any one question—skip it, and then come back to it if time permits.
3. If a question near the end of a math section seems easy, beware—you've probably fallen into a trap!
4. If you're not aiming for a perfect or nearly perfect score, forget about the last couple of questions in each section. Concentrate on getting as many of the earlier questions right as possible. Remember that a correct answer to an easy test question counts the same as a correct answer to a hard question.
5. Guess when you can eliminate some of the answer choices and when you can come up with a reasonable answer to a grid-in question.

# Calculators and the SAT I

## OVERVIEW

*Plan to bring a calculator to the exam room. Don't panic, though, if you forget it. A calculator may be useful when you need to do ordinary arithmetic calculations, but no solution will require a calculator.*

## WHAT TYPE OF CALCULATOR IS ALLOWED?

Along with a few sharpened pencils, an eraser, and a watch, bring to the exam room a calculator that you know how to use. The exam proctor does not have any calculators to lend, and you will not be allowed to share a calculator. Any one of the following types of calculators may be used:

- A basic four-function calculator
- A scientific calculator
- A graphing calculator

Scientific and graphing calculators offer no practical advantage on this test. A solar-powered basic calculator with an easy-to-read display will serve you well.

## WHEN SHOULD YOU USE A CALCULATOR?

Although calculators don't tell you *how* to solve problems, they can be helpful when doing routine arithmetic computations. A calculator may come in handy if it is necessary to:

- add, subtract, multiply, or divide whole numbers and decimals;
- find or estimate a square root;
- change a fraction into decimal form.

Since SAT I math questions never require complicated arithmetic, don't expect to use a calculator on every question. Here is a grid-in question for which using a calculator is appropriate.

**EXAMPLE:** If the sales tax on a $10.00 scarf is $0.80, what would a $25 shirt cost with tax?

**SOLUTION:**

- Find the sales tax rate by answering this question:

What percent of 10 is 0.80?

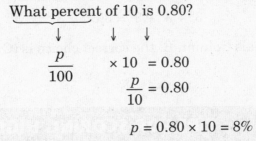

$$\frac{p}{100} \times 10 = 0.80$$

$$\frac{p}{10} = 0.80$$

$$p = 0.80 \times 10 = 8\%$$

- To add a percent $p$ of a given amount (the price of the shirt) to that amount, _use a calculator_ to multiply the given amount by 1.$p$, as in

$$\$25 \times 1.08 = \$27$$

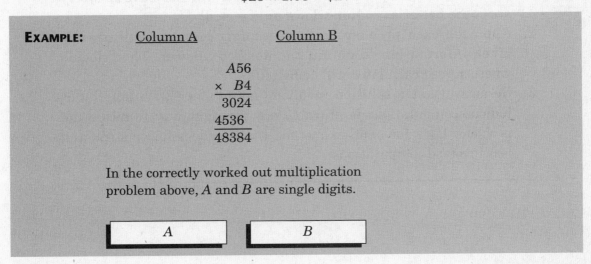

**EXAMPLE:**   Column A          Column B

A56
× B4
3024
4536
48384

In the correctly worked out multiplication problem above, $A$ and $B$ are single digits.

|  A  |  |  B  |

**SOLUTION:**   Write the two partial products obtained by multiplying each digit of $A56$ by $B4$.

- Since $4 \times A56 = 3024$, $A56 = \frac{3024}{4} = 756$. To divide 3024 by 4, a calculator is helpful. Since $A56 = 756$, $A = 7$.
- Similarly, $B \times 756 = 4536$. Using a calculator to divide gives $B = \frac{4536}{756} = 6$.

Since $7 > 6$, Column A is bigger than Column B, and the correct choice is (A).

A calculator would not be helpful in answering the QC question that follows.

**EXAMPLE:**     Column A          Column B

| $3^9 + 3^9 + 3^9$ |  | $3^{10}$ |

**SOLUTION:**   You could use a calculator to perform repeated multiplication of 3 in both columns, but tedious computation is not what the SAT I is about. Instead, you should recognize that

$$3^9 + 3^9 + 3^9 = 3(3^9) = 3^{1+9} = 3^{10}$$

Since Column A equals Column B, the correct choice is (C).

## TIPS FOR SCORING HIGH

1. Before the day of the test, check that your calculator is in good working order. If the calculator uses batteries, insert new ones. If you bring a solar-powered calculator, you will not have to worry about the batteries running down unexpectedly.

2. Approach each problem by first deciding how you will use the given information to obtain the desired answer. Then decide whether your calculator will be helpful.

3. Be aware that a solution involving many steps with lengthy or tedious computations is probably not the right way to tackle the problem. Look for another method that involves fewer steps and less involved computations.

# Math Strategies You Need to Know

After reading this chapter, you will know key math strategies that can help you get the correct answers to test questions that at first glance may seem too difficult or too time consuming. Of course, if you know how to solve a problem without relying on a special strategy, then do so!

---

## LESSONS IN THIS CHAPTER

---

---

# General Math Strategies

## OVERVIEW

*This lesson deals with math strategies that do not depend on the format of the test question.*

**STRATEGY 1**

### USE THE TEST BOOKLET AS A SCRATCH PAD

Each page of questions in the test booklet contains blank areas where you can do your calculations. Also, feel free to mark up any diagrams that are given. Sometimes it may help to draw your own diagram.

**EXAMPLE:**   Points *H*, *J*, *K*, and *M* lie on the same line, in that order. If *HK* = 8, *JM* = 11, and *JK* = 5, what is the ratio of *HJ* to *HM*?

**SOLUTION:**   Draw a diagram that fits the conditions of the problem:

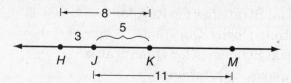

Since *HJ* = *HK* − *JK* = 3, then *HM* = *HJ* + *JM* = 3 + 11 = 14. Thus,

$$\frac{HJ}{HM} = \frac{3}{14}$$

The correct answer is $\frac{3}{14}$. Grid in as 3/14.

**STRATEGY 2**

### FIND THE QUANTITY THE QUESTION ASKS FOR

When you read a test question, circle the quantity that you need to find. After you solve the problem, make sure that your answer represents the quantity that you circled in the question.

**26**

**EXAMPLE:**   If $(5-3)(x-1) = 4$, then $x^2 =$
(A) 3
(B) 6
(C) 8
(D) 9
(E) 12

**SOLUTION:**   As you read the question, circle the quantity you need to find:

$$\text{If } (5-3)(x-1) = 4, \text{ then } \boxed{x^2} =$$

Although the question asks you to find the value of $x^2$, you must first solve the equation for $x$:

$$(5-3)(x-1) = 4$$

• Remove the first parentheses:                                                   $2(x-1) = 4$
  Multiply each term inside the remaining parentheses by 2:          $2x - 2 = 4$
• Add 2 to both sides of the equation:                                              $2x = 6$
• Divide both sides of the equation by 2:                                       $x = \dfrac{6}{2} = 3$

A regular multiple-choice question usually includes answer choices that anticipate that you will make a mistake in reading the question or in doing the computations. For example, if you misread $x^2$ and $x$, you will choose (A) as the answer. If you incorrectly interpret $x^2$ as two times $x$, you will get 6, choice (B), as an answer. Since $x^2 = 3^2 = 9$, the correct choice is (D).

**STRATEGY**

## 3   PLUG IN NUMBERS TO FIND A PATTERN

You can solve some problems by substituting a few numbers for a given variable until you discover a pattern.

**EXAMPLE:**   If $n$ is an even integer, which of the following expressions always represent(s) an odd integer?

    I. $n - 3$
    II. $n^2 + 3n$
    III. $\dfrac{n}{2} + 3$

(A) I only
(B) I and II only
(C) I and III only
(D) I, II, and III
(E) None

**SOLUTION:** Pick a small even number for $n$. Evaluate each Roman numeral statement using this number.

- I. If $n = 2$, then $n - 3 = 2 - 3 = -1$, which is an odd integer. Test two more even numbers. If $n = 4$, then $n - 3 = 4 - 3 = 1$, which is also an odd integer. If $n = 6$, then $n - 3 = 3$, again an odd integer. Since statement I appears to be true, eliminate choice (E).
- II. If $n = 2$, then $n^2 + 3n = 2^2 + 3(2) = 4 + 6 = 10$. Since statement II is false, eliminate choices (B) and (D).
- III. If $n = 2$, $\frac{n}{2} + 3 = \frac{2}{2} + 3 = 1 + 3 = 4$. Since statement III is false, eliminate choice (C).

Since only Roman numeral statement I is true, the correct choice is (A).

> **EXAMPLE:** When a positive integer $k$ is divided by 5, the remainder is 3. What is the remainder when $3k$ is divided by 5?

**SOLUTION:** Find a few positive integers that, when divided by 5, give 3 as a remainder. Any positive integer that is the sum of 5 and 3 or of a multiple of 5 (i.e., 10, 15, etc.) and 3 will have this property. For example, when 8, 13, and 18 are each divided by 5, the remainder is 3:

| $k$ | $k \div 5$ | |
|---|---|---|
| 8 | $8 \div 5 = 1$ | Remainder 3 |
| 13 | $13 \div 5 = 2$ | Remainder 3 |
| 18 | $18 \div 5 = 3$ | Remainder 3 |

Now, using the same values for k, divide $3k$ by 5 and find the remainders.

| $k$ | $3k$ | $3k \div 5$ | |
|---|---|---|---|
| 8 | 24 | $24 \div 5 = 4$ | Remainder 4 |
| 13 | 39 | $39 \div 5 = 7$ | Remainder 4 |
| 18 | 54 | $54 \div 5 = 10$ | Remainder 4 |

The correct answer is 4.

**STRATEGY**

## 4 CHOOSE A CONVENIENT BASE VALUE

If you need to figure out how a quantity changes without knowing its value, choose any starting value that makes the arithmetic easy. Carry out the computations using this value. Then compare the final answer with the base value that you chose.

**EXAMPLE:** The current value of a stock is 20% less than its value when it was purchased. By what percent must the current value of the stock rise in order for the stock to have its original value?
(A) 20%
(B) 25%
(C) 30%
(D) $33\frac{1}{3}$ %
(E) 50%

**SOLUTION:** Choose a convenient starting value of the stock. When working with percents, 100 is usually a good starting value.

- Assume the original value of the stock was $100.
- Find the current value of the stock. Since the current value is 20% less than the original value, the current value is

$$\$100 - 0.20(\$100) = \$100 - \$20 = \$80$$

- Find the amount by which the current value of the stock must increase in order to regain the original value. The value must rise from $80 to $100, which is a change of $20.
- To find the *percent* by which the current value of the stock must rise in order for the stock to have its original value, find what percent $20 is of $80. Since

$$\frac{\$20}{\$80} \times 100\% = \frac{1}{4} \times 100\% = 25\%$$

the current value of the stock must increase 25% in order to regain its original value.

The correct choice is (B).

**EXAMPLE:** Fred gives $\frac{1}{3}$ of his compact discs to Andy and then gives $\frac{3}{4}$ of the remaining compact discs to Jerry. Fred now has what fraction of the original number of discs?
(A) $\frac{1}{12}$

(B) $\frac{1}{6}$

(C) $\frac{1}{3}$

(D) $\frac{5}{12}$

(E) $\frac{1}{2}$

**SOLUTION:** Since Fred gives $\frac{1}{3}$ and then $\frac{3}{4}$ of his compact discs away, pick any number that is divisible by both 3 and 4 for the original number of compact discs. Assume Fred starts with 12 compact discs.

- After Fred gives $4 \left(= \frac{1}{3} \times 12\right)$ discs to Andy, he is left with 8 (= 12 − 4) compact discs.
- Since $\frac{3}{4}$ of 8 is $\frac{3}{4} \times 8$ or 6, Fred gives 6 of the remaining compact discs to Jerry. This leaves Fred with 2 (= 8 − 6) of the original 12 compact discs.
- Since $\frac{2}{12} = \frac{1}{6}$, Fred now has $\frac{1}{6}$ of the original number of discs.

The correct choice is (B).

**STRATEGY 5**

## COMBINE OR MULTIPLY EQUATIONS

Some questions can be answered quickly if you recognize that you need to combine or multiply equations.

**EXAMPLE:** If $8a - b^2 = 24$ and $8b + b^2 = 56$, then $a + b =$
(A) 3
(B) 7
(C) 8
(D) 10
(E) 80

**SOLUTION:** Pairs of equations usually need to be added or subtracted.

- Write the given pair of equations, placing the first on top of the second:

$$8a - b^2 = 24$$

$$8b + b^2 = 56$$

- Notice that the left side of the first equation contains $-b^2$ and the left side of the second equation contains $b^2$. Eliminate $b^2$ by adding the corresponding sides of the two equations:

$$
\begin{array}{r}
8a - b^2 = 24 \\
+ \quad 8b + b^2 = 56 \\
\hline
8a + 8b + 0 = 80
\end{array}
$$

- Divide each member of the new equation by 8:

$$\frac{8a}{8} + \frac{8b}{8} = \frac{80}{8}$$

$$a + b = 10$$

The correct choice is (D).

---

**EXAMPLE:** If $x - 2y = 1$, what is the value of $2^{3x-6y}$?

**SOLUTION:**

- Factor the exponent of $2^{3x-6y}$:

$$3x - 6y = 3(x - 2y)$$

- Since it is given that $x - 2y = 1$, substitute 1 for $x - 2y$ in the factored expression:

$$3x - 6y = 3(1) = 3$$

- Find the value of $2^{3x-6y}$ by replacing $3x - 6y$ with 3:

$$2^{3x-6y} = 2^3 = 2 \times 2 \times 2 = 8$$

The correct answer is 8.

# LESSON
# 2-2

# Special Math Strategies for Regular Multiple-Choice Questions

## OVERVIEW

*The correct answer to a regular multiple-choice question is given to you, along with four other choices that are designed to distract you. Sometimes you can use a "backdoor" strategy that will allow you to identify the correct answer without going through a standard mathematical solution.*

STRATEGY

# 6

## TEST NUMERICAL ANSWER CHOICES IN THE QUESTION

When each of the answer choices for a regular multiple-choice question is an integer, you may be able to find the correct answer by plugging each of the possible choices back into the question until you find the number that gives the correct answer. This method of solution is called "backsolving."

> **EXAMPLE:** If $2^{2x-1} = 32$, then $x =$
> (A) 2
> (B) 3
> (C) 4
> (D) 5
> (E) 6

**SOLUTION:** Since the answer choices give the possible values of $x$, replace $x$ with each of these values until you find the one that makes $2^{2x-1}$ equal to 32. As the value of $x$ increases, so does the value of $2^{2x-1}$. Therefore, you may be able to save time by starting with choice (C), the middle $x$-value. If this value of $x$ makes $2^{2x-1}$ too large, then the correct answer must be *smaller* than the middle $x$-value. In this case, you can eliminate choices (C), (D), and (E). Similarly, you can eliminate choices (A), (B), and (C) if the middle $x$-value makes $2^{2x-1}$ too *small*.

- Plug in choice (C) for $x$. If $x = 4$, then

$$2^{2x-1} = 2^{2(4)-1}$$
$$= 2^7$$
$$= 2 \times 2 \times 2 \times 2 \times 2 \times 2 \times 2$$
$$= 128$$

Since the correct value of $x$ must be smaller than 4, try choice (B). If choice (B) doesn't work, the correct answer must be choice (A).

- Plug in choice (B) for $x$. If $x = 3$, then

$$2^{2x-1} = 2^{2(3)-1}$$
$$= 2^5$$
$$= 2 \times 2 \times 2 \times 2 \times 2$$
$$= 32$$

Hence, choice (B) is correct.

**STRATEGY**

**7**

### CHANGE VARIABLE ANSWER CHOICES INTO NUMERICAL ANSWER CHOICES

If you don't know how to find the answer to a regular multiple-choice question in which the answer choices contain letters, substitute numbers for the letters.

> **EXAMPLE:**   If $m + n$ is an odd number when $m$ and $n$ are positive integers, which expression always represents an even number?
> (A) $m - n$
> (B) $m^2 + n^2$
> (C) $(m + n)^2$
> (D) $m^2 - n^2$
> (E) $(mn)^2$

**SOLUTION:**   Pick simple numbers for $m$ and $n$ such that $m + n$ is an odd number. Let $m = 2$ and $n = 1$; then $m + n = 3$, and 3 is an odd number. Try the answer choices in turn:

- (A) $m - n = 2 - 1 = 1$
- (B) $m^2 + n^2 = 2^2 + 1^2 = 4 + 1 = 5$
- (C) $(m + n)^2 = (2 + 1)^2 = 3^2 = 9$
- (D) $m^2 - n^2 = 2^2 - 1^2 = 4 - 1 = 3$
- (E) $(mn)^2 = (2 \times 1)^2 = 2^2 = 4$

Since the only even number among the transformed answer choices is 4, the correct choice is (E).

STRATEGY

# 8

## RULE OUT IMPOSSIBLE ANSWER CHOICES

Sometimes you can simplify a problem or guess the correct answer by using mathematical reasoning to eliminate choices that cannot possibly be correct answers.

**EXAMPLE:** If $-1 < a < 0$, which of the following statements is always true?
(A) $a^3 < a^2 < a$
(B) $a < a^2 < a^3$
(C) $a^3 < a < a^2$
(D) $a < a^3 < a^2$
(E) $a^2 < a^3 < a$

**SOLUTION:** If $a$ is negative, then $a^2$ is positive and $a^3$ is negative. Since $a^2$ is positive, it is larger than $a$ and $a^3$. Since the correct answer must have the form $\_\_\_ < \_\_\_ < a^2$, you can eliminate answer choices (A), (B), and (E). You can guess from the two choices that remain, or you can try plugging in numbers to see whether (C) or (D) is correct. If $a = -\frac{1}{2}$, then

$$a^3 = \left(-\frac{1}{2}\right)^3 = \left(-\frac{1}{2}\right) \times \left(-\frac{1}{2}\right) \times \left(-\frac{1}{2}\right) = -\frac{1}{8}$$

Since $-\frac{1}{2} < -\frac{1}{8}$, $a$ is less than $a^3$, so $a < a^3 < a^2$. The correct choice is (D).

## TIPS FOR SCORING HIGH

1. When choice (E) is "It cannot be determined from the information given," look at the question number. This choice is rarely the correct answer to a question that appears near the end of a test section, but *may* be the correct answer to a question early in the section.

2. Before starting your solution, look at the five answer choices. The types of numbers or the forms of the answer choices may help you decide on a solution strategy. For example, if there are whole numbers in the question, and decimal or fractional answers appear as choices, the solution may involve division.

3. Keep in mind that, if one of the five answer choices does not look similar to the other four, the dissimilar choice is rarely the correct answer. For example, if the five choices are these:
(A) $\pi(r - \sqrt{2})$ (B) $\pi(r + \sqrt{2})$ (C) $\pi(r^2 + 2)$ (D) $\pi(r^2 - 2)$ (E) $\frac{1}{2}r$
choice (E) is probably not correct since it is the only choice that does not involve $\pi$ and does not include the sum or difference of two terms.

4. If you think you have solved a problem correctly, but do not find your answer among the five choices, try writing your answer in an equivalent form. For example, if you don't find $\frac{3}{2}$, look for 1.5; if you don't find $2a + 4$, look for $2(a + 2)$; if you don't find $\frac{3x-y}{x}$, look for $3 - \frac{y}{x}$ since

$$\frac{3x-y}{x} = \frac{3x}{x} - \frac{y}{x} = 3 - \frac{y}{x}$$

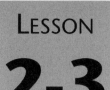

# Special Math Strategies for Quantitative Comparison Questions

## OVERVIEW

*You may be able to figure out the answer to a QC question by:*

- *doing easy arithmetic;*
- *eliminating the same quantity in both columns;*
- *trying to make the two columns look alike;*
- *performing the same operation on both columns;*
- *using centered information to transform the columns;*
- *replacing variables with particular test values.*

STRATEGY

9

### WHEN THE ARITHMETIC IS EASY, DO IT

Some QC questions near the beginning of a section can sometimes be answered quickly by doing simple arithmetic.

**EXAMPLE:**

| Column A | Column B |
|---|---|
| 25% of 48 | $\frac{1}{3}$ of 36 |

**SOLUTION:** Do the arithmetic in each column. Quantity A is 12 since

$$25\% = \frac{1}{4} \quad \text{and} \quad \frac{1}{4} \times 48 = 12$$

Quantity B is also 12 since

$$\frac{1}{3} \times 36 = 12$$

Hence, the correct choice is (C).

**36**

**STRATEGY 10**

## CANCEL THE SAME QUANTITIES IN BOTH COLUMNS

Think of a QC question as a balanced scale on which quantity A and quantity B are being weighed on either side of the scale. If the same number is on both sides, then removing it from both sides has no effect on the scale balance. You can simplify a comparison by:

- Canceling any quantity that is added or subtracted in both columns.
- Canceling any *positive* quantity that multiplies or divides in both columns.

**EXAMPLE:**

| Column A | Column B |
|---|---|
| $\frac{13}{14} + \frac{3}{7}$ | $\frac{13}{14} + \frac{4}{8}$ |

**SOLUTION:** Since $\frac{13}{14}$ appears in both columns, it does not affect the comparison and therefore can be canceled.

| Column A | Column B |
|---|---|
| $\frac{1\cancel{3}}{\cancel{14}} + \frac{3}{7}$ | $\frac{1\cancel{3}}{\cancel{14}} + \frac{4}{8}$ |
| $\frac{3}{7}$ | $\frac{4}{8}$ |

Using a calculator, you find that $\frac{3}{7} = 3 \div 7 = 0.428. \ldots$ Since $\frac{4}{8} = \frac{1}{2} = 0.5$, quantity B is greater than quantity A. Hence, the correct choice is (B).

**STRATEGY 11**

## TRY TO MAKE THE TWO COLUMNS LOOK ALIKE

You may be able to compare the quantities in the two columns term by term if you can make these quantities look alike.

**EXAMPLE:**

| Column A | Column B |
|---|---|
| $x(x + 2)$ | $(x + 1)^2$ |

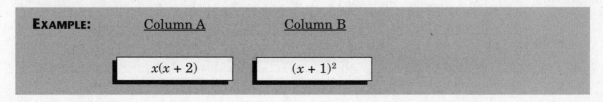

**SOLUTION:**   Try to make the quantities in the two columns look alike by writing each as a polynomial. Then compare the two polynomials term by term.

| Column A | Column B |
|:---:|:---:|
| $x(x+2)$ | $(x+1)^2$ |
| $x(x) + x(2)$ | $(x+1)(x+1)$ |
| $x^2 + 2x$ | $x^2 + 2x + 1$ |

Since both polynomials include $x^2$ and $2x$, quantity B is always greater than quantity A by 1. Hence, the correct choice is (B).

## PERFORM THE SAME OPERATION ON BOTH COLUMNS

Sometimes the size relationship between the quantities in the two columns becomes more apparent after the same arithmetic operation is performed on both columns.

- The same number can be added or subtracted in both columns.
- The same *positive* number can multiply or divide in both columns.
- The quantity in each column can be raised to the second power to eliminate any square root radicals, provided that each quantity is positive.

**EXAMPLE:**

| Column A | Column B |
|:---:|:---:|
| $3\sqrt{2}$ | $2\sqrt{3}$ |

**SOLUTION:**   Eliminate the square root radicals by raising the quantity in each column to the second power. Then compare the quantities that result.

- In Column A, $(3\sqrt{2})^2 = (3\sqrt{2})(3\sqrt{2}) = 9 \cdot 2 = 18$
- In Column B, $(2\sqrt{3})^2 = (2\sqrt{3})(2\sqrt{3}) = 4 \cdot 3 = 12$
- Since Column A > Column B, the correct choice is (A).

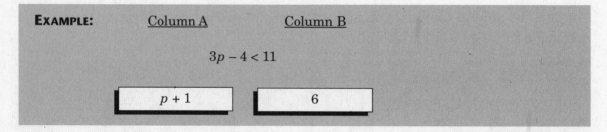

**EXAMPLE:**

| Column A | Column B |
|:---:|:---:|
| | |

$$3p - 4 < 11$$

| Column A | Column B |
|:---:|:---:|
| $p + 1$ | $6$ |

**SOLUTION:**   Simplify the comparison by subtracting 1 from the quantity in each column. You are left with $p$ in Column A and 5 in Column B. To find out how $p$ compares with 5, solve the given inequality:

$$3p - 4 < 11$$

- Add 4 on both sides of the inequality:   $3p - 4 + (4) < 11 + (4)$
- Simplify:   $3p < 15$
- Divide both sides of the inequality by 3:   $\dfrac{3p}{3} < \dfrac{15}{3}$

$$p < 5$$

Since $p$ is less than 5, quantity B is greater than quantity A. The correct choice is (B).

## USE CENTERED INFORMATION TO TRANSFORM THE COLUMNS

Sometimes centered information, which applies to both columns, can be used to transform the columns so that the comparison of values becomes easier.

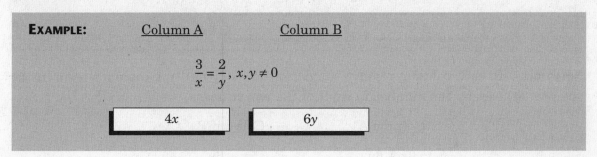

**EXAMPLE:**    Column A        Column B

$$\frac{3}{x} = \frac{2}{y},\ x, y \neq 0$$

4x        6y

**SOLUTION:**   Use the given information, $\frac{3}{x} = \frac{2}{y}$, to transform one of the columns so that it contains the same variable as the other column.

- Cross-multiplying in the equation $\frac{3}{x} = \frac{2}{y}$ gives $2x = 3y$.
- In Column A, $4x$ can now be expressed in terms of $y$. Since

$$4x = 2(2x) = 2(3y) = 6y,$$

quantity A is $6y$.

Since quantity B is also $6y$, the two columns always have the same value. The correct choice is (C).

## REPLACE VARIABLES WITH TEST VALUES

If variable quantities are being compared, then replacing these variables with particular numbers may enable you to tell which column, if either, contains the bigger quantity. Use a variety of test values, such as

- 0 and 1,
- a number between 0 and 1,
- an integer greater than 1,
- a negative integer.

Be sure, however, not to use any test value that has been ruled out by any centered information. If you find that two different test values make different answer choices true, there is no need to continue. The correct answer for the question must be choice (D).

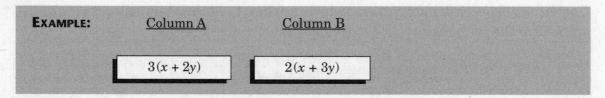

**EXAMPLE:**

| Column A | Column B |
|----------|----------|
| $3(x + 2y)$ | $2(x + 3y)$ |

**SOLUTION:**   In each column, remove the parentheses by multiplying each term inside the parentheses by the number in front of the parentheses.

| Column A | Column B |
|----------|----------|
| $3x + 6y$ | $2x + 6y$ |
| $3x + \cancel{6y}$ | $2x + \cancel{6y}$ |
| $3x$ | $2x$ |

Since there is no restriction on the value of $x$, you can assume that $x$ can be positive, zero, or negative.

- If $x = 1$, then $3x = 3$ and $2x = 2$. Since quantity A > quantity B, the correct answer must be either choice (A) or choice (D).
- If $x = 0$, then $3x = 0$ and $2x = 0$. Since quantity A = quantity B, the correct answer must be either choice (C) or choice (D).

Since the size relationship between $3x$ and $2x$ can change for particular values of $x$, the correct choice is (D).

## STRATEGY 15

## PAY ATTENTION TO ANY RESTRICTIONS ON THE VALUES OF VARIABLES

Centered information, which applies to both columns, may restrict the value(s) of one or more variables. This may affect the size relationship between the quantities to be compared. Consider the following QC question, in which the value of $x$ is limited to numbers greater than 1.

Column A        Column B

$x > 1$

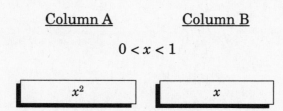

Since $x$ is greater than 1, increasing powers of $x$ grow larger in value. Hence, $x^2 > x$. The correct choice is (A).

Now compare the same two quantities, but with the restriction that $x$ is some number between 0 and 1.

Column A        Column B

$0 < x < 1$

Since $x$ is between 0 and 1, increasing powers of $x$ become *smaller* in value. For example, if $x = \frac{1}{2}$, then

$$x^2 = \left(\frac{1}{2}\right)^2 = \frac{1}{4}$$

Since $x > x^2$, quantity B is now larger than quantity A. The correct choice is (B).

In the QC question below there is no restriction on the value of $x$.

Column A        Column B

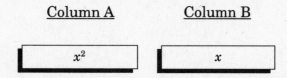

If a number greater than 1 is chosen for $x$, the answer is choice (A). If $x$ is replaced by a number between 0 and 1, the answer becomes choice (B). If $x = 1$ or $x = 0$, the answer is choice (C). Since more than one answer choice is possible, the size relationship between the columns depends on the particular number that replaces $x$. In this case the correct choice is (D).

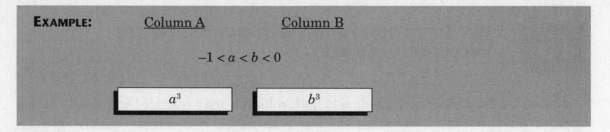

**EXAMPLE:**

Column A          Column B

$$-1 < a < b < 0$$

$a^3$          $b^3$

**SOLUTION:** According to the given information, $a$ and $b$ are between –1 and 0 with $a$ less than $b$. Pick convenient numbers for $a$ and $b$ that satisfy these conditions. Since $-\frac{1}{2} < -\frac{1}{4}$, let $a = -\frac{1}{2}$ and $b = -\frac{1}{4}$. Then, in Column A,

$$a^3 = \left(-\frac{1}{2}\right)^3 = \left(-\frac{1}{2}\right) \times \left(-\frac{1}{2}\right) \times \left(-\frac{1}{2}\right) = -\frac{1}{8}$$

In Column B,

$$b^3 = \left(-\frac{1}{4}\right)^3 = \left(-\frac{1}{4}\right) \times \left(-\frac{1}{4}\right) \times \left(-\frac{1}{4}\right) = -\frac{1}{64}$$

Since $-\frac{1}{64}$ is closer to 0 ("less negative") than $-\frac{1}{8}$, quantity B is greater than quantity A. The correct choice is (B).

# PART II

## REVIEWING KEY MATH SKILLS AND CONCEPTS

# ANSWER SHEET FOR TUNE-UP EXERCISES
## LESSON # _____ :

Before you work through the SAT I Tune-Up exercises found at the end of each review lesson, you may wish to reproduce the answer form below. This may contain more ovals and grids than you will need for a particular exercise section. Mark only those you need, and leave the rest blank.

## Multiple Choice

1. Ⓐ Ⓑ Ⓒ Ⓓ Ⓔ
2. Ⓐ Ⓑ Ⓒ Ⓓ Ⓔ
3. Ⓐ Ⓑ Ⓒ Ⓓ Ⓔ
4. Ⓐ Ⓑ Ⓒ Ⓓ Ⓔ
5. Ⓐ Ⓑ Ⓒ Ⓓ Ⓔ

6. Ⓐ Ⓑ Ⓒ Ⓓ Ⓔ
7. Ⓐ Ⓑ Ⓒ Ⓓ Ⓔ
8. Ⓐ Ⓑ Ⓒ Ⓓ Ⓔ
9. Ⓐ Ⓑ Ⓒ Ⓓ Ⓔ
10. Ⓐ Ⓑ Ⓒ Ⓓ Ⓔ

11. Ⓐ Ⓑ Ⓒ Ⓓ Ⓔ
12. Ⓐ Ⓑ Ⓒ Ⓓ Ⓔ
13. Ⓐ Ⓑ Ⓒ Ⓓ Ⓔ
14. Ⓐ Ⓑ Ⓒ Ⓓ Ⓔ
15. Ⓐ Ⓑ Ⓒ Ⓓ Ⓔ

16. Ⓐ Ⓑ Ⓒ Ⓓ Ⓔ
17. Ⓐ Ⓑ Ⓒ Ⓓ Ⓔ
18. Ⓐ Ⓑ Ⓒ Ⓓ Ⓔ
19. Ⓐ Ⓑ Ⓒ Ⓓ Ⓔ
20. Ⓐ Ⓑ Ⓒ Ⓓ Ⓔ

## Quantitative Comparison

1. Ⓐ Ⓑ Ⓒ Ⓓ Ⓔ
2. Ⓐ Ⓑ Ⓒ Ⓓ Ⓔ
3. Ⓐ Ⓑ Ⓒ Ⓓ Ⓔ
4. Ⓐ Ⓑ Ⓒ Ⓓ Ⓔ
5. Ⓐ Ⓑ Ⓒ Ⓓ Ⓔ

6. Ⓐ Ⓑ Ⓒ Ⓓ Ⓔ
7. Ⓐ Ⓑ Ⓒ Ⓓ Ⓔ
8. Ⓐ Ⓑ Ⓒ Ⓓ Ⓔ
9. Ⓐ Ⓑ Ⓒ Ⓓ Ⓔ
10. Ⓐ Ⓑ Ⓒ Ⓓ Ⓔ

11. Ⓐ Ⓑ Ⓒ Ⓓ Ⓔ
12. Ⓐ Ⓑ Ⓒ Ⓓ Ⓔ
13. Ⓐ Ⓑ Ⓒ Ⓓ Ⓔ
14. Ⓐ Ⓑ Ⓒ Ⓓ Ⓔ
15. Ⓐ Ⓑ Ⓒ Ⓓ Ⓔ

16. Ⓐ Ⓑ Ⓒ Ⓓ Ⓔ
17. Ⓐ Ⓑ Ⓒ Ⓓ Ⓔ
18. Ⓐ Ⓑ Ⓒ Ⓓ Ⓔ
19. Ⓐ Ⓑ Ⓒ Ⓓ Ⓔ
20. Ⓐ Ⓑ Ⓒ Ⓓ Ⓔ

## Grid In

1. 2. 3. 4. 5.

# Arithmetic Skills and Concepts

This chapter reviews the arithmetic operations and arithmetic reasoning skills that are tested by many of the easy to medium-difficultly questions that appear on the SAT I.

---

## LESSONS IN THIS CHAPTER

---

---

# Numbers, Symbols, and Variables

## OVERVIEW

*All numbers that appear on the SAT I are real numbers.* **Real numbers** *include all the different types of numbers, both positive and negative, you encountered in arithmetic and beginning algebra classes.*

*A* **variable** *is a symbol that serves as a placeholder for an unknown member of a given set of numbers. Letters of the alphabet such as a, b, and x are used as variables to help make general statements about how numbers behave.*

## TYPES OF NUMBERS

The set of real numbers is comprised of these sets of numbers:

- *Whole numbers* = 0, 1, 2, 3, . . .

- *Integers* = . . . $\underbrace{-4, -3, -2, -1,}_{\text{negative integers}}$ 0, $\overbrace{1, 2, 3, 4,}^{\text{positive integers}}$ . . .

- *Rational numbers* = $\begin{cases} \bullet \text{ Fractions such as } \frac{2}{3} \text{ that have integers above and} \\ \text{ below the fraction bar.} \\ \bullet \text{ Terminating decimals such as } 0.75. \\ \bullet \text{ Nonending, repeating decimals such as } 0.33333 \dots \end{cases}$

- *Irrational numbers* = $\begin{cases} \text{Numbers such as } \pi \text{ and } \sqrt{2} \text{ that do not have exact} \\ \text{decimal equivalents.} \end{cases}$

## COMPARISON SYMBOLS

Table 3.1 summarizes the symbols used to compare two numbers.

TABLE 3.1  COMPARISON SYMBOLS

| Symbol | Translation | Example |
|---|---|---|
| = | is equal to | $5 = 5$ |
| ≠ | is *not* equal to | $5 \neq 3$ |
| > | is greater than | $5 > 3$ |
| ≥ | is greater than *or* equal to | $x \geq 5$ means that $x$ can be 5 *or* any number greater than 5. |
| < | is less than | $3 < 5$ |
| ≤ | is less than *or* equal to | $x \leq 3$ means that $x$ can be 3 *or* any number less than 3. |

Since 7 is between 6 and 8, you can write $6 < 7 < 8$. The inequality $6 < 7 < 8$ means that 6 is less than 7 and, at the same time, 7 is less than 8. In general, if $x$ is between $a$ and $b$ with $a < b$, then $a < x < b$. The inequality $a \leq x \leq b$ means that $x$ is between $a$ and $b$, or may be equal to either $a$ or $b$, as shown in Figure 3.1.

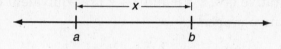

**Figure 3.1**  $a \leq x \leq b$

## INDICATING MULTIPLICATION

When a number and a variable are written consecutively, as in $2y$, the operation of multiplication is understood. The number 2 in front of $y$ is called the **coefficient** of $y$. Thus, $3n$ means 3 times $n$, and 3 is the coefficient of $n$. If $n = 4$, then

$$3n = 3(4) = 12$$

The expression $2ab$ means 2 times $a$ times $b$. If $a = 5$ and $b = 7$, then

$$2ab = 2(5)(7) = 10(7) = 70$$

When terms such as $3n$ and $2n$ differ only in their numerical coefficients, they are called **like terms**.

## LAWS OF ARITHMETIC

Real numbers behave in predictable ways.

- *Commutative law*: The order in which *two* numbers are added or multiplied does not matter. For example:

$$3 + 4 = 4 + 3 \quad \text{and} \quad 3 \times 4 = 4 \times 3$$

- *Associative law*: The order in which *three* numbers are added or multiplied does not matter. For example:

$$2 \times (3 \times 4) = (2 \times 3) \times 4$$

where the parentheses indicate the pair of numbers that are multiplied first.

- *Distributive law*: To multiply a number by a sum of two values, you multiply each value in the sum by that number and then add the two products. For example:

$$3 \times (2 + 4) = 3 \times 2 + 3 \times 4$$
$$= \quad 6 \quad + \quad 12$$
$$= 18$$

## APPLYING THE LAWS OF ARITHMETIC TO ALGEBRAIC EXPRESSIONS

The commutative, associative, and distributive laws can be applied to expressions that contain variables since these expressions represent real numbers. For example:

- By the commutative law, $x + 3$ and $3 + x$ are equivalent expressions.
- By the distributive property,

$$3(2x + 5) = 3(2x) + 3(5) = 6x + 15$$

- By the reverse of the distributive property,

$$4x + 5x = (4 + 5)x = 9x$$

Thus, like terms are combined by combining their numerical coefficients. Here are three more examples:

**EXAMPLE:**      $3y + 4y = 7y$

**EXAMPLE:**      $9p + 3p + 2p = 14p$

**EXAMPLE:**      $5x - x = 4x \quad$ since $5x - x = 5x - 1x = 4x$

# LESSON 3-1 TUNE-UP EXERCISES

## Multiple Choice

**1** If $a = 9 \times 23$ and $b = 9 \times 124$, what is the value of $b - a$?

(A) 901
(B) 903
(C) 906
(D) 909
(E) 911

**2** By how much does the product of 8 and 25 exceed the product of 15 and 10?

(A) 25
(B) 50
(C) 75
(D) 100
(E) 125

**3** How many containers, each holding 16 fluid ounces of milk, are needed to hold 5 quarts of milk? (1 quart = 32 fluid ounces.)

(A) 6
(B) 8
(C) 10
(D) 12
(E) 14

**4** If the current odometer reading of a car is 31,983 miles, what is the LEAST number of miles that the car must travel before the odometer displays four digits that are the same?

(A) 17
(B) 239
(C) 350
(D) 650
(E) 1350

**5** In store $A$ a scarf costs \$12, and in store $B$ the same scarf is on sale for \$8. How many scarfs can be bought in store $B$ with the amount of money, excluding sales tax, needed to buy 10 scarfs in store $A$?

(A) 4
(B) 12
(C) 15
(D) 18
(E) 21

**6** Let $*$ represent one of the four basic arithmetic operations such that, for any nonzero real number $r$,

$$r * 0 = r \quad \text{and} \quad r * r = 0$$

Which arithmetic operation(s) does the symbol $*$ represent?

(A) + only
(B) − only
(C) + and −
(D) × only
(E) +, −, or ×

**7** Kurt has saved \$160 to buy a stereo system that costs \$400 including taxes. If he earns \$8 an hour after all payroll deductions have been made, how many hours will he have to work in order to have exactly enough money to buy the stereo system?

(A) 20
(B) 24
(C) 25
(D) 30
(E) 40

**8**   If the present time is exactly 1:00 P.M., what was the time exactly 39 hours ago?

(A) 4:00 P.M.
(B) 4:00 A.M.
(C) 9:00 P.M.
(D) 9:00 A.M.
(E) 10:00 P.M.

**9**   Let # represent one of the four basic arithmetic operations such that, for any nonzero real numbers $r$ and $s$,

$$r \# 0 = r \text{ and } r \# s = s \# r$$

Which arithmetic operation(s) does the symbol # represent?

(A) + only
(B) × only
(C) − only
(D) ≠ and ×
(E) + and ÷

**10**   If $w = (6)(6)(6)$, $x = (5)(6)(7)$, and $y = (4)(6)(8)$, which inequality statement is true?

(A) $x < y < w$
(B) $w < x < y$
(C) $y < w < x$
(D) $y < x < w$
(E) $w < y < x$

## Quantitative Comparison

Each question consists of two quantities in boxes, one in Column A and one in Column B. You are to compare the two quantities and on the answer sheet fill in

A  if the quantity in Column A is greater;
B  if the quantity in Column B is greater;
C  if the two quantities are equal;
D  if the relationship cannot be determined from the information given

AN E RESPONSE WILL NOT BE SCORED

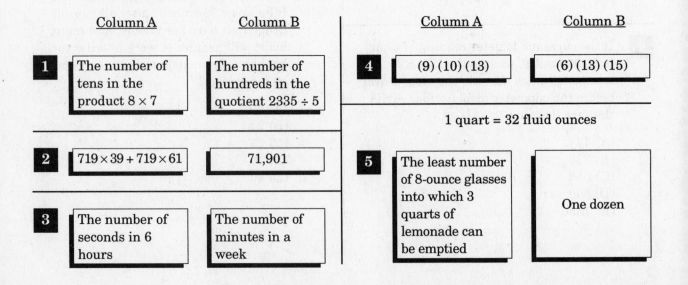

| | Column A | Column B |
|---|---|---|
| **1** | The number of tens in the product $8 \times 7$ | The number of hundreds in the quotient $2335 \div 5$ |
| **2** | $719 \times 39 + 719 \times 61$ | 71,901 |
| **3** | The number of seconds in 6 hours | The number of minutes in a week |

| | Column A | Column B |
|---|---|---|
| **4** | (9) (10) (13) | (6) (13) (15) |

1 quart = 32 fluid ounces

| | Column A | Column B |
|---|---|---|
| **5** | The least number of 8-ounce glasses into which 3 quarts of lemonade can be emptied | One dozen |

| Column A | Column B |          | Column A | Column B |
|----------|----------|----------|----------|----------|

**6** $3(8x + 4)$     $4(3 + 6x)$

**7** $\dfrac{1}{3}$     $0.333$

**8** $2(x + 1) + 3x - 2$     $5x$

$a$ and $b$ are positive integers.

**9** $3(a + 2b)$     $3a + 2b$

**Questions 10 and 11.**

$$2 < x < 7$$
$$3 < y < 8$$

**10** $y - x$     $6$

**11** $4x$     $y$

---

$$4 < ab < 12$$
$a$ and $b$ are positive integers.

**12** $ab$     $a + b$

$$
\begin{array}{rr}
18 & \quad 18 \\
59 & 59 \\
34 & 34 \\
x & x \\
y & z \\
+\ 27 & +\ 27 \\
\hline
221 & 208 \\
\end{array}
$$

**13** $y - z$     $13$

---

## Grid In

**1** The houses in a certain community are numbered consecutively from 2019 to 2176. How many houses are in the community?

**2** If 1 kilobyte of computer memory is equivalent to 1024 bytes and 1 byte is equivalent to 8 bits, how many kilobytes are equivalent to 40,960 bits?

**3** A television set will cost, including taxes and finance charges, $495 if the buyer puts $129 down and then pays off the balance in eight equal monthly payments. Under this purchase plan, what will each of the monthly payments be?

**4** If $13 \le k \le 21$, $9 \le p \le 19$, $2 < m < 6$, and $k, p,$ and $m$ are integers, what is the largest possible value of $\dfrac{k-p}{m}$?

**5** For some fixed value of $x$, $9(x + 2) = y$. After the value of $x$ is increased by 3, $9(x + 2) = w$. What is the value of $w - y$?

# Powers and Roots

### OVERVIEW

*Each of the numbers that are multiplied together to obtain a product is called a **factor** of that product. Since $5 \times 6 = 30$, 5 and 6 are each factors of 30. The product of identical factors may be indicated by writing the number of times the factor is repeated a half-line above and to the right of the factor. For example:*

$$\underbrace{2 \times 2 \times 2 \times 2 \times 2}_{\text{2 is used as a factor 5 times.}} \quad = \quad 2\overset{\text{⌐ Exponent}}{\underset{\text{⌐ Base}}{5}}$$

## MEANING OF EXPONENT

Repeated multiplication of the same number may be indicated in a more compact form by using an **exponent** that tells the number of times the number, called the **base**, appears as a factor in the product. The notation $b^n$ is read as "$b$ raised to the $n$th power." In $b^n$, the number $b$ is the base and $n$ is the exponent.

When a number or variable appears without an exponent, the exponent is understood to be 1. For example, $3 = 3^1$ and $y = y^1$.

**EXAMPLE:**     What is the value of $3^4 - 4^3$?

**SOLUTION:**   Since $3^4 = 3 \times 3 \times 3 \times 3 = 81$ and $4^3 = 4 \times 4 \times 4 = 64$,

$$3^4 - 4^3 = 81 - 64 = 17$$

**EXAMPLE:**     If $2^x = 8$, what is the value of $x^3$?

**SOLUTION:**   Since $2^x = 8 = 2 \times 2 \times 2 = 2^3$, then $x = 3$. Hence,

$$x^3 = 3^3 = 3 \times 3 \times 3 = 27$$

## LAWS OF EXPONENTS

You should know these rules for working with exponents:

- To multiply powers with the *same* nonzero base, *add* their exponents. For example:

$$2^4 \times 2^3 = 2^{4+3} = 2^7 \quad \text{and} \quad b^2 \cdot b^3 = b^{2+3} = b^5$$

Don't try to multiply powers that have different bases. For example:

$$2^5 \times 3^4 \neq (2 \times 3)^{5+4}$$

- To divide powers with the *same* nonzero base, *subtract* their exponents. For example:

$$2^7 \div 2^3 = \frac{2^7}{2^3} = 2^{7-3} = 2^4 \quad \text{and} \quad \frac{b^5}{b} = \frac{b^5}{b^1} = b^{5-1} = b^4$$

- Don't try to divide powers that have different bases. For example:

$$2^5 \div 3^4 \neq \left(\frac{2}{3}\right)^{5-4}$$

- To raise a power to another power, *multiply* the exponents. For example:

$$(2^3)^7 = 2^{3 \times 7} = 2^{21} \quad \text{and} \quad (b^4)^3 = b^{4 \times 3} = b^{12}$$

These rules are summarized in Table 3.2, where $b$, $m$, and $n$ stand for positive integers except that $b$ cannot be 0.

### TABLE 3.2  LAWS OF EXPONENTS  ($b \neq 0$)

| Product Law | Quotient Law | Power Law |
|---|---|---|
| $b^m \times b^n = b^{m+n}$ | $b^m \div b^n = b^{m-n}$ | $(b^m)^n = b^{m \times n}$ |

## ORDER OF OPERATIONS

Arithmetic operations are not necessarily performed in the order in which they are encountered. Instead,

P:  First evaluate expressions within *P*arentheses.
E:  Then evaluate *E*xponents.
M:  Next, *M*ultiply and
D:  *D*ivide, working from left to right.
A:  Finally, *A*dd and
S:  *S*ubtract, again working from left to right.

The first letters of the key words in the sequence of operations spell "PEMDAS," which can be remembered by memorizing the sentence "*Please Excuse My Dear Aunt Sally.*"

**EXAMPLE:**       Find the value of $5 + 18 \div (4 - 1)^2$.

**Solution:**
$$
\begin{aligned}
5 + 18 \div (4 - 1)^2 &= 5 + 18 \div 3^2 \\
&= 5 + 18 \div 9 \\
&= 5 + \quad 2 \\
&= 7
\end{aligned}
$$

## SQUARE AND CUBE ROOT RADICALS

The **square** of a number is the product of a number and itself. Thus, $3^2$ represents the square of 3 since $3^2 = 3 \times 3 = 9$. Finding the square root of a number reverses the process of squaring a number. The **square root** of a nonnegative number $N$ is one of two identical numbers whose product is $N$. The symbol for square root is $\sqrt{\phantom{x}}$. Thus, $\sqrt{9} = 3$ since $3 \times 3 = 9$. In $\sqrt{9}$, the symbol $\sqrt{\phantom{x}}$ is called a **radical**, and 9, the number underneath the radical sign, is called the **radicand**.

The symbol $\sqrt[3]{\phantom{x}}$ represents the **cube root** of the number that is written underneath the radical. Thus $\sqrt[3]{8} = 2$ since $2 \times 2 \times 2 = 8$.

## PERFECT SQUARES

The square root of a whole number may not be a whole number. For example, $\sqrt{7}$ does not equal a whole number since it is not possible to find two identical whole numbers whose product is 7. A whole number is a **perfect square** if its square root is also a whole number. The numbers 1, 4, 9, 16, 25, 36, 49, 64, 81, and 100 are perfect squares.

## PROPERTIES OF SQUARE ROOT RADICALS

Here are some properties of square root radicals that you should know:

- $\sqrt{a} \times \sqrt{b} = \sqrt{a \times b}$. The product of the square roots of two numbers is the square root of the product. For example:

$$
\sqrt{7} \times \sqrt{3} = \sqrt{7 \times 3} = \sqrt{21}
$$

- $\dfrac{\sqrt{a}}{\sqrt{b}} = \sqrt{\dfrac{a}{b}} \quad (b \neq 0)$. The quotient of the square roots of two numbers is the square root of the quotient. For example:

$$
\frac{\sqrt{21}}{\sqrt{7}} = \sqrt{\frac{21}{7}} = \sqrt{3}
$$

- $(\sqrt{N})^2 = N$. The square of a square root radical is the radicand. For example:

$$
(\sqrt{5})^2 = \sqrt{5} \times \sqrt{5} = 5
$$

- $a\sqrt{b} + c\sqrt{b} = (a + c)\sqrt{b}$. To combine square root radicals with the same radicands, combine their rational coefficients and keep the same radical factor, as in

$$5\sqrt{3} + 2\sqrt{3} = 7\sqrt{3}$$

- $\sqrt{a+b} \neq \sqrt{a} + \sqrt{b}$. Radicals cannot be distributed over the operations of addition and subtraction. For example:

$$\sqrt{4+9} \neq \sqrt{4} + \sqrt{9}$$

- To simplify a square root radical, write the radicand as the product of two numbers, one of which is the highest perfect square factor of the radicand. Then write the radical over each factor and simplify. For example:

$$\sqrt{20} = \sqrt{4 \times 5} = \sqrt{4} \times \sqrt{5} = 2\sqrt{5}$$

# LESSON 3-2  TUNE-UP EXERCISES

---

**Multiple Choice**

---

**1**  If $x = 3$ and $y = \sqrt{1 + 7x - 2x^2}$, then $y =$

(A) 1
(B) 2
(C) 3
(D) 4
(E) 5

**2**  If $x = 32 - 16 \div 2 \times 4$, then $x =$

(A) 0
(B) 2
(C) 4
(D) 8
(E) 32

**3**  What is the greatest number of integer values of $x$ for which $1 < x^2 < 50$?

(A) Five
(B) Six
(C) Seven
(D) Eight
(E) Nine

**4**  If $2 = p^3$, then $8p$ must equal

(A) $p^6$
(B) $p^8$
(C) $p^{10}$
(D) $8\sqrt{2}$
(E) 16

**5**  If $5 = a^x$, then $\dfrac{5}{a} =$

(A) $a^{x+1}$
(B) $a^{x-1}$
(C) $a^{1-x}$
(D) $a^{\frac{x}{5}}$
(E) $a^{\frac{5}{x}}$

**6**  If $b^3 = 4$, then $b^6 =$

(A) 2
(B) 8
(C) 12
(D) 16
(E) 64

**7**  If $w$ is a positive number and $w^2 = 2$, then $w^3 =$

(A) $\sqrt{2}$
(B) $2\sqrt{2}$
(C) $3\sqrt{2}$
(D) 4
(E) 6

**8**  If $8^{x+1} = 64$, what is the value of $3^{2x+1}$?

(A) 1
(B) 3
(C) 9
(D) 27
(E) 81

**9**  If $0 < y < x$, which statement MUST be true?

(A) $\sqrt{x} - \sqrt{y} = \sqrt{x-y}$
(B) $\sqrt{x} + \sqrt{x} = \sqrt{2x}$
(C) $x\sqrt{y} = y\sqrt{x}$
(D) $\sqrt{xy} = (\sqrt{x})(\sqrt{y})$
(E) $\sqrt{x+y} = \sqrt{x} + \sqrt{y}$

**10**  Given $y = wx^2$ and $y$ is not 0. If the values of $x$ and $w$ are each doubled, then the value of $y$ is multiplied by

(A) 1
(B) 2
(C) 4
(D) 6
(E) 8

**11** If $\sqrt{n}$ is a positive integer, how many values of $n$ are in the interval $100 < n < 199$?

(A) Three
(B) Four
(C) Five
(D) Six
(E) Seven

**12** If $(2^3)^2 = 4^p$, then $3^p =$

(A) 3
(B) 6
(C) 9
(D) 27
(E) 81

**13** If $y = 2$ and $z = \sqrt[3]{2+y+y^2}$, then $z^2 =$

(A) 1
(B) 4
(C) 9
(D) 16
(E) 25

**14** If $x$ is a positive integer, which of the following statements must be true?

I. $\left(\dfrac{x}{x}\right)^{99} = \left(\dfrac{x+1}{x+1}\right)^{100}$

II. $(x^x)^2 = x^{x^2}$

III. $\dfrac{x^{100}}{x^{99}} = 1^x$

(A) I only
(B) II only
(C) I and III
(D) II and III
(E) I, II, and III

**15** If $y = 25 - x^2$ and $1 \le x \le 5$, what is the smallest possible value of $y$?

(A) 0
(B) 1
(C) 5
(D) 10
(E) 15

## Quantitative Comparison

Each question consists of two quantities in boxes, one in Column A and one in Column B. You are to compare the two quantities and on the answer sheet fill in

A  if the quantity in Column A is greater;
B  if the quantity in Column B is greater;
C  if the two quantities are equal;
D  if the relationship cannot be determined from the information given

AN E RESPONSE WILL NOT BE SCORED

| | Column A | Column B | | Column A | Column B |
|---|---|---|---|---|---|
| **1** | $2^4$ | $4^2$ | **4** | $3n^2$ | $(3n)^2$ |
| **2** | $2^6 - 2^5$ | $2$ | **5** | $5 \times 6^{10}$ | $6 \times 5^{10}$ |
| **3** | $2\sqrt{3}$ | $3\sqrt{2}$ | **6** | $2\sqrt{75}$ | $3\sqrt{27}$ |

| | Column A | Column B | | Column A | Column B |
|---|---|---|---|---|---|
| **7** | $2^{10}$ | $2^9 + 2^9$ | | | |

$n$ is a positive integer.

| | Column A | Column B |
|---|---|---|
| **11** | $2n$ | $2^n$ |

| | Column A | Column B |
|---|---|---|
| **8** | $8^4$ | $2^{11}$ |

| | Column A | Column B |
|---|---|---|
| **12** | $\sqrt{160 \times 10^5}$ | $4 \times 10^3$ |

$$50 \times 40 \times 30 = 6 \times 10^N$$

| | Column A | Column B |
|---|---|---|
| **9** | $N$ | $3$ |

$$7^2 \times 7^w = 7^8$$
$$8^8 \div 8^z = 8^3$$
$w$ and $z$ are positive integers.

| | Column A | Column B |
|---|---|---|
| **13** | $w$ | $z$ |

$$9^4 = 3^n$$

| | Column A | Column B |
|---|---|---|
| **10** | $n$ | $6$ |

## Grid In

**1**  If $2^4 \times 4^2 = 16^x$, then $x =$

**3**  If $(y - 1)^3 = 8$, what is the value of $(y + 1)^2$?

**2**  If $3^{x-1} = 9$ and $4^{y+2} = 64$, what is the value of $\frac{x}{y}$?

**4**  If $\frac{p+p+p}{p \cdot p \cdot p} = 12$ and $p > 0$, what is the value of $p$?

# LESSON 3-3

# Divisibility and Factors

## OVERVIEW

*When 12 is divided by 3, the remainder is 0. Thus, 12 is **divisible by** 3, and, as a result, 3 is called a factor of 12. The numbers 1, 2, 4, 6, and 12 are also factors of 12 since 12 is divisible by each of these numbers.*

## THE PARTS OF A DIVISION EXAMPLE

The parts of a division example have special names. For example, $7 \div 3$ can be written in fractional form as $\frac{7}{3}$ or in long-division form as

$$
\begin{array}{r}
2 \quad \leftarrow \text{ Quotient} \\
\text{Divisor} \rightarrow \quad 3\overline{)\,7} \quad \leftarrow \text{ Dividend} \\
-\,6 \\
\hline
1 \quad \leftarrow \text{ Remainder}
\end{array}
$$

Thus, $7 \div 3 = \frac{7}{3} = 2 + \frac{1}{3}$. In general,

$$\frac{\text{Dividend}}{\text{Divisor}} = \text{Quotient} + \frac{\text{Remainder}}{\text{Divisor}}$$

## ODD AND EVEN NUMBERS

An **even number** is a number that is divisible by 2. The numbers 0, 2, 4, 6, and 8 are even numbers. An **odd number** is a number that is *not* divisible by 2. The numbers 1, 3, 5, 7, and 9 are odd numbers. Consecutive even numbers and consecutive odd numbers always differ by 2.

You can predict whether the sum or product of two numbers will be even or odd by using these rules:

$$
\begin{array}{lll}
\text{even} + \text{even} = \text{even} & \text{and} & \text{even} \times \text{even} = \text{even} \\
\text{odd} + \text{odd} = \text{even} & \text{and} & \text{odd} \times \text{odd} = \text{odd} \\
\text{odd} + \text{even} = \text{odd} & \text{and} & \text{even} \times \text{odd} = \text{even}
\end{array}
$$

## FACTORS AND MULTIPLES

Here are some terms you should know:

- The **factors** of a number $N$ are the numbers that can be divided into $N$ with a remainder of 0. For example, the factors of 32 are

$$1, 2, 4, 8, 16, \text{ and } 32$$

  since 32 is divisible by each of these numbers.
- The **common factors** of two numbers are the factors that the numbers have in common. For example:

Factors of 12:    1,    2,    3,    4,    6,    12

Factors of 21:    1,    3,    7,    21

  The common factors of 12 and 21 are 1 and 3.
- Any number that can be obtained by multiplying a number $N$ by a positive integer is called a **multiple** of $N$. For example, some multiples of 6 are 12, 18, 24, 30, and 36. The multiples of a number are also multiples of any factor of that number. Two factors of 6 are 2 and 3. The numbers 12, 18, 24, 30, and 36 are multiples of 6, and they are also multiples of 2 and 3.

## PRIME AND COMPOSITE NUMBERS

A positive integer greater than 1 is either prime or composite, but not both.

- A **prime number** is a number that has exactly two different factors, itself and 1. The set of prime numbers includes

$$2, 3, 5, 7, 11, 13, 17, 19, 23, \ldots$$

  The only even number that is prime is 2.
- A **composite number** is a number that has more than two different factors. For example:

$$4, 6, 8, 9, 10, 12, 14, 15, 16, \ldots,$$

  are composite numbers.
- The number 1 is neither prime or composite.

## PRIME FACTORIZATION

The process of breaking down a number into the product of two or more other numbers is called **factoring.** Factoring reverses multiplication:

$$\text{Multiplying:} \quad 7 \times 3 = 21$$

$$\text{Factoring:} \quad 21 = 7 \times 3$$

Since $21 = 7 \times 3$, 7 and 3 are factors of 21.

The **prime factorization** of a number breaks down the number into the product of prime numbers. For example:

- The prime factorization of 18 is

$$18 = 3 \times 3 \times 2$$

- The prime factorization of 30 is

$$30 = 5 \times 3 \times 2$$

# LESSON 3-3 TUNE-UP EXERCISES

## Multiple Choice

**1** Which number CANNOT be the remainder when an integer $N$ is divided by 4?

(A) 0
(B) 1
(C) 2
(D) 3
(E) 4

**2** When a whole number $N$ is divided by 5, the quotient is 13 and the remainder is 4. What is the value of $N$?

(A) 55
(B) 59
(C) 65
(D) 69
(E) 79

**3** If $p$ is divisible by 3 and $q$ is divisible by 4, then $pq$ must be divisible by each of the following EXCEPT

(A) 3
(B) 4
(C) 6
(D) 9
(E) 12

**4** Which number has the most factors?

(A) 12
(B) 18
(C) 25
(D) 70
(E) 100

**5** Which number is divisible by 2 and by 3?

(A) 112
(B) 4308
(C) 6122
(D) 701,456
(E) 23,451

**6** All numbers that are divisible by both 3 and 10 are also divisible by

(A) 4
(B) 9
(C) 15
(D) 12
(E) 20

**7** If $x$ represents any even number and $y$ represents any odd number, which of the following numbers is even?

(A) $y + 2$
(B) $x - 1$
(C) $(x + 1)(y - 1)$
(D) $y(y + 2)$
(E) $x + y$

**8** For how many different positive integers $p$ is $\frac{105}{p}$ greater than an integer?

(A) Five
(B) Six
(C) Seven
(D) Eight
(E) Nine

**9**  If $n$ is an odd integer, which expression always represents an odd integer?

(A) $(2n - 1)^2$

(B) $n^2 + 2n + 1$

(C) $(n - 1)^2$

(D) $\dfrac{n + 1}{2}$

(E) $3^n + 1$

**10**  If $k - 1$ is a multiple of 4, what is the next larger multiple of 4?

(A) $k + 1$

(B) $4k$

(C) $k - 5$

(D) $k + 3$

(E) $4(k - 1)$

**11**  After $m$ marbles are put into $n$ jars, each jar contains the same number of marbles, with two marbles remaining. In terms of $m$ and $n$, how many marbles were put into each jar?

(A) $\dfrac{m}{n} + 2$

(B) $\dfrac{m}{n} - 2$

(C) $\dfrac{m + 2}{n}$

(D) $\dfrac{m - 2}{n}$

(E) $\dfrac{mn}{n + 2}$

**12**  When $p$ is divided by 4, the remainder is 3; and when $p$ is divided by 3, the remainder is 0. What is a possible value of $p$?

(A) 8

(B) 11

(C) 15

(D) 18

(E) 21

**13**  If $n$ is an integer and $n^2 + 5$ is an odd integer, then which statement(s) must be true?

  I. $n^2 - 1$ is even.

 II. $n$ is even.

III. $5n$ is even.

(A) I only

(B) I and II only

(C) I and III only

(D) II and III only

(E) I, II, and III

**14**  When the number of people who contribute equally to a gift decreases from four to three, each person must pay an additional \$10. What is the cost of the gift?

(A) \$30

(B) \$60

(C) \$90

(D) \$120

(E) \$180

**15**  If $n$ is any even integer, what is the remainder when $(n + 1)^2$ is divided by 4?

(A) 0

(B) 1

(C) 2

(D) 3

(E) 4

**16**  A jar contains between 40 and 50 marbles. If the marbles are taken out of the jar three at a time, two marbles will be left in the jar. If the marbles are taken out of the jar five at a time, four marbles will be left in the jar. How many marbles are in the jar?

(A) 41

(B) 43

(C) 44

(D) 47

(E) 49

## Quantitative Comparison

Each question consists of two quantities in boxes, one in Column A and one in Column B. You are to compare the two quantities and on the answer sheet fill in

A if the quantity in Column A is greater;
B if the quantity in Column B is greater;
C if the two quantities are equal;
D if the relationship cannot be determined from the information given

AN E RESPONSE WILL NOT BE SCORED

| Column A | Column B | | Column A | Column B |
|---|---|---|---|---|

$N$ is a positive integer.

**1** | The remainder when $N$ is divided by 4 | The remainder when $N$ is divided by 8

**2** | The number of divisors of 21 | The number of divisors of 12

**3** | The number of prime numbers greater than 10 and less than 21 | The number of prime numbers less than 10

**4** | The number of common factors of 24 and 30 | The number of common factors of 12 and 18

**5** | The remainder when 416,879 is divided by 7 | The remainder when 793,804 is divided by 5

**6** | The remainder when $1 \times 2 \times 3 \times 4 \times 5$ is divided by 90 | The remainder when $5 \times 6 \times 7 \times 8 \times 9$ is divided by 90

When $n$ is divided by 5, the remainder is 2

**7** | The remainder when $n + 3$ is divided by 5 | The remainder when $n - 2$ is divided by 5

$k$ is a prime number.

**8** | The number of distinct prime factors of $12k$ | The number of distinct prime factors of $21k$

When $k$ is divided by 4, the remainder is 2

**9** | The remainder when $2k$ is divided by 4 | The remainder when $k + 2$ is divided by 4

---

**Grid In**

**1** When $a$ is divided by 7, the remainder is 5; and when $b$ is divided by 7, the remainder is 4. What is the remainder when $a + b$ is divided by 7?

**2** How many integers from −3000 to 3000, inclusive, are divisible by 3?

**3** When a positive integer $k$ is divided by 6, the remainder is 1. What is the remainder when $5k$ is divided by 3?

# Number Lines and Signed Numbers

## OVERVIEW

*This lesson reviews the rules for working with positive and negative numbers.*

## ORDERING NUMBERS ON A NUMBER LINE

The size order of real numbers can be pictured by drawing a ruler-like line called a **number line**. As shown in Figure 3.2, positive numbers are located in increasing size order to the right of 0, and their opposites appear in descending size order to the left of 0.

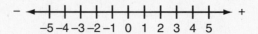

**Figure 3.2**  *The Real Number Line*

On the number line:

- Each number is *less than* the number that is located to its right. For example, $-3 < -2$.
- The larger of two negative numbers is the number that is closer to 0. For example, $-1 > -100$, since $-1$ is closer to 0 than is $-100$.
- If $a > b$, then $-b > -a$. For example, since $7 > 5$, $-5 > -7$.

## ABSOLUTE VALUE

The **absolute value** of a number $x$, denoted by $|x|$, is the distance on the number line from that number to 0. Since distance can never be negative, the absolute value of a number is always nonnegative.

- The absolute value of a positive number is that number. For example, $|2| = 2$.
- The absolute value of 0 is 0. For example, $|0| = 0$.
- the absolute value of a negative number is the opposite of that number. For example, $|-2| = 2$.

## MULTIPLYING AND DIVIDING SIGNED NUMBERS

The product or quotient of two numbers with the same sign is positive, and the product or quotient of two numbers with different signs is negative.

$$(+) \times (+) = + \qquad (+) \div (+) = +$$
$$(-) \times (-) = + \qquad (-) \div (-) = +$$
$$(+) \times (-) = - \qquad (+) \div (-) = -$$
$$(-) \times (+) = - \qquad (-) \div (+) = -$$

Thus, the square of either a positive or a negative number is positive. If $x^2 = 9$, $x$ may be equal to either 3 *or* –3 since

$$(3)^2 = (-3)^2 = 9$$

## MULTIPLYING MORE THAN TWO SIGNED NUMBERS

When more than two signed numbers are multiplied together, the sign of the product can be determined by using these rules:

- The product of an *even* number of negative factors is *positive*. For example:

$$(-2)^4 = \underbrace{(-2) \times (-2) \times (-2) \times (-2)}_{\text{4 factors}} = 16$$

- The product of an *odd* number of negative factors is *negative*. For example:

$$(-2)^3 = \underbrace{(-2) \times (-2) \times (-2)}_{\text{3 factors}} = -8$$

## ADDING AND SUBTRACTING SIGNED NUMBERS

The rules for adding and subtracting signed numbers are summarized in Table 3.3.

**TABLE 3.3**

**RULES FOR ADDING AND SUBTRACTING SIGNED NUMBERS**

| Operation | Sign of Numbers | Procedure |
|---|---|---|
| Addition | Same<br>(+) + (+) = +<br>(−) + (−) = − | Add numbers while ignoring their signs. Write the sum using the *common* sign.<br>(a) (+5) + (+8) = **+13**<br>(b) (−5) + (−8) = **−13** |
|  | Different | Subtract numbers while ignoring their signs. The answer has the same sign as the number having the *larger absolute value*.<br>(a) (+5) + (−8) = **−3**<br>(b) (−5) + (+8) = **+3** |
| Subtraction | Same | (a) (+5) − (+8) = (+5) + (−8) = **−3**<br><br>Take the opposite and *add*. |
|  | Different | (b) (+5) − (−8) = (+5) + (+8) = **+13**<br><br>Take the opposite and *add*. |

# LESSON 3-4   TUNE-UP EXERCISES

---

### Multiple Choice

---

**1**  If $2b = -3$, what is the value of $1 - 4b$?

(A) $-7$
(B) $-5$
(C) 5
(D) 6
(E) 7

---

**2**  If $a$ is a negative integer and $b$ is a positive integer, which of the following statements must be true?

I. $b + a > 0$

II. $\dfrac{b-a}{a} < 0$

III. $a^b < 0$

(A) None
(B) I only
(C) II only
(D) III only
(E) I and II only

---

**3**  What is the value of $\dfrac{(-2)^3 - (-4)^2}{(-1)(-2)(-3)}$

(A) $-4$

(B) $\dfrac{-4}{3}$

(C) $\dfrac{4}{3}$

(D) 4

(E) 6

---

**4**  For which value of $k$ is the value of $k(k-2)(k+1)$ negative?

(A) $-2$
(B) $-1$
(C) 0
(D) 2
(E) 3

---

**5**  $p^2(2-5) + (-p)^2 =$

(A) $-4p^2$
(B) $-p^2$
(C) $-2p^2$
(D) $2p^2$
(E) $4p^2$

---

**6**  If $n + 5$ is an odd integer, then $n$ could be which of the following?

(A) 1
(B) $-1$
(C) $-2$
(D) $-3$
(E) $-7$

---

**7**  Which of the following statements must be true when $a < 0$ and $b > 0$?

I. $a + b > 0$

II. $b - a > 0$

III. $a\left[\dfrac{a}{b}\right] > 0$

(A) I only
(B) II only
(C) I and III only
(D) II and III only
(E) I, II, and III

---

**8**  $(-2)^3 + (-3)^2 =$

(A) $-12$
(B) $-1$
(C) 0
(D) 1
(E) 2

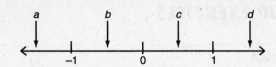

Questions 9 and 10 refer to the diagram above.

**9** Which of the following statements must be true?

I. $c^2 < c$
II. $a^2 > c$
III. $b < \dfrac{1}{b}$

(A) I only
(B) I and II only
(C) I and III only
(D) I, II, and III
(E) None

**10** Which of the following statements must be true?

I. $ad > b$
II. $ab > ad$
III. $\dfrac{1}{a} > \dfrac{1}{b}$

(A) II only
(B) I and II only
(C) II and III only
(D) I, II, and III
(E) None

**11** If $(y - 3)^2 = 16$, what is the smallest possible value of $y^2$?

(A) −4
(B) 1
(C) 7
(D) 16
(E) 49

**12** A set of five integers contains at least one positive integer. If the product of these five integers is negative, then, at most, how many of the five integers can be negative?

(A) None
(B) One
(C) Two
(D) Three
(E) Four

**13** If $a^2b^3c > 0$, which of the following statements must be true?

I. $bc > 0$
II. $ac > 0$
III. $ab > 0$

(A) I only
(B) I and II only
(C) I and III only
(D) II and III only
(E) I, II, and III

**14** If $a^2b^3c^5$ is negative, which product is always negative?

(A) $bc$
(B) $b^2c$
(C) $ac$
(D) $ab$
(E) $bc^2$

**15** If $a \neq 0$, which of the following statements must be true?

I. $(-a)^2 = -a^2$
II. $a - b = -(b - a)$
III. $a > -a$

(A) I only
(B) II only
(C) III only
(D) I and III only
(E) II and III only

## Quantitative Comparison

Each question consists of two quantities in boxes, one in Column A and one in Column B. You are to compare the two quantities and on the answer sheet fill in

   A  if the quantity in Column A is greater;
   B  if the quantity in Column B is greater;
   C  if the two quantities are equal;
   D  if the relationship cannot be determined from the information given

AN E RESPONSE WILL NOT BE SCORED

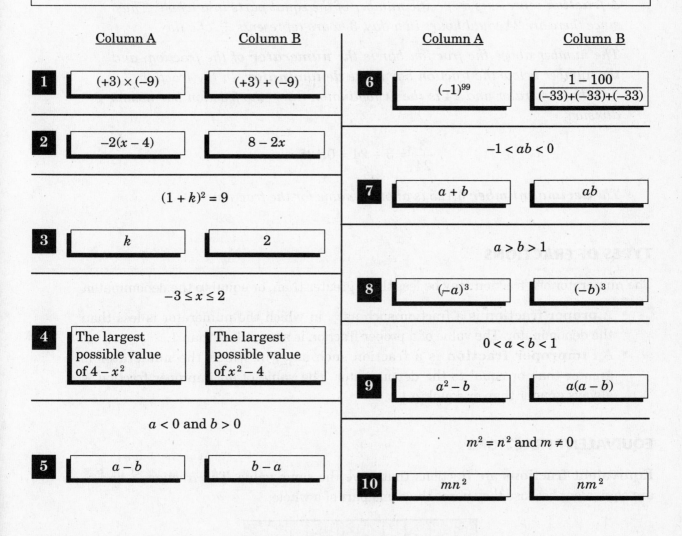

Column A     Column B        Column A     Column B

**1**   $(+3) \times (-9)$    $(+3) + (-9)$

**6**   $(-1)^{99}$    $\dfrac{1 - 100}{(-33)+(-33)+(-33)}$

**2**   $-2(x - 4)$    $8 - 2x$

$$-1 < ab < 0$$

**7**   $a + b$    $ab$

$$(1 + k)^2 = 9$$

**3**   $k$    $2$

$$a > b > 1$$

**8**   $(-a)^3$    $(-b)^3$

$$-3 \le x \le 2$$

**4**   The largest possible value of $4 - x^2$    The largest possible value of $x^2 - 4$

$$0 < a < b < 1$$

**9**   $a^2 - b$    $a(a - b)$

$$a < 0 \text{ and } b > 0$$

$$m^2 = n^2 \text{ and } m \ne 0$$

**5**   $a - b$    $b - a$

**10**   $mn^2$    $nm^2$

## Grid In

**1**   If $p^2 = 16$ and $q^2 = 36$, what is the largest possible value of $q - p$?

**2**   If $-4 \le x \le 2$ and $y = 1 - x^2$, what number is obtained when the *smallest* possible value of $y$ is subtracted from the *largest* possible value of $y$?

# Fractions and Decimals

### OVERVIEW

A **fraction** represents a specific number of the equal parts in a whole. Thus, since there are 24 equal hours in a day, 3 hours represents $\frac{3}{24}$ of a day.

The number above the fraction bar is the **numerator** of the fraction, and the number below the fraction bar is the **denominator**. In the fraction $\frac{3}{24}$, 3 is the numerator and 24 is the denominator. Since the fraction bar means division,

$$\frac{3}{24} = 3 \div 24 = 0.125$$

The **decimal number** 0.125 is another name for the fraction $\frac{3}{24}$.

## TYPES OF FRACTIONS

The numerator of a fraction may be less than, greater than, or equal to the denominator.

- A **proper fraction** is a fraction such as $\frac{2}{3}$, in which the numerator is less than the denominator. The value of a proper fraction is always less than 1.
- An **improper fraction** is a fraction such as $\frac{3}{2}$, in which the numerator is greater than or equal to the denominator. The value of an improper fraction is always greater than or equal to 1.

## EQUIVALENT FRACTIONS

**Equivalent fractions** are fractions that have the same value. The fractions $\frac{1}{2}$ and $\frac{5}{10}$ are equivalent because they name the same part of a whole:

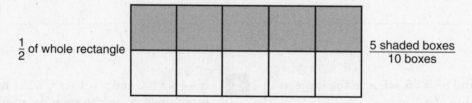

$\frac{1}{2}$ of whole rectangle      $\frac{5 \text{ shaded boxes}}{10 \text{ boxes}}$

Multiplying or dividing the numerator and denominator of a fraction by the same nonzero number always produces an equivalent fraction:

$$\frac{1}{2} = \frac{1 \times 5}{2 \times 5} = \frac{5}{10} \quad \text{and} \quad \frac{5}{10} = \frac{5 \div 5}{10 \div 5} = \frac{1}{2}$$

## REDUCING FRACTIONS TO LOWEST TERMS

A fraction is in **lowest terms** when its numerator and denominator do not have any common factors other than 1. To write a fraction in lowest terms, divide the numerator and denominator by their largest common factor.

**EXAMPLE:**      Write $\frac{16}{24}$ in lowest terms.

**SOLUTION:**    The largest number by which 16 and 24 are both divisible is 8.

$$\frac{16}{24} = \frac{16 \div 8}{24 \div 8} = \frac{2}{3}$$

You may find it easier to perform the division mentally by thinking "How many times does 8 go into 16?" and "How many times does 8 go into 24?" Write the answers above the numerator and denominator of the fraction as shown below:

$$\frac{\overset{2}{\cancel{16}}}{\underset{3}{\cancel{24}}} = \frac{2}{3}$$

## MIXED NUMBERS

A **mixed number** is a number such as $2\frac{3}{8}$, which represents the sum of a whole number and a proper fraction. Thus, $2\frac{3}{8}$ means $2 + \frac{3}{8}$.

- To change an improper fraction into a mixed number, divide the denominator of the improper fraction into the numerator. Write the quotient with the remainder expressed as a fraction in lowest terms. For example:

$$\frac{19}{7} = 19 \div 7 = \begin{array}{r} 2 \\ 7{\overline{)}\,19} \\ -14 \\ \hline 5 \end{array} \leftarrow \text{Remainder}$$

Since the quotient is 2 and the remainder is 5, $\frac{19}{7} = 2\frac{5}{7}$.

- To rewrite a mixed number such as $3\frac{5}{9}$ as an improper fraction, multiply 3 by 9, add the product to 5, and then write the sum over 9:

$$3\frac{5}{9} = \frac{(3 \times 9) + 5}{9} = \frac{27 + 5}{9} = \frac{32}{9}$$

## PLACE VALUE IN DECIMAL NUMBERS

The **place value** of each digit of a decimal number is 10 times as great as the place value of the digit to its right.

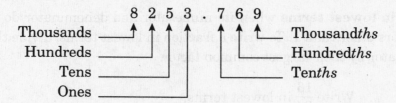

Thus, the decimal number 8253.769 is equivalent to

$$8253.769 = 8 \times 1000 + 2 \times 100 + 5 \times 10 + 3 \times 1 + 7 \times 0.1 + 6 \times 0.01 + 9 \times 0.001$$

## COMPARING DECIMALS

To compare two decimal numbers:

- Write one decimal underneath the other so that the decimal points and digits with the same place value are aligned.
- If one decimal is shorter than the other, add zeros to the right of the last digit of the shorter decimal until the decimal numbers have the same number of digits.
- Start from the left and compare the digits that are in the same column. Stop when you find unlike digits. The greater digit indicates the decimal with the greater value.

**EXAMPLE:**   In which of the following lists are the numbers written in order from least to greatest?
(A)  0.0361, 0.3061, 0.306
(B)  0.3061, 0.306, 0.0361
(C)  0.0361, 0.306, 0.3061
(D)  0.306, 0.3061, 0.0301
(E)  0.306, 0.0361, 0.3061

**SOLUTION:**   Write the three decimal numbers that are being compared, one underneath the other, so that digits with the same place value are aligned in the same vertical columns:

0.0361
0.3061
0.3060
⌣———— Add 0.

Then compare the digits in the same columns. The least decimal is 0.0361. Also, 0.306<u>0</u> < 0.306<u>1</u> since 0 < 1 in the last decimal position.

The correct arrangement, from least to greatest, of the decimal numbers is as follows:

$$0.0361, \quad 0.306, \quad 0.3061$$

Thus, the correct choice is (C).

## REWRITING FRACTIONS AS DECIMALS

To rewrite a fraction as a decimal number, use a calculator to divide the denominator of the fraction into the numerator of the fraction.

**EXAMPLE:** Rewrite $\dfrac{11}{40}$ as a decimal.

**SOLUTION:** Using a calculator, divide 40 into 11:

$$\frac{11}{40} = 11 \div 40 = 0.275$$

## COMPARING FRACTIONS

To compare two fractions, use these facts:

- If two fractions have the same denominator but different numerators, then the larger fraction is the fraction with the greater numerator. For example, $\frac{5}{9} > \frac{4}{9}$.
- If two fractions have the same numerator but different denominators, then the larger fraction is the fraction with the smaller denominator. For example, $\frac{4}{5} > \frac{4}{7}$.
- If two fractions have different numerators and different denominators, then the larger fraction can be determined by comparing the decimal values of the two fractions. For example, $\frac{11}{23} > \frac{7}{15}$ since

$$\frac{11}{23} = 0.47826\ldots \quad \text{and} \quad \frac{7}{15} = 0.46666\ldots$$

## MULTIPLYING AND DIVIDING DECIMALS BY POWERS OF 10

Multiplying or dividing a decimal number by a power of 10 (10, 100, 1000, etc.) affects only the position of the decimal point of the number.

- To multiply a decimal number by a power of 10, move the decimal point one place to the *right* for each 0 in the power of 10.
- To divide a decimal number by a power of 10, move the decimal point one place to the *left* for each 0 in the power of 10.

**EXAMPLE:** Multiply 6.25 by 1000.

**SOLUTION:** Since there are three 0s in 1000, move the decimal point in 6.25 three places to the *right*:

$$6.25 \times 1000 = 6\underset{\frown}{.250} \times 1000 = 6250$$

**EXAMPLE:**   Divide 8.24 by 100.

**SOLUTION:**   Since there are two 0s in 100, move the decimal point two places to the *left*.

$$8.24 \div 100 = \underset{\smile}{08}.24 \div 100 = 0.0824$$

# LESSON 3-5  TUNE-UP EXERCISES

| **Multiple Choice** |
|---|

**1**  What part of an hour elapses from 4:56 P.M. to 5:32 P.M.?

(A) $\dfrac{1}{4}$

(B) $\dfrac{1}{2}$

(C) $\dfrac{2}{3}$

(D) $\dfrac{3}{5}$

(E) $\dfrac{3}{4}$

**2**  If each of the fractions $\dfrac{3}{k}$, $\dfrac{4}{k}$, $\dfrac{5}{k}$ is in lowest terms, which of the following could be the value of $k$?

(A) 48
(B) 49
(C) 50
(D) 51
(E) 52

**3**  Which number has the greatest value?

(A) 0.2093
(B) 0.2908
(C) 0.2893
(D) 0.2938
(E) 0.2909

**4**  The elapsed time from 11:00 A.M. to 3:00 P.M. on Wednesday of the same day is what fraction of the elapsed time from 11:00 A.M. on Wednesday to 3:00 P.M. on Friday of the same week?

(A) $\dfrac{1}{15}$

(B) $\dfrac{1}{14}$

(C) $\dfrac{4}{51}$

(D) $\dfrac{1}{13}$

(E) $\dfrac{1}{12}$

**5**  In which number is the digit 3 in the hundredths place?

(A) 300.000
(B)  30.000
(C)   0.300
(D)   0.030
(E)   0.003

**6**  In which arrangement are the fractions listed from least to greatest?

(A) $\dfrac{9}{19}, \dfrac{1}{2}, \dfrac{8}{15}$

(B) $\dfrac{1}{2}, \dfrac{8}{15}, \dfrac{9}{19}$

(C) $\dfrac{9}{19}, \dfrac{8}{15}, \dfrac{1}{2}$

(D) $\dfrac{1}{2}, \dfrac{9}{19}, \dfrac{8}{15}$

(E) $\dfrac{8}{15}, \dfrac{1}{2}, \dfrac{9}{19}$

**7**  After the formula $V = \dfrac{4}{3}\pi r^3$ has been evaluated for some positive value of $r$, the formula is again evaluated using one-half of the original value of $r$. The new value of $V$ is what fractional part of the original value of $V$?

(A) $\dfrac{1}{16}$

(B) $\dfrac{1}{9}$

(C) $\dfrac{1}{8}$

(D) $\dfrac{1}{4}$

(E) $\dfrac{1}{2}$

**8** Each inch on ruler $A$ is marked in equal $\frac{1}{8}$-inch units, and each inch on ruler $B$ is marked in equal $\frac{1}{12}$-inch units. When ruler $A$ is used, a side of a triangle measures 12 of the $\frac{1}{8}$-inch units. When ruler $B$ is used, how many $\frac{1}{12}$-inch units will the same side measure?

(A) 8
(B) 12
(C) 18
(D) 20
(E) 24

**9** $60 + 2 + \frac{4}{8} + \frac{3}{500} =$

(A) 60.256
(B) 62.43
(C) 62.506
(D) 62.53
(E) 62.560

**10** If $N \times \frac{7}{12} = \frac{7}{12} \times \frac{3}{14}$, then $\frac{1}{N} =$

(A) 8

(B) $\frac{14}{3}$

(C) $\frac{12}{7}$

(D) $\frac{1}{6}$

(E) $\frac{3}{14}$

**11** If $n = 2.5 \times 10^{25}$, then $\sqrt{n} =$

(A) $0.5 \times 10^5$
(B) $0.5 \times 10^{12}$
(C) $5 \times 10^5$
(D) $5 \times 10^{\sqrt{24}}$
(E) $5 \times 10^{12}$

**12** If $y$ is a real number and $y = \frac{x-2}{x+3}$, then $x$ CANNOT equal which of the following numbers?

(A) −3
(B) −2
(C) 0
(D) 2
(E) 3

**13** If eight pencils cost $0.42, how many pencils can be purchased with $2.10?

(A) 16
(B) 24
(C) 30
(D) 36
(E) 40

**14** A store sells 8-ounce containers of orange juice at $0.69 each and 12-ounce containers of orange juice at $0.95 each. How much money will be saved by purchasing a total of 48 ounces of orange juice in 12-ounce rather than 8-ounce containers?

(A) $0.24
(B) $0.32
(C) $0.34
(D) $0.48
(E) $0.56

**15** If $b \neq 0$, then $\frac{(-2b)^3}{-8b^3} =$

(A) −1

(B) $-\frac{1}{4}$

(C) $\frac{1}{4}$

(D) $\frac{3}{4}$

(E) 1

## Quantitative Comparison

Each question consists of two quantities in boxes, one in Column A and one in Column B. You are to compare the two quantities and on the answer sheet fill in

A  if the quantity in Column A is greater;
B  if the quantity in Column B is greater;
C  if the two quantities are equal;
D  if the relationship cannot be determined from the information given

AN E RESPONSE WILL NOT BE SCORED

| Column A | Column B |
|---|---|
| **1** 0.4648 rounded to the nearest hundredth | 0.4551 rounded to the nearest hundredth |
| **2** $4\frac{3}{7}$ | $\frac{30}{7}$ |
| **3** $\frac{63}{18}$ | $\frac{7}{2}$ |
| **4** $\frac{11}{15}$ | $\frac{5}{7}$ |
| **5** $\left(\frac{1}{2}\right)^3$ | $\left(\frac{1}{3}\right)^2$ |

| Column A | Column B |
|---|---|
| **6** The number of whole numbers between $\frac{23}{5}$ and $\frac{99}{5}$ | The number of whole numbers between $\frac{11}{5}$ and $\frac{59}{3}$ |

$x$ is a positive integer.

| Column A | Column B |
|---|---|
| **7** $\frac{x+1}{x+2}$ | $\frac{1}{2}$ |

$k$ is a positive integer.

| Column A | Column B |
|---|---|
| **8** $3.12k \times 10^3$ | $0.00312k \times 10^7$ |

## Grid In

**1** If $\frac{4\Delta}{6\Delta} + \frac{5}{17} = 1$, what digit does $\Delta$ represent?

**2** On a certain map, 1.5 inches represents a distance of 45 miles. If two points on the map are 0.8 inch apart, how many miles apart are these two points?

**3** Four lemons cost $0.68. At the same rate, 1 pound of lemons costs $1.19. How many lemons typically weigh 1 pound?

**4** If the charge for a taxi ride is $2.50 for the first $\frac{1}{2}$ mile and $0.75 for each additional $\frac{1}{8}$ mile, how many miles did the taxi travel for a ride that cost $10.75?

**5** One cubic foot of a certain metal weighs 8 pounds and costs $4.20 per pound. If 1 cubic foot is equivalent to 1728 cubic inches, what is the cost of 288 cubic inches of the same metal?

# Operations with Fractions

## OVERVIEW

*Fractions are multiplied, divided, and combined (added or subtracted) according to the following rules, where a, b, c, and d stand for real numbers with b, d ≠ 0:*

- $\dfrac{a}{b} \times \dfrac{c}{d} = \dfrac{a \times c}{b \times d}$

- $\dfrac{a}{b} \pm \dfrac{c}{b} = \dfrac{a \pm c}{b}$

- $\dfrac{a}{b} \div \dfrac{c}{d} = \dfrac{a}{b} \times \dfrac{d}{c} = \dfrac{a \times d}{b \times c}$

- $\dfrac{a}{b} \pm \dfrac{c}{d} = \dfrac{ad \pm bc}{bd}$

## MULTIPLYING FRACTIONS

To multiply fractions, write the product of the numerators over the product of the denominators. Then write the result in lowest terms. For example:

$$\frac{4}{9} \times \frac{5}{8} = \frac{20}{72} = \frac{20 \div 4}{72 \div 4} = \frac{5}{18}$$

Sometimes it is easier to divide out pairs of common factors of any numerator and denominator *before* multiplying. For example:

$$\frac{\overset{2}{\cancel{8}}}{\underset{3}{\cancel{21}}} \times \frac{\overset{1}{\cancel{7}}}{\underset{5}{\cancel{20}}} = \frac{2}{15}$$

## FINDING A FRACTION OF A NUMBER

To find a fraction of another number, replace the word *of* with the times symbol. Then multiply.

**EXAMPLE:**     What is $\dfrac{3}{5}$ of 20?

**SOLUTION:**  $\frac{3}{5}$ of $20 = \frac{3}{5} \times 20 = 3 \times 4 = 12$

**EXAMPLE:**  After John spends $\frac{1}{4}$ of his salary on food and $\frac{2}{3}$ of what remains on clothes, what part of his original salary does he have left?

**SOLUTION 1:**

- After John spends $\frac{1}{4}$ of his original salary on food, $\frac{3}{4}\left(=1-\frac{1}{4}\right)$ of his salary remains.
- John spends $\frac{2}{3}$ of what remains on clothes. Since

$$\frac{2}{3} \text{ of } \frac{3}{4} = \frac{2}{3} \times \frac{3}{4} = \frac{1}{2}$$

John spends $\frac{1}{2}$ of his original salary on clothes.
- After John spends $\frac{1}{4}$ of his original salary on food and $\frac{1}{2}$ of his original salary on clothes, he has $\frac{1}{4}\left(=1-\frac{1}{4}-\frac{1}{2}\right)$ of his original salary left.

**Solution 2:**  Pick a convenient amount as John's original salary. Choose a number that is divisible by 4, for example, $100.

- John spends $\frac{1}{4}$ of his original salary on food. Since $\frac{1}{4} \times \$100 = \$25$, he spends $25 on food, so he has $75 left.
- John spends $\frac{2}{3}$ of $75 on clothes. Since $\frac{2}{3} \times \$75 = \$50$, he spends $50 on clothes, so $25 remains from his original salary.
- Since $\frac{\$25}{\$100} = \frac{1}{4}$, John has $\frac{1}{4}$ of his original salary left.

## DIVIDING FRACTIONS

To divide a fraction by another fraction, invert the second fraction by interchanging its numerator and denominator. Then multiply. For example:

$$\frac{12}{35} \div \frac{3}{14} = \frac{\overset{4}{\cancel{12}}}{\underset{5}{\cancel{35}}} \times \frac{\overset{2}{\cancel{14}}}{\underset{1}{\cancel{3}}} = \frac{4 \times 2}{5 \times 1} = \frac{8}{5}$$

**EXAMPLE:**  Divide: $3 \div \dfrac{12}{7}$.

**SOLUTION:**  $3 \div \dfrac{12}{7} = \dfrac{3}{1} \times \dfrac{7}{12} = \dfrac{7}{4}$

## RECIPROCALS

The **reciprocal** of any nonzero number $x$ is $\frac{1}{x}$. For example, the reciprocal of 3 is $\frac{1}{3}$. The reciprocal of any nonzero fraction $\frac{a}{b}$ ($b \neq 0$) is $\frac{b}{a}$. For example, the reciprocal of $\frac{2}{5}$ is $\frac{5}{2}$.

The product of any nonzero number and its reciprocal is 1.

## POWERS, SQUARE ROOTS, AND RECIPROCALS OF PROPER FRACTIONS

If a number is between 0 and 1, then:

- The square of the number is *less than* the original number. For example:

$$\left(\frac{1}{3}\right)^2 < \frac{1}{3} \text{ since } \frac{1}{9} < \frac{1}{3}$$

  In general, as the exponent of a proper fraction increases, the value of that expression decreases.
- The square root of the number is *greater than* the original number;

$$\sqrt{0.25} > 0.25 \text{ since } \sqrt{0.25} = 0.5 \text{ and } 0.5 > 0.25$$

- The reciprocal of the number is *greater than* the original number. For example, the reciprocal of $\frac{1}{2}$ is 2, which is greater than $\frac{1}{2}$.

In each of the above cases, taking powers, roots, or reciprocals of numbers that are *greater than* 1 gives results that are exactly opposite to those obtained here.

## ADDING AND SUBTRACTING FRACTIONS

To add or subtract fractions with like denominators:

- Write the sum or difference of the numerators over the common denominator.
- Express the result in lowest terms, if possible.

**EXAMPLE:**  Add: $\frac{7}{12} + \frac{1}{12}$.

**SOLUTION:**  $\frac{7}{12} + \frac{1}{12} = \frac{7+1}{12} = \frac{8}{12} = \frac{2}{3}$

**EXAMPLE:**  Write $\frac{3x}{8} - \frac{x}{8}$ as a single fraction.

**SOLUTION:**  $\frac{3x}{8} - \frac{x}{8} = \frac{3x-x}{8}$

$$= \frac{2x}{8}$$

$$= \frac{x}{4}$$

To combine fractions with unlike denominators:

- Find the *Least Common Denominator* (LCD) of the fractions. For example, the LCD of $\frac{3}{8}$ and $\frac{5}{12}$ is 24 since 24 is the smallest number that is divisible by both 8 and 12.
- Change each fraction into an equivalent fraction that has the LCD as its denominator.
- Write the sum or difference of the numerators over the common denominator.
- Express the result in lowest terms, if possible.

**EXAMPLE:**    Add $\frac{1}{3} + \frac{2}{5}$.

**SOLUTION:**    The LCD is 15. Change each fraction into an equivalent fraction that has 15 as its denominator. Multiply the first fraction by 1 in the form of $\frac{5}{5}$, and multiply the second fraction by 1 in the form of $\frac{3}{3}$.

$$\frac{1}{3} + \frac{2}{5} = \frac{5}{5}\left(\frac{1}{3}\right) + \frac{3}{3}\left(\frac{2}{5}\right)$$

$$= \frac{5}{15} + \frac{6}{15}$$

$$= \frac{11}{15}$$

## FRACTIONS WITH RADICALS IN THEIR DENOMINATORS

To eliminate a radical of the form $\sqrt{b}$ from the denominator of a fraction, multiply the numerator and denominator of the fraction by $\sqrt{b}$. For example:

$$\frac{6}{\sqrt{3}} = \frac{6}{\sqrt{3}} \cdot \frac{\sqrt{3}}{\sqrt{3}} = \frac{6\sqrt{3}}{3} = 2\sqrt{3}$$

## OPERATIONS WITH MIXED NUMBERS

Mixed numbers have a whole number and a fractional part.

- To multiply or divide mixed numbers, first change the mixed numbers to improper fractions. Then follow the rules for multiplying and dividing fractions.

**EXAMPLE:**    Multiply: $1\frac{7}{8} \times 3\frac{1}{3}$.

**SOLUTION:**    $1\frac{7}{8} \times 3\frac{1}{3} = \frac{15}{8} \times \frac{10}{3} = \frac{25}{4}$

**EXAMPLE:**        Divide: $5\frac{1}{2} \div 1\frac{3}{8}$.

**SOLUTION:**   $5\frac{1}{2} \div 1\frac{3}{8} = \frac{11}{2} \div \frac{11}{8} = \frac{11}{2} \times \frac{8}{11} = 4$

- To add or subtract mixed numbers, write the second mixed number under the first mixed number. If the denominators of the fractions are different, find the LCD and change the fractions into equivalent fractions with the LCD as denominator. Add or subtract the fractions. Then add or subtract the whole numbers.

**EXAMPLE:**        Add: $2\frac{1}{4} + 5\frac{2}{3}$.

**SOLUTION:**

$$2\frac{1}{4} = 2\frac{3}{12}$$

$$+ \ 5\frac{2}{3} = 5\frac{8}{12}$$

$$= 7\frac{11}{12}$$

**EXAMPLE:**        Subtract: $6\frac{2}{9} - 4\frac{5}{9}$.

**SOLUTION:**

$$6\frac{2}{9} = 5\frac{9+2}{9} = 5\frac{11}{9}$$

$$- \ 4\frac{4}{9} \qquad\quad = -4\frac{4}{9}$$

$$= 1\frac{7}{9}$$

## SIMPLIFYING COMPLEX FRACTIONS

A fraction that has another fraction in its numerator or denominator is called a **complex fraction**. To simplify a complex fraction:

- Find the LCD of the denominators of the fractions that are contained in the numerator or denominator of the complex fraction.
- Multiply the numerator and denominator of the complex fraction by the LCD.
- Simplify the resulting fraction.

**EXAMPLE:**        Simplify: $\dfrac{4}{\dfrac{3}{2}}$.

**SOLUTION 1:** Multiply the numerator and the denominator by 2:

$$\frac{2}{2} \times \frac{4}{\frac{3}{2}} = \frac{2 \times 4}{2 \times \frac{3}{2}} = \frac{8}{3}$$

**SOLUTION 2:** Divide the numerator by the denominator:

$$\frac{4}{\frac{3}{2}} = 4 \div \frac{3}{2} = 4 \times \frac{2}{3} = \frac{8}{3}$$

## LESSON 3-6   TUNE-UP EXERCISES

---

### Multiple Choice

---

**1** John completes a race in $9\frac{1}{3}$ minutes, and Steve finishes the same race in $7\frac{3}{4}$ minutes. How many seconds after Steve finishes the race does John complete the race?

(A) 72
(B) 90
(C) 92
(D) 95
(E) 100

**2** Multiplying a number by $\frac{1}{2}$ and then dividing the result by $\frac{3}{4}$ is equivalent to performing which of the following operations on the number?

(A) Multiplying by $\frac{2}{3}$
(B) Dividing by $\frac{2}{3}$
(C) Multiplying by $\frac{3}{8}$
(D) Dividing by $\frac{3}{8}$
(E) Multiplying by 3

**3** $\left[\frac{1}{10}\right]^3 + \left[\frac{2}{10}\right]^3 + \left[\frac{3}{10}\right]^3 =$

(A) 0.018
(B) 0.025
(C) 0.032
(D) 0.036
(E) 0.040

**4** What is $\frac{3}{4}$ of $\frac{2}{3}$ of 12?

(A) 2
(B) 3
(C) 4
(D) 6
(E) 8

**5** What is $\frac{2}{21}$ of $3 \times 5 \times 7$?

(A) 6
(B) 9
(C) 10
(D) 12
(E) 15

**6** For which value of $n$ is $\frac{1}{4} < n < \frac{1}{3}$ ?

(A) $\frac{5}{24}$
(B) $\frac{6}{24}$
(C) $\frac{7}{24}$
(D) $\frac{8}{24}$
(E) $\frac{9}{24}$

**7** If $p = \dfrac{\frac{4}{\sqrt{3}} - \frac{1}{\sqrt{3}}}{\sqrt{3}}$, what is the value of $p^2$?

(A) 1
(B) 3
(C) 4
(D) 9
(E) 16

**8** If $k = \dfrac{\frac{c}{a}}{b}$, which expression equals $\frac{1}{k}$ ?

(A) $\frac{ac}{b}$
(B) $\frac{b}{ac}$
(C) $\frac{a}{bc}$
(D) $\frac{bc}{a}$
(E) $\frac{1}{abc}$

**9** An item is on sale on Monday at $\frac{1}{3}$ off the list price of $d$ dollars. On Friday, the sale price of the item is reduced by $\frac{1}{4}$ of the current price. What is the number of dollars in the final sale price of the item?

(A) $\frac{d}{2}$

(B) $\frac{d}{3}$

(C) $\frac{d}{4}$

(D) $\frac{d}{6}$

(E) $\frac{d}{8}$

**10** If $\frac{a}{b}$ is a fraction less than 1, which of the following fractions MUST be greater than 1?

(A) $\left[\frac{a}{b}\right]^2$

(B) $2\left[\frac{a}{b}\right]$

(C) $\frac{a}{2b}$

(D) $\frac{a+2}{b+2}$

(E) $\frac{2}{\frac{a}{b}}$

**11** Juan gives $\frac{2}{3}$ of his $p$ pencils to Roger and then gives $\frac{1}{4}$ of the pencils that he has left to Maria. In terms of $p$, how many pencils does Juan now have?

(A) $\frac{1}{6}p$

(B) $\frac{1}{4}p$

(C) $\frac{1}{3}p$

(D) $\frac{1}{2}p$

(E) $\frac{5}{6}p$

**12** The statement $a^3 < a^2 < a$ is true when

I. $a > 1$
II. $0 < a < 1$
III. $-1 < a < 0$

(A) I and III only
(B) II and III only
(C) II only
(D) III only
(E) None

## Quantitative Comparison

Each question consists of two quantities in boxes, one in Column A and one in Column B. You are to compare the two quantities and on the answer sheet fill in

A  if the quantity in Column A is greater;
B  if the quantity in Column B is greater;
C  if the two quantities are equal;
D  if the relationship cannot be determined from the information given

AN E RESPONSE WILL NOT BE SCORED

| Column A | Column B | | Column A | Column B |
|---|---|---|---|---|
| | | | | |

**1**

| $2x$ divided by $\frac{1}{4}$ | $x$ multiplied by 8 |
|---|---|

**2**

| $\frac{7}{12}$ of an hour | Time that elapses from 1:54 P.M. to 2:28 P.M. |
|---|---|

| Column A | Column B | Column A | Column B |
|---|---|---|---|

**3** | $\dfrac{1}{\frac{4}{4}}$ | $\dfrac{4}{\frac{1}{4}}$

$$0 < x < 1$$

**11** | $\dfrac{1}{\sqrt{x}}$ | $x$

**4** | $\dfrac{3}{8}$ of $\dfrac{8}{3}$ | $\dfrac{719}{720}$

$$-1 < b < 0$$

**12** | $b$ | $\dfrac{1}{b}$

**5** | $\dfrac{3}{9} + \dfrac{4}{12} + \dfrac{5}{15}$ | $1$

$$a + \dfrac{1}{5} = b$$

**13** | $a - 1$ | $b - 1$

$$a = \dfrac{7}{10} \text{ and } b = 1\dfrac{3}{4}$$

**6** | $\dfrac{a}{b}$ | $\dfrac{2}{5}$

$$0 < x < 1$$

**14** | $1 - x^2$ | $1 - x^3$

$$\dfrac{1}{n} > 1$$

**7** | $n$ | $\sqrt{n}$

### Questions 15 and 16.

$A$ is $\dfrac{2}{3}$ of $B$.

$B$ is $\dfrac{3}{4}$ of $C$.

$A, B, C > 0$.

$$x > 0$$

**8** | $\dfrac{x}{0.2}$ | $4x$

**15** | $\dfrac{A}{C}$ | $\dfrac{C}{A}$

$$0 < a < 1$$

**9** | $(a^3)^2$ | $a^5$

**16** | $8B$ | $6A + 3C$

**10** | $\dfrac{1}{1 - \left(\frac{1}{2}\right)^2}$ | $\dfrac{4}{3}$

---

| **Grid In** |
| :---: |

**1** If $y = \sqrt{\dfrac{x+4}{2}}$, what is the value of $y$ when $x = \dfrac{1}{2}$ ?

**2** What fraction of $\dfrac{10}{9}$ is $\dfrac{5}{6}$ ?

**3** If $25\left(\dfrac{x}{y}\right) = 4$, what is the value of $100\left(\dfrac{y}{x}\right)$ ?

# Fraction Word Problems

## OVERVIEW

*Many types of fraction problems can be solved either by:*

- *forming a fraction in which the number in the numerator is being compared to the number in the denominator; or*
- *finding the whole when a fractional part of it is given.*

## FINDING WHAT FRACTION ONE NUMBER IS OF ANOTHER

To find what fraction a given number "$\frac{is}{of}$" another number, form the fraction

$$\frac{\text{PART } is}{of \text{ WHOLE}}$$

In the fraction above, the "part" is given number and the "whole" is the number to which the part is being compared.

**EXAMPLE:** There are 15 boys and 12 girls in a class. The number of boys is what fraction of the total number of students in the class?

**SOLUTION:** Since the number of boys is being compared to the total number of students in the class, form the fraction in which the number of boys is the numerator and the total number of students in the class, boys plus girls, is the denominator:

$$\frac{\text{boys}}{\text{whole class}} = \frac{15}{15 + 12} = \frac{15}{27} = \frac{5}{9}$$

## FINDING THE WHOLE WHEN A PART OF IT IS KNOWN

If $\frac{4}{7}$ of some unknown number is 12, then multiplying 12 (the "part") by the reciprocal of the given fraction gives the unknown number (the "whole"). Thus,

$$\text{whole} = \overset{3}{\cancel{12}} \times \frac{7}{\underset{1}{\cancel{4}}} = 3 \times 7 = 21$$

You can also find the unknown number by following these steps:

- Find $\frac{1}{7}$ of the number. Since $\frac{4}{7}$ of the unknown number is 12, $\frac{1}{7}$ of the unknown number is $\frac{1}{4}$ of 12, which is 3.
- Multiply to find the whole number. Since $\frac{1}{7}$ of the unknown number is 3, $\frac{7}{7}$ of the unknown number is $7 \times 3 = 21$.

Hence, the unknown number is 21.

The above procedure can be quickly applied using the following shorthand notation:

$$\frac{4}{7} \Rightarrow 12 \qquad\qquad \text{[You are told that } \frac{4}{7} \text{ of the unknown number is 12.]}$$

$$\frac{1}{7} \Rightarrow \frac{1}{4}(12) = 3 \qquad [\, \frac{1}{7} \text{ of the unknown number is 3.]}$$

$$\frac{7}{7} \Rightarrow 7(3) = 21 \qquad [\, \frac{7}{7} \text{ of the unknown number is 21.]}$$

**EXAMPLE:**  Of the books that are on a shelf, $\frac{1}{3}$ are math books, $\frac{1}{4}$ are science books, and the remaining books are history books. If the shelf contains 10 history books, how many books are on the shelf?

**SOLUTION 1:**

- Math and science books account for $\frac{7}{12}$ of the books on the shelf since

$$\frac{1}{3} + \frac{1}{4} = \frac{4}{12} + \frac{3}{12} = \frac{7}{12}$$

- History books make up $\frac{5}{12}\left(= 1 - \frac{7}{12}\right)$ of the total number of books on the shelf.
- Since there are 10 history books, the problem reduces to answering the question "$\frac{5}{12}$ of what number is 10?"

$$\frac{5}{12} \Rightarrow 10 \qquad\qquad [\, \frac{5}{12} \text{ of the unknown number is 10.]}$$

$$\frac{1}{12} \Rightarrow \frac{1}{5}(10) = 2 \qquad [\, \frac{1}{12} \text{ of the unknown number is 2.]}$$

$$\frac{12}{12} \Rightarrow 12(2) = 24 \qquad [\, \frac{12}{12} \text{ or the whole number is 24.]}$$

The shelf contains a total of 24 books.

**SOLUTION 2:**

- Use algebra by letting $x$ represent the total number of books on the shelf.
- Write an equation that states that the sum of the numbers of math, science, and history books is $x$:

$$\frac{x}{3} + \frac{x}{4} + 10 = x$$

- To eliminate the fractions from this equation, multiply each member of the equation by 12, which is the LCD of 3 and 4:

$$12\left[\frac{x}{3}\right] + 12\left[\frac{x}{4}\right] + 12(10) = 12x$$

$$4x \quad + \quad 3x \quad + \quad 120 \ = 12x$$

$$120 = 12x - 7x$$

$$120 = 5x$$

$$\frac{120}{5} = x$$

$$24 = x$$

# LESSON 3-7  TUNE-UP EXERCISES

| Multiple Choice |
| --- |

**1** If $\frac{3}{8}$ of a number is 6, what is $\frac{7}{8}$ of the same number?

(A) 8
(B) 12
(C) 14
(D) 16
(E) 24

**2** After Claire has read the first $\frac{5}{8}$ of a book, there are 120 pages left to read. How many pages of the book has Claire read?

(A) 160
(B) 200
(C) 240
(D) 300
(E) 320

**3** $\frac{4}{9}$ of 27 is $\frac{6}{5}$ of what number?

(A) 5
(B) 10
(C) 12
(D) 15
(E) 20

**4** If $\frac{3}{5}$ of a class that includes 10 girls are boys, how many students are in the class?

(A) 15
(B) 20
(C) 21
(D) 25
(E) 30

**5** What is the sum of all two-digit whole numbers in which one digit is $\frac{3}{4}$ of the other digit?

(A) 77
(B) 102
(C) 129
(D) 154
(E) 231

**6** Boris has $D$ dollars. If he lends $\frac{1}{4}$ of this amount of money and then spends $\frac{1}{3}$ of the money that he has left, how many dollars in terms of $D$, does Boris now have?

(A) $\frac{D}{2}$

(B) $\frac{D}{3}$

(C) $\frac{D}{4} - \frac{1}{3}$

(D) $\frac{3}{4}D - \frac{1}{3}$

(E) $\frac{11}{12}D$

**7** The value obtained by increasing $a$ by $\frac{1}{5}$ of its value is numerically equal to the value obtained by decreasing $b$ by $\frac{1}{2}$ of its value. Which equation expresses this fact?

(A) $1.2a = 0.5b$
(B) $0.2a = 0.5b$
(C) $0.8a = 1.5b$
(D) $1.2a = 1.5b$
(E) $a + 0.2 = b - 0.5$

**8** How many times must a jogger run around a circular $\frac{1}{4}$-mile track in order to have run $3\frac{1}{2}$ miles?

(A) 9
(B) 12
(C) 14
(D) 15
(E) 16

**9** A chocolate bar that weighs $\frac{9}{16}$ of a pound is cut into seven equal parts. How much do three parts weigh?

(A) $\frac{21}{112}$ pound

(B) $\frac{27}{112}$ pound

(C) $\frac{16}{63}$ pound

(D) $\frac{47}{63}$ pound

(E) $\frac{85}{112}$ pound

**10** At a high school basketball game, $\frac{3}{5}$ of the students who attended were seniors, $\frac{1}{3}$ of the other students who attended were juniors, and the remaining 80 students who attended were all sophomores. How many seniors attended this game?

(A) 120
(B) 175
(C) 180
(D) 210
(E) 300

**11** Of the 75 people in a room, $\frac{2}{5}$ are college graduates. If $\frac{4}{9}$ of the students who are not college graduates are seniors in high school, how many people in the room are neither college graduates nor high school seniors?

(A) 15
(B) 20
(C) 25
(D) 36
(E) 40

**12** If $\frac{2}{3}$ of $\frac{3}{4}$ of a number is 24, what is $\frac{1}{4}$ of the same number?

(A) 8
(B) 12
(C) 16
(D) 20
(E) 24

**13** A man paints $\frac{3}{4}$ of a house in 2 days. If he continues to work at the same rate, how much more time will he need to paint the rest of the house?

(A) $\frac{1}{4}$ day

(B) $\frac{1}{2}$ day

(C) $\frac{2}{3}$ day

(D) 1 day

(E) $\frac{4}{3}$ days

**14** A water tank is $\frac{3}{5}$ full. After 12 gallons are poured out, the tank is $\frac{1}{3}$ full. When the tank is full, how many gallons of water does it hold?

(A) 25
(B) 32
(C) 35
(D) 42
(E) 45

**15** In a school election, Susan received $\frac{2}{3}$ of the ballots cast, Mary received $\frac{1}{5}$ of the remaining ballots, and Bill received all of the other votes. If Bill received 48 votes, how many votes did Susan receive?

(A) 75
(B) 90
(C) 120
(D) 150
(E) 180

**16** After Arlene pumps gas into the gas tank of her car, the gas gauge moves from exactly $\frac{1}{8}$ full to exactly $\frac{7}{8}$ full. If the gas costs $1.50 per gallon and Arlene is charged $18.00 for the gas, what is the capacity, in gallons, of the gas tank?

(A) 24
(B) 20
(C) 18
(D) 16
(E) 15

**17** At the beginning of the day, the prices of stocks $A$ and $B$ are the same. At the end of the day, the price of stock $A$ has increased by $\frac{1}{10}$ of its original price and the price of stock $B$ has decreased by $\frac{1}{10}$ of its original price. The new price of stock $A$ is what fraction of the new price of stock $B$?

(A) $\frac{2}{10}$
(B) $\frac{9}{11}$
(C) $\frac{9}{10}$
(D) $\frac{11}{10}$
(E) $\frac{11}{9}$

**18** After $\frac{3}{4}$ of the people in a room leave, three people enter the same room. The number of people who are now in the room, assuming no other people enter or leave, is $\frac{1}{3}$ of the original number of people who were in the room. How many people left the room?

(A) 9
(B) 18
(C) 24
(D) 27
(E) 36

---

## Grid In

**1** After a number is increased by $\frac{1}{3}$ of its value, the result is 24. What was the original number?

**2** In an election, $\frac{1}{2}$ of the male voters and $\frac{2}{3}$ of the female voters cast their ballots for candidate $A$. If the number of female voters was $1\frac{1}{2}$ times the number of male voters, what fraction of the total number of votes cast did candidate $A$ receive?

# Percent

## OVERVIEW

*Since there are 100 cents in 1 dollar, 15 cents represents 15 percent of a dollar.* **Percent** *means the number of* hundred*ths or the number of parts out of 100. Instead of writing the word percent, you can use the symbol %. Thus,*

$$P\% = \frac{P}{100}$$

## REWRITING FRACTIONS AND DECIMALS AS PERCENTS

Since fractions, decimals, and percents all represent parts of a whole, you can change from one of these types of numbers to another. Fractions and decimals greater than 1 are equivalent to percents greater than 100.

- To rewrite a decimal as a percent, move the decimal point two places to the right and add the percent sign (%). For example:

$$0.45 = 45\% \quad 0.08 = 8\% \quad 1.5 = 1.50 = 150\%$$

- To rewrite a fraction as a percent, first change the fraction to a decimal. Then change the decimal to a percent. For example:

$$\frac{3}{4} = 0.75 = 75\% \quad \frac{5}{2} = 2.50 = 250\% \quad 4 = 4.00 = 400\%$$

## REWRITING PERCENTS AS DECIMALS AND AS FRACTIONS

You may need to express a percent as a decimal or as a fraction.

- To rewrite a percent as a decimal, drop the percent symbol and divide the number that remains by 100. For example:

$$85\% = \frac{85}{100} = 0.85 \quad \frac{3}{4}\% = \frac{\frac{3}{4}}{100} = \frac{0.75}{100} = 0.0075$$

- To rewrite a percent as a fraction, drop the percent symbol and make the number that remains the numerator of a fraction whose denominator is 100. For example:

$$27\% = \frac{27}{100} \qquad 7\frac{1}{2}\% = \frac{7\frac{1}{2}}{100} = \frac{7.5}{100} = \frac{7.5 \times 10}{100 \times 10} = \frac{75}{1000}$$

## SOME FRACTION, DECIMAL, AND PERCENT EQUIVALENTS

Some calculations can go faster if you memorize the conversions in Table 3.4 and their multiples.

| TABLE 3.4   SOME CONVERSIONS YOU SHOULD KNOW | | | | | | | |
|---|---|---|---|---|---|---|---|
| Fraction | $\frac{1}{10}$ | $\frac{1}{8}$ | $\frac{1}{6}$ | $\frac{1}{5}$ | $\frac{1}{4}$ | $\frac{1}{3}$ | $\frac{1}{2}$ |
| Decimal | 0.10 | 0.125 | 0.1666... | 0.20 | 0.25 | 0.3333... | 0.50 |
| Percent | 10% | 12.5% | $16\frac{2}{3}\%$ | 20% | 25% | $33\frac{1}{3}\%$ | 50% |

**EXAMPLE:**   Write 350% as a fraction.

**SOLUTION 1:**   $350\% = 7 \times 50\% = 7 \times \frac{1}{2} = \frac{7}{2}$

**SOLUTION 2:**   $350\% = \frac{350}{100} = \frac{350 \div 50}{100 \div 50} = \frac{7}{2}$

**EXAMPLE:**   Write 37.5% as a fraction.

**SOLUTION 1:**   $37.5\% = 3 \times 12.5\% = 3 \times \frac{1}{8} = \frac{3}{8}$

**SOLUTION 2:**   $37.5\% = \frac{37.5}{100} = \frac{375}{1000} = \frac{375 \div 125}{1000 \div 125} = \frac{3}{8}$

## FINDING THE PERCENT OF A NUMBER

To find the percent of a given number:

- Rewrite the percent as a decimal.
- Use your calculator to multiply the given number by the decimal form of the percent.

**EXAMPLE:**   What is 15% of 80?

**SOLUTION:**   Rewrite 15% as 0.15. Then multiply:

$$15\% \text{ of } 80 = 0.15 \times 80 = 12$$

## FINDING THE PERCENT OF INCREASE OR DECREASE

When a quantity goes up or down in value, the percent of change can be calculated by comparing the amount of the change to the original amount.

$$\text{Percent of change} = \frac{\text{Amount of change}}{\text{Original amount}} \times 100\%$$

**EXAMPLE:** If the price of an item increases from $70 to $84, what is the percent of increase in price?

**SOLUTION:** The original amount is $70, and the amount of increase is $84 − $70 = $14.

$$\text{Percent of increase} = \frac{\text{Amount of increase}}{\text{Original amount}} \times 100\%$$

$$= \frac{14}{70} \times 100\%$$

$$= \frac{1}{5} \times 100\%$$

$$= 20\%$$

## FINDING AN ORIGINAL AMOUNT AFTER A PERCENT CHANGE

If you know the number that results after a given number is increased by $P\%$, you can find the original number by dividing the new amount by $1 + P\%$.

$$\text{Original amount} = \frac{\text{New amount after an increase of } P\%}{1 + \dfrac{P}{100}}$$

Similarly, if you know the number that results after a given number is decreased by $P\%$, you can find the original number by dividing the new amount by $1 - P\%$.

$$\text{Original amount} = \frac{\text{New amount after a decrease of } P\%}{1 - \dfrac{P}{100}}$$

**EXAMPLE:** A pair of tennis shoes cost $48.60 including sales tax. If the sales tax rate is 8%, what is the cost of the tennis shoes before the tax is added?

**SOLUTION:** 
$$\text{Cost of tennis shoes} = \frac{\text{Cost with tax included}}{1 + \text{tax rate}}$$

$$= \frac{48.60}{1 + 0.08}$$

$$= \frac{48.60}{1.08}$$

Use a calculator to divide:     $= 45$

The tennis shoes cost $45 without tax.

# LESSON 3-8   TUNE-UP EXERCISES

| **Multiple Choice** |
| --- |

**1**  What is 20% of $\frac{2}{3}$ of 15?

(A) 1
(B) 2
(C) 3
(D) 4
(E) 5

**2**  Which expression is equivalent to $\frac{2}{5}\%$?

(A) 0.40
(B) 0.04
(C) 0.004
(D) 0.0004
(E) 0.00004

**3**  The fraction $\dfrac{0.25 + 0.15}{0.50}$ is equivalent to what percent?

(A) 20%
(B) 25%
(C) 40%
(D) 60%
(E) 80%

**4**  In a movie theater, 480 of the 500 seats were occupied. What percent of the seats were NOT occupied?

(A)  0.4%
(B)  2%
(C)  4%
(D)  20%
(E)  40%

**5**  After 2 months on a diet, John's weight dropped from 168 pounds to 147 pounds. By what percent did John's weight drop?

(A) $12\frac{1}{2}\%$

(B) $14\frac{2}{7}\%$

(C) 21%

(D) 25%

(E) $28\frac{4}{7}\%$

**6**  Which expression is equivalent to 0.01%?

(A) $\sqrt{\dfrac{1}{100}}$

(B) $\dfrac{0.01}{10}$

(C) $\dfrac{1}{100}$

(D) $\dfrac{\frac{1}{100}}{\frac{1}{10}}$

(E) $\dfrac{1}{10^4}$

**7**  If the result of increasing $a$ by 300% of $a$ is $b$, then $a$ is what percent of $b$?

(A) 20%
(B) 25%
(C) $33\frac{1}{3}\%$
(D) 40%
(E) $66\frac{2}{3}\%$

**8** After a 20% increase, the new price of a radio is $78.00. What was the original price of the radio?

(A) $15.60
(B) $60.00
(C) $62.40
(D) $65.00
(E) $97.50

**9** After a discount of 15%, the price of a shirt is $51. What was the original price of the shirt?

(A) $44.35
(B) $58.65
(C) $60.00
(D) $64.00
(E) $65.00

**10** Three students use a computer for a total of 3 hours. If the first student uses the computer 28% of the total time, and the second student uses the computer 52% of the total time, how many minutes does the third student use the computer?

(A) 24
(B) 30
(C) 36
(D) 42
(E) 50

**11** What is 20% of 25% of $\frac{4}{5}$ ?

(A) 0.0025
(B) 0.004
(C) 0.005
(D) 0.04
(E) 0.05

| Minimum Age Requirement (years) | Number of States |
|---|---|
| 14 | 7 |
| 15 | 12 |
| 16 | 28 |
| 17 | 2 |
| 18 | 1 |

**12** The table above shows the minimum age requirement for obtaining a driver's license. In what percent of the states can a person obtain a driver's license before the age of 16?

(A) 94%
(B) 47%
(C) 38%
(D) 19%
(E) 6%

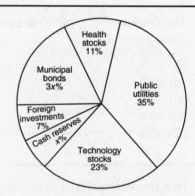

Investment Portfolio Valued at $250,000

**Questions 13 and 14.**

The graph above shows how $250,000 is invested.

**13** How much money is invested in municipal bonds?

(A) $45,000
(B) $37,500
(C) $35,000
(D) $30,000
(E) $15,000

**14** After 20% of the amount that is invested in technology stocks is reinvested in health stocks, how much money is invested in health stocks?

(A) $77,500
(B) $65,000
(C) $50,000
(D) $45,000
(E) $39,000

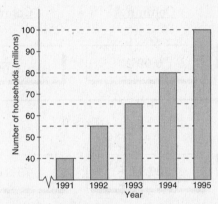

Growth of Computers in U.S. Households

**Questions 15 and 16.**

The graph above shows the number of U.S. households with computers for the years 1991 to 1995.

**15** What was the percent of increase in the number of households with computers from 1991 to 1995?

(A) 60%
(B) 75%
(C) 80%
(D) 120%
(E) 150%

**16** The greatest percent of increase in the number of households with computers occurred in which 2 consecutive years?

(A) 1991 to 1992
(B) 1992 to 1993
(C) 1993 to 1994
(D) 1994 to 1995
(E) It cannot be determined from the information given.

**17** In a factory that manufactures light bulbs, 0.04% of the bulbs manufactured are defective. It is expected that there will be one defective light bulb in what number of bulbs that are manufactured?

(A) 2500
(B) 1250
(C) 1000
(D) 500
(E) 250

**18** A discount of 25% on the price of a pair of shoes, followed by another discount of 8% on the new price of the shoes, is equivalent to a single discount of what percent of the original price?

(A) 17%
(B) 29%
(C) 31%
(D) 33%
(E) 35%

**19** If 8 kilograms of alcohol are added to 17 kilograms of pure water, what percent by weight of the resulting solution is alcohol?

(A) 8%
(B) 17%
(C) 32%
(D) 68%
(E) 75%

**20** If $a$ is 40% of $b$, then $b$ exceeds $a$ by what percent of $a$?

(A) 60%
(B) 100%
(C) 140%
(D) 150%
(E) 250%

## Quantitative Comparison

Each question consists of two quantities in boxes, one in Column A and one in Column B. You are to compare the two quantities and on the answer sheet fill in

A  if the quantity in Column A is greater;
B  if the quantity in Column B is greater;
C  if the two quantities are equal;
D  if the relationship cannot be determined from the information given

AN E RESPONSE WILL NOT BE SCORED

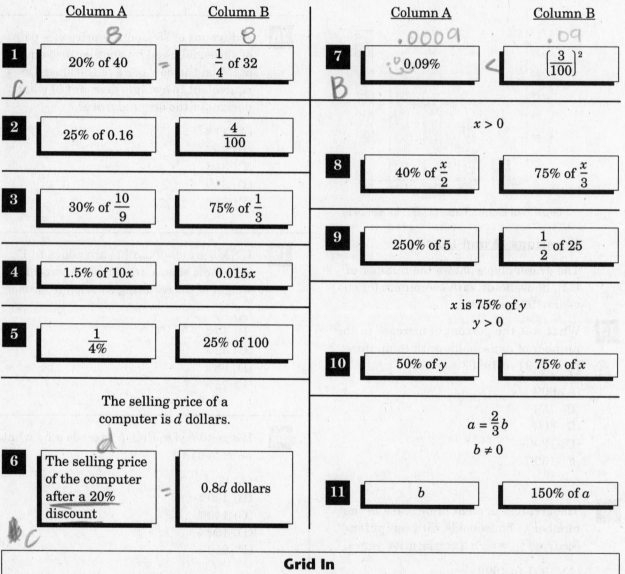

| Column A | Column B |
|---|---|
| **1** 20% of 40 | $\frac{1}{4}$ of 32 |
| **2** 25% of 0.16 | $\frac{4}{100}$ |
| **3** 30% of $\frac{10}{9}$ | 75% of $\frac{1}{3}$ |
| **4** 1.5% of 10$x$ | 0.015$x$ |
| **5** $\frac{1}{4\%}$ | 25% of 100 |

The selling price of a computer is $d$ dollars.

| **6** The selling price of the computer after a 20% discount | 0.8$d$ dollars |

| Column A | Column B |
|---|---|
| **7** 0.09% | $\left(\frac{3}{100}\right)^2$ |

$x > 0$

| **8** 40% of $\frac{x}{2}$ | 75% of $\frac{x}{3}$ |
| **9** 250% of 5 | $\frac{1}{2}$ of 25 |

$x$ is 75% of $y$
$y > 0$

| **10** 50% of $y$ | 75% of $x$ |

$a = \frac{2}{3}b$
$b \neq 0$

| **11** $b$ | 150% of $a$ |

## Grid In

**1** A store offers a 4% discount if a consumer pays cash rather than paying by credit card. If the cash price of an item is $84.00, what is the credit-card purchase price of the same item?

**2** During course registration, 28 students enroll in a certain college class. After three boys are dropped from the class, 44% of the class consists of boys. What percent of the original class did girls comprise?

# ANSWERS TO CHAPTER 3 TUNE-UP EXERCISES

## LESSON 3-1 *(Multiple Choice)*

**1. D** 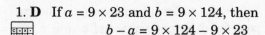 If $a = 9 \times 23$ and $b = 9 \times 124$, then
$$b - a = 9 \times 124 - 9 \times 23$$
You can use your calculator to do the arithmetic on the right side of the equation, but it's easier to use the reverse of the distributive law:
$$b - a = 9(124 - 23)$$
$$= 9(101)$$
$$= 909$$

**2. B** To figure out by what amount quantity $A$ exceeds quantity $B$, calculate $A - B$:
$$(8 \times 25) - (15 \times 10) = 200 - 150 = 50$$

**3. C** Since 1 quart = 32 fluid ounces, 5 quarts = $5 \times 32 = 160$ fluid ounces. If each container holds 16 fluid ounces, then $\frac{160}{16} = 10$ containers are needed to hold 5 quarts of milk.

**4. B** *Solution 1*: If the current odometer reading of a car is 31,983 miles, then the next mileage reading in which at least four digits are the same will be 32,222. Hence, the least number of miles that the car must travel before the odometer displays four digits that are the same is $32,222 - 31,983 = 239$.
*Solution 2*: Find the sum of 31,983 and the number given in each answer choice beginning with (A). Stop when you get a sum (31,983 + 239) in which at least four of the digits are the same. Hence, the correct choice is (B).

**5. C** In store $A$ 10 scarfs cost $10 \times \$12 = \$120$. Since the same scarf costs $8 in store $B$, $\frac{120}{8} = 15$ scarfs can be bought in store $B$ with $120.

**6. B** If $r * 0 = r$, then $*$ represents either addition or subtraction or both. It is also given that $r * r = 0$. Since $r - r = 0$ and $r + r \neq 0$ when $r$ is not 0, $*$ can represent only subtraction.

**7. D** Since Kurt has saved $160 to buy a stereo system that costs $400, he needs to earn an additional $400 - \$160 = \$240$. Earning $8 an hour, he will need to work $\frac{240}{8} = 30$ hours to have enough money to buy the stereo system.

**8. E** Since 24 hours + 12 hours + 3 hours = 39 hours, break down the problem by figuring out the time 24 hours ago, 12 hours before that time, and then 3 hours earlier. If the present time is exactly 1:00 P.M., then 24 hours ago it was also 1:00 P.M., and 12 hours before that time it was 1:00 A.M. Three hours before 1:00 A.M. the time was 10:00 P.M.

**9. A** If $r$ # $0 = r$, then the symbol # can represent either addition or subtraction, but not multiplication or division. Since it is also given that $r$ # $s = s$ # $r$, the operation # is commutative. Since addition is commutative and subtraction is not commutative, the symbol # represents only addition.

**10. D** Each product contains 6, so 6 can be ignored. Then $w = 36$, $x = 35$, and $y = 32$, so $y < x < w$.

*(Quantitative Comparison)*

**1. A** In Column A, $8 \times 7 = 56$. In Column B, use your calculator to divide: $2335 \div 5 = 467$. Since Column A contains 5 tens and Column B contains 4 hundreds, Column A > Column B.

**2. B** Do the arithmetic in Column A by reversing the distributive law:
$$719 \times 39 + 719 \times 61 = 719(39 + 61)$$
$$= 719(100)$$
$$= 71,900$$
Since $71,901 > 71,900$, Column B > Column A.

**3. A** Since there are 60 seconds in a minute and 60 minutes in an hour, the number of seconds in 6 hours is $6 \times 60 \times 60 = 21,600$ (Column A). Since there are 60 minutes in an hour, 24 hours in a day, and 7 days in a week, the number of minutes in a week is $60 \times 24 \times 7 = 10,080$ (Column B). Since $21,600 > 10,080$, Column A > Column B.

**4. C** To make the comparison between $(9)(10)(13)$ and $(6)(13)(15)$ easier, cross out 13 in each column. Since the product of the two remaining factors in Column A is $9 \times 10 = 90$ and the product of the two

*(Lesson 3-1   Quantitative Comparison continued)*

remaining factors in Column B is $6 \times 15 = 90$, Column A = Column B.

5. **C**   Since 1 quart = 32 fluid ounces, 3 quarts = $3 \times 32 = 96$ fluid ounces. The least number of 8-ounce glasses into which 3 quarts of lemonade can be emptied is $\frac{96}{8} = 12$. Since one dozen is another name for 12, Column A = Column B.

6. **C**   Remove the parentheses in each column by multiplying each term inside the parentheses by the number in front of the parentheses. In Column A,
$$3(8x + 4) = 3(8x) + 3(4)$$
$$= 24x + 12$$
In Column B,
$$4(3 + 6x) = 4(3) + 4(6x)$$
$$= 12 + 24x$$
or $24x + 12$. Since Column A and Column B can be made to look exactly alike, Column A = Column B.

7. **A**   Since the fraction $\frac{1}{3}$ equals the nonending, repeating decimal 0.3333..., and 0.3333... is greater than 0.333, Column A > Column B.

8. **C**   Simplify the expression in Column A and then compare the two columns:
$$2(x + 1) + 3x - 2 = 2x + 2 + 3x - 2$$
$$= (2x + 3x) + 2 - 2$$
$$= 5x$$
Since Column A can be made to look exactly like Column B, Column A = Column B.

9. **A**   To compare $3(a + 2b) = 3a + 6b$ in Column A to $3a + 2b$ in Column B, ignore $3a$ since it is being added in both columns. Since $b$ is a positive integer, $6b$ is always greater than $2b$, so Column A > Column B.

10. **B**   The largest value of $y$ in $3 < y < 8$ is smaller than 8, and the smallest value of $x$ in $2 < x < 7$ is greater than 2. Since the difference $y - x$ must be less than 6, Column B > Column A.

11. **A**   If $2 < x < 7$, then $8 < 4x < 28$. Since $4x$ is always greater than 8 and $y$ is always less than 8, $4x > y$, so Column A > Column B.

12. **D**   Pick positive integers for $a$ and $b$ that satisfy the given inequality, $4 < ab < 12$.

- If $a = 1$ and $b = 5$, then $ab = 5$ and $a + b = 6$, so Column B > Column A.
- If $a = 2$ and $b = 3$, then $ab = 6$ and $a + b = 5$, so Column A > Column B.

Since the different answers can be correct for different values of $a$ and $b$, the correct choice is (D).

13. **C**   Write the two sums horizontally, aligning like terms in the same column. Then subtract:
$$\begin{array}{r} 18 + 59 + 34 + x + y + 27 = 221 \\ - \quad 18 + 59 + 34 + x + z + 27 = 208 \\ \hline y - z \qquad = 13 \end{array}$$
Hence, Column A = Column B.

*(Grid In)*

1. **158**   In general, if $A$ and $B$ are positive integers, then the number of integers from $A$ to $B$ is $(B - A) + 1$. If the number of houses in a certain community are numbered consecutively from 2019 to 2176, there are $(2176 - 2019) + 1 = 157 + 1 = 158$ houses in the community.

2. **5**   Since 1 kilobyte is equivalent to $1024 \times 8$ or 8192 bits, 40,960 bits are equivalent to $\frac{40,960}{8192}$ or 5 kilobytes.

3. **45.8**   The balance that needs to be paid off is $495 - $129 or $366. Since eight equal monthly payments will be made, each monthly payment is $\frac{\$366}{8}$ or $45.75. Since 45.75 will not fit the grid, grid in 45.8.

4. **4**   The fraction $\frac{k-p}{m}$ will have its largest value when $k - p$ has its greatest value and $m$ has its least value. The largest value of $k - p$ is $21 - 9$ or 12. The inequality $2 < m < 6$ means that $m$ is greater than 2 but less than 6. Since $m$ is an integer, the least value of $m$ is 3. Hence, the largest possible value of $\frac{k-p}{m}$ is $\frac{12}{3}$ or 4.

5. **27**   For some fixed value of $x$, $9(x + 2) = y$. If the value of $x$ is increased by 3, then the value of $y$ is increased by $9 \times 3 = 27$. Since after $x$ is increased by 3, $9(x + 2) = w$, the value of $w - y$ is 27.

## LESSON 3-2   *(Multiple Choice)*

1. **B**   If $x = 3$ and $y = \sqrt{1 + 7x - 2x^2}$, then
$$y = \sqrt{1 + 7(3) - 2(3)^2}$$
$$= \sqrt{1 + 21 - 2(9)}$$

*(Lesson 3-2   Multiple Choice continued)*

$$= \sqrt{1+21-18}$$
$$= \sqrt{1+3}$$
$$= \sqrt{4}$$
$$= 2$$

2. **A**   Given $x = 32 - 16 \div 2 \times 4$, find the value of $x$ by working from left to right, doing any divisions and multiplications before any additions or subtractions:

$$x = 32 - 16 \div 2 \times 4$$
$$= 32 - 8 \times 4$$
$$= 32 - 32$$
$$= 0$$

3. **B**   If $1 < x^2 < 50$, then $\sqrt{1} < \sqrt{x^2} < \sqrt{50}$, which can be written as $1 < x < \sqrt{50}$. Since $\sqrt{50}$ is between 7 and 8, there are six integer values of $x$ that are greater than 1 but less than $\sqrt{50}$: 2, 3, 4, 5, 6, and 7.

4. **C**   If $2 = p^3$, then

$$(2)^3 = (p^3)^3$$
$$8 = p^{3 \times 3}$$
$$= p^9$$

Hence, $8p = p^9 \times p = p^{10}$.

5. **B**   To divide powers with the *same* base, keep the base and *subtract* the exponents. If $5 = a^x$, then

$$\frac{5}{a} = \frac{a^x}{a}$$
$$= a^{x-1}$$

6. **D**   If $b^3 = 4$, then

$$b^6 = (b^3)^2 = (4)^2 = 16$$

7. **B**   If $w$ is a positive number and $w^2 = 2$, then $w = \sqrt{2}$, so

$$w^3 = w^2 \cdot w = 2\sqrt{2}$$

8. **D**   If powers of the same base are equal, then their exponents must be equal. Since $8^{x+1} = 64 = 8^2$, then $x + 1 = 2$, so $x = 1$. To find the value of $3^{2x+1}$, replace $x$ with 1:

$$3^{2x+1} = 3^{2(1)+1}$$
$$= 3^3$$
$$= 27$$

9. **D**   Test each choice in turn, using particular values for $x$ and $y$. Since it is given that $0 < y < x$, let $y = 1$ and $x = 4$. The only statement that is true is (D), $\sqrt{xy} = (\sqrt{x})(\sqrt{y})$, since

$$\sqrt{xy} = \sqrt{4 \cdot 1} \quad = \sqrt{4} \quad = 2$$

and

$$(\sqrt{x})(\sqrt{y}) = (\sqrt{4})(\sqrt{1}) = (2)(1) = 2$$

10. **E**   Given $y = wx^2$ and $y$ is not 0. Since the values of $x$ and $w$ are each doubled, replace $w$ with $2w$ and $x$ with $2x$ in the original equation:

$$y_{\text{new}} = (2w)(2x)^2$$
$$= (2w)(4x^2)$$
$$= 8(wx^2)$$
$$= 8y$$

Hence, the original value of $y$ is multiplied by 8.

11. **B**   If $\sqrt{n}$ is a positive integer, then $n$ must be a perfect square integer. The perfect square integers in the interval $100 < n < 199$ are as follows:

$$121(=11^2), \ 144(=12^2), \ 169(=13^2), \ 196(=14^2)$$

Hence, there are four perfect square integers in the given interval.

12. **D**   If $(2^3)^2 = 4^p$, you can find $p$ by expressing each side of the equation as a power of the same base and then setting the exponents of the two bases equal:

$$(2^3)^2 = 4^p$$
$$2^6 = 2^{2p}$$
$$6 = 2p$$
$$3 = p$$

Since $p = 3$, then

$$3^p = 3^3 = 3 \times 3 \times 3 = 27$$

13. **B**   If $y = 2$, then

$$z = \sqrt[3]{2 + y + y^2}$$
$$= \sqrt[3]{2 + 2 + (2)^2}$$
$$= \sqrt[3]{2 + 2 + 4}$$
$$= 2$$

Since $z = 2$, then $z^2 = 2^2 = 4$.

14. **A**   Determine whether each Roman numeral statement is true or false when $x$ is a positive integer.

- I. Since $\left[\frac{x}{x}\right]^{99} = (1)^{99} = 1$ and $\left[\frac{x+1}{x+1}\right]^{100} = (1)^{99} = 1$, statement I is true.
- II. Since $(x^x)^2 = x^{2x}$ and $x^{x^2}$ is not equal to $x^{2x}$ for all positive integer values of $x$, statement II is false.
- III. Since $\frac{x^{100}}{x^{99}} = x^{100-99} = x^1 = x$ and $1^x = 1$, statement III is false.

Since only Roman numeral statement I is true, the correct choice is (A).

15. **A**   If $y = 25 - x^2$, the smallest possible value of $y$ is obtained by subtracting the largest possible value of $x^2$ from 25. Since $1 \le x \le 5$, the largest possible value of $x^2$ is $5^2 = 25$. When $x^2 = 25$, then $y = 25 - 25 = 0$.

*(Lesson 3-2  Quantitative Comparison)*

1. **C**  Evaluate and then compare each column. In Column A,
$$2^4 = 2 \times 2 \times 2 \times 2 = 16$$
In Column B,
$$4^2 = 4 \times 4 = 16$$
Hence, Column A = Column B.

2. **A**  Evaluate Column A. Since $2^5 = 2 \times 2 \times 2 \times 2 \times 2 = 32$,
$$2^6 - 2^5 = 2(2^5) - 2^5$$
$$= 2(32) - 32$$
$$= 64 - 32$$
$$= 32$$
Hence, Column A > Column B.

3. **B**  Since the quantities in the two columns are both greater than 1, compare their squares. Squaring Column A gives
$$(2\sqrt{3})^2 = (2^2)(\sqrt{3})^2 = 4 \times 3 = 12$$
Squaring Column B gives
$$(3\sqrt{2})^2 = (3^2)(\sqrt{2})^2 = 9 \times 2 = 18$$
Since the square of Column B is greater than the square of Column A, Column B > Column A.

4. **B**  In Column B,
$$(3n)^2 = (3)^2 (n)^2 = 9n^2$$
Since $9n^2$ is always greater than $3n^2$, Column B > Column A.

5. **A**  Rewrite Column A as
$$5 \times 6^{10} = 5 \times 6 \times 6^9 = 30 \times 6^9$$
and rewrite Column B as
$$6 \times 5^{10} = 6 \times 5 \times 5^9 = 30 \times 5^9$$
To compare $30 \times 6^9$ with $30 \times 5^9$, ignore 30 since it appears in both expressions. Since $6^9 > 5^9$, Column A > Column B.

6. **A**  Simplify the radical in each column. In Column A,
$$2\sqrt{75} = 2\sqrt{25}\,\sqrt{3} = 10\sqrt{3}$$
In Column B,
$$3\sqrt{27} = 3\sqrt{9}\,\sqrt{3} = 9\sqrt{3}$$
Since $10\sqrt{3} > 9\sqrt{3}$, Column A > Column B.

7. **C**  In Column B,
$$2^9 + 2^9 = 2 \times 2^9 = 2^1 \times 2^9 = 2^{1+9} = 2^{10}$$
Since Column A is also $2^{10}$, Column A = Column B.

8. **A**  In Column A,
$$8^4 = (2^3)^4 = 2^{3\times4} = 2^{12}$$
Since Column B is $2^{11}$, Column A > Column B.

9. **A**  Use the centered information to find the value of $N$. Rewrite the left side of the given equation, $50 \times 40 \times 30 = 6 \times 10^N$, as an integer multiplied by a power of 10:
$$(5 \times 10) \times (4 \times 10) \times (3 \times 10) = 6 \times 10^N$$
$$60 \times 10^3 = 6 \times 10^N$$
$$6 \times 10^4 = 6 \times 10^N$$
Comparing the two sides of the equations tells you that $N$ must be equal to 4. Since Column B = 3, Column A > Column B.

10. **A**  Use the centered information to find the value of $n$. You are told that $9^4 = 3^n$. Writing the left side of the equation as a power of 3 gives
$$9^4 = (3^2)^4 = 3^{2\times4} = 3^8 = 3^n$$
so $n = 8$. Since Column B = 6, Column A > Column B.

11. **D**  To compare $2n$ with $2^n$ for positive integer values of $n$, pick a few different values for $n$.
- If $n = 1$, then $2n = 2(1) = 2$ and $2^n = 2^1 = 2$, so Column A = Column B.
- If $n = 3$, then $2n = 2(3) = 6$ and $2^n = 2^3 = 2 \times 2 \times 2 = 8$, so Column B > Column A.

Since two different answers are possible, the correct choice is (D).

12. **C**  Simplify the radical in Column A:
$$\sqrt{160 \times 10^5} = \sqrt{16 \times 10 \times 10^5}$$
$$= \sqrt{16 \times 10^6}$$
$$= \sqrt{16} \times \sqrt{10^6}$$
$$= 4 \times 10^3$$
Hence, Column A = Column B.

13. **A**  Use the centered information to find the values of $w$ and $z$.
- Since $7^2 \times 7^w = 7^8$, then $7^{2+w} = 7^8$, and $w = 6$.
- Since $8^8 \div 8^z = 8^3$, then $8^{8-z} = 8^3$, and $z = 5$.

Hence, Column A > Column B.

*(Grid In)*

1. **2**  Given $2^4 \times 4^2 = 16^x$, find the value of $x$ by expressing each side of the equation as a power of the same base.
$$2^4 \times (2^2)^2 = (2^4)^x$$
$$2^4 \times 2^4 = 2^{4x}$$
$$2^{4+4} = 2^{4x}$$
$$2^8 = 2^{4x}$$
$$8 = 4x, \text{ so } x = 2$$

2. **3**  To find the value of $\frac{x}{y}$, given that $3^{x-1} = 9$ and $4^{y+2} = 64$, use the two equations to find the values of $x$ and $y$.
- Since $3^{x-1} = 9 = 3^2$, then $x - 1 = 2$, so $x = 3$.

*(Lesson 3-2 Grid In continued)*

- Since $4^{y+2} = 64 = 4^3$, then $y + 2 = 3$, so $y = 1$.
  Hence,
  $$\frac{x}{y} = \frac{3}{1} = 3$$

3. **16** Since $2^3 = 2 \times 2 \times 2 = 8$ and $(y-1)^3 = 8$, then $y - 1 = 2$, so $y = 3$. Hence,
   $$(y+1)^2 = (3+1)^2 = 4^2 = 4 \times 4 = 16$$

4. **1/2** Since
   $$\frac{p+p+p}{p \cdot p \cdot p} = \frac{3p}{p \cdot p \cdot p} = \frac{3}{p^2} = 12$$
   then $\frac{p^2}{3} = \frac{1}{12}$, so $p^2 = \frac{3}{12} = \frac{1}{4}$. Hence, $p = \frac{1}{2}$ since $\frac{1}{2} \times \frac{1}{2} = \frac{1}{4}$. Grid in as 1/2.

## LESSON 3-3 *(Multiple Choice)*

1. **E** If a whole number is divided by a positive integer $N$, the remainder must be a whole number *less than N*. Since the divisor is 4, the remainder can be 0, 1, 2, or 3, but not 4.

2. **D** In any division example, the divisor times the quotient plus the remainder should equal the dividend. If the quotient of $N$ divided by 5 is 13 and the remainder is 4 $\left[\frac{N}{5} = 13 + \frac{4}{5}\right]$, then
   $$N = (5 \times 13) + 4 = 65 + 4 = 69$$

3. **D** *Solution 1*: If $p$ is divisible by 3 and $q$ is divisible by 4, then $pq$ must be divisible by any combination of prime factors of 3 and 4. Since $3 = 3 \times 1$ and $4 = 2 \times 2$, $pq$ is divisible by each of the following: 3, 4, $3 \times 2$ or 6, and $3 \times 2 \times 2$ or 12. Since no product of prime factors of 3 and 4 equals 9, $pq$ cannot be divisible by 9.
   *Solution 2*: Pick numbers for $p$ and $q$. Then test each choice until you find a number that does not divide $pq$ evenly. For example, if $p = 6$ and $q = 8$, then $pq = 48$. Testing each answer choice, you find that 48 is divisible by 3, 4, and 6, but not by 9.

4. **E** When figuring out how many factors a number has, be sure to include the number itself and 1. Try each choice in turn:
   - (A) There are 5 factors of 12: 1, 3, 4, 6, and 12.
   - (B) There are 6 factors of 18: 1, 2, 3, 6, 9, and 18.
   - (C) There are 3 factors of 25: 1, 5, and 25.
   - (D) There are 8 factors of 70: 1, 2, 5, 7, 10, 14, 35, and 70.
   - (E) There are 9 factors of 100: 1, 2, 4, 5, 10, 20, 25, 50, and 100.
   Hence, 100 has the most factors. The correct choice is (E).

5. **B** *Solution 1*: A number is divisible by 3 if the sum of its digits is divisible by 3. In choice (B), the sum of the digits of 4308 is $4 + 3 + 0 + 8 = 15$, which is divisible by 3. In choice (E), the sum of the digits of 23,451 is $2 + 3 + 4 + 5 + 1 = 15$, so (E) is also divisible by 3. A number is divisible by 2 only if its last digit is even. Since the last digit of 4308 is even but the last digit of 23,451 is odd, the correct choice must be (B).
   *Solution 2*: Using a calculator, test each choice in turn to find a number that gives a 0 remainder when divided by 2 and by 3.

6. **C** *Solution 1*: Since 3 and 10 do not have any common factors other than 1, any number that is divisible by both 3 and 10 must be divisible by their product, 30. Any number that is divisible by 30 must also be divisible by any factor of 30. Since 15 is the only answer choice that is a factor of 30, any number that is divisible by both 3 and 10 must also be divisible by 15.
   *Solution 2*: Pick an easy number that is divisible by both 3 and 10, say 30. Then divide 30 by each of the answer choices in turn. Stop when you find a number (15) that divides evenly into 30. The correct choice is (C).

7. **C** Since $x$ represents any even number and $y$ represents any odd number, let $x = 2$ and $y = 3$. Evaluate the expression in each of the answer choices until you find one that produces an even number.
   - (A) $y + 2 = 3 + 2 = 5$
   - (B) $x - 1 = 2 - 1 = 1$
   - (C) $(x + 1)(y - 1) = (2 + 1)(3 - 1) = (3)(2) = 6$. There is no need to go further. The correct choice is (C).

8. **C** Break down 105 into its prime factors:
   $$\frac{105}{p} = \frac{1 \times 5 \times 21}{p} = \frac{1 \times 3 \times 5 \times 7}{p}$$

*(Lesson 3-3 Multiple Choice continued)*

Thus, when $p$ equals any of the seven positive integers 1, 3, 5, 7, 15 ($3 \times 5$), 21 ($3 \times 7$), or 35 ($5 \times 7$), $\frac{105}{p}$ is an integer.

9. **A** Since $n$ is an odd integer, let $n = 3$. Evaluate each answer choice in turn until you find an odd number. For choice (A),

$$(2n - 1)^2 = (2(3) - 1)^2 = (6-1)^2 = 25$$
$$= 2(3) - 1 = 5$$

There is no need to go further. The correct choice is (A).

10. **D** Consecutive multiples of 4, such as 4, 8, and 12, always differ by 4. If $k - 1$ is a multiple of 4, then the next larger multiple of 4 is obtained by adding 4 to $k - 1$, which gives $k - 1 + 4$ or $k + 3$.

11. **D** You are told that, after $m$ marbles are put into $n$ jars, each jar contains the same number of marbles, with two marbles remaining. If $x$ represents the number of marbles put into each jar, $m$ divided by $n$ equals $x$ with a remainder of 2. This statement can be written as

$$\frac{m}{n} = x + \frac{2}{n}$$

Since

$$x = \frac{m}{n} - \frac{2}{n} = \frac{m-2}{n},$$

$\frac{m-2}{n}$ marbles were put into each jar.

12. **C** You are given that, when $p$ is divided by 4, the remainder is 3, and when $p$ is divided by 3, the remainder is 0. Identify the answer choices that are divisible by 3. Then substitute each of these choices for $p$ until you find the choice that produces a remainder of 3 when divided by 4.

Choices (C), (D), and (E) are each divisible by 3. For choice (C), let $p = 15$. When 15 is divided by 4, the remainder is 3.

There is no need to go further. The correct choice is (C).

13. **D** Determine whether each Roman numeral statement is true or false.

- I. Subtracting any multiple of 2 from an odd integer always produces another odd integer. For example, $7 - 4 = 3$. Since $n^2 + 5$ is an odd integer and $(n^2 + 5) - 6 = n^2 - 1$, $n^2 - 1$ is also an odd integer. Hence, statement I is false.
- II. The sum of an even integer and an odd integer is an odd integer. Since

$n^2 + 5$ is an odd integer, $n^2$ must be even, so $n$ must also be even. Hence, statement II is true.

- III. The product of an odd integer and an even integer is an even integer. Since you have determined that $n$ is even, $5n$ is also even. Hence, statement III is true.

Since only Roman numeral statements II and III are true, the correct choice is (D).

14. **D** You are told that, if the number of people who contribute equally to a gift decreases from four to three, each person must pay an additional $10. You can eliminate choices (A) and (C), which are not divisible by both 3 and 4. Check each of the remaining choices until you find the right one.

- (B) 60 divided by 3 is 20, and 60 divided by 4 is 15. The difference between 20 and 15 is 5.
- (D) 120 divided by 3 is 40, and 120 divided by 4 is 30. The difference between 40 and 30 is 10.

There is no need to test the last choice. The correct choice is (D).

15. **B** Since $n$ is any even integer, pick a simple number for $n$. If $n = 4$, then $(n+1)^2 = (4+1)^2 = 25$. When 25 is divided by 4, the remainder is 1.

16. **C** You need to find the answer choice that produces remainders of 2 and 4 when divided by 3 and 5, respectively.

- (A) The remainder when 41 is divided by 3 is 2, and the remainder when 41 is divided by 5 is 1.
- (B) The remainder when 43 is divided by 3 is 1.
- (C) The remainder when 44 is divided by 3 is 2, and the remainder when 44 is divided by 5 is 4.

There is no need to continue. The correct choice is (C).

*(Quantitative Comparison)*

1. **D** Try different values for $N$.

- Let $N = 8$. If $N$ is divided by 4 or by 8, the remainder is 0, so Column A = Column B.
- Let $N = 12$. If $N$ is divided by 4, the remainder is 0. If $N$ is divided by 8, the remainder is 4. In this case, Column B > Column A.

*(Lesson 3-3   Quantitative Comparison continued)*

Since two answer choices are possible, the correct choice is (D).

2. **B**  The divisors of 21 are 1, 3, 7, and 21. The divisors of 12 are 1, 2, 3, 4, 6, and 12. Since 12 has six divisors and 21 has four divisors, Column B > Column A.

3. **C**  The prime numbers greater than 10 and less than 21 are 11, 13, 17, and 19. The prime numbers less than 10 are 2, 3, 5, and 7. Hence, Column A = Column B.

4. **C**  The factors common to 24 and 30 are 1, 2, 3, and 6. The factors common to 12 and 18 are 1, 2, 3, and 6. Hence, Column A = Column B.

5. **B**   In Column A,
$$416{,}879 \div 7 = 59{,}554.143$$
To convert the decimal part of the quotient into the remainder, multiply it by the divisor. Since $0.143 \times 7$ is approximately 1, the remainder is 1.
In Column B,
$$793{,}804 \div 5 = 15{,}876.80$$
Since $0.80 \times 5 = 4$, the remainder is 4. Thus, Column B > Column A.

6. **A**  For Column A, write the factors of 90 and then cancel like factors:

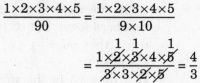

$$\frac{1\times2\times3\times4\times5}{90} = \frac{1\times2\times3\times4\times5}{9\times10}$$

$$= \frac{1\times\overset{1}{\cancel{2}}\times\overset{1}{\cancel{3}}\times4\times\overset{1}{\cancel{5}}}{\cancel{3}\times3\times\cancel{2}\times\cancel{5}} = \frac{4}{3}$$

Hence, dividing $1\times2\times3\times4\times5$ by 90 is equivalent to dividing 4 by 3, which gives a remainder of 1.
For Column B,

$$\frac{5\times6\times7\times8\times9}{90} = \frac{\overset{1}{\cancel{5}}\times\overset{3}{\cancel{6}}\times7\times8\times\overset{1}{\cancel{9}}}{3\times3\times2\times5} = 168$$

Hence, dividing $5\times6\times7\times8\times9$ by 90 gives a quotient of 168 and a remainder of 0. Thus, Column A > Column B.

7. **C**  Pick an easy number for $n$ that gives a remainder of 2 when divided by 5. Let $n = 7$.
- In Column A, $n + 3 = 10$. When 10 is divided by 5, the remainder is 0.
- In Column B, $n - 2 = 5$. When 5 is divided by 5, the remainder is 0.

Thus Column A = Column B.

8. **D**  If $k = 2$, $12k = 24 = 1 \times 2^3 \times 3$ in Column A and $21k = 42 = 1 \times 2 \times 3 \times 7$ in Column B. Since 1 is not a prime number, Column A has two different prime factors and Column B contains 3 different prime factors so Column B > Column A. If $k = 3$, $12k = 36 = 1 \times 2^2 \times 3^2$ and $21k = 63 = 1 \times 3^2 \times 7$. Since each column now has the same number of prime factors, the answer depends on the value of $k$ so the correct choice is (D).

9. **C**  Pick an easy number for $k$ that gives a remainder of 2 when $k$ is divided by 4. Let $k = 6$.
- In Column A, $2k = 12$. When 12 is divided by 4, the remainder is 0.
- In Column B, $k + 2 = 8$. When 8 is divided by 4, the remainder is 0.

Hence, Column A = Column B.

*(Grid In)*

1. **2**  When $a$ is divided by 7, the remainder is 5, so let $a = 12$. When $b$ is divided by 7, the remainder is 4, so let $b = 11$. Then $a + b = 23$. When 23 is divided by 7, the remainder is 2.

2. **2001**  All multiples of 3 from $-3000$ to 3000 are divisible by 3. Since
$$3 = 1\times3,\ 6 = 2\times3,\ 9 = 3\times3,\ \ldots,$$
$$3000 = 1000\times3$$
there are 1000 multiples of 3 from 3 to 3000, inclusive. Similarly, there are 1000 multiples of 3 from $-3000$ to $-3$. Since 0 is also divisible by 3, there are $1000 + 1000 + 1$ or 2001 integers from $-3000$ to 3000, inclusive, that are divisible by 3.

3. **2**  When a positive integer $k$ is divided by 6, the remainder is 1, so let $k = 7$. Then $5k = 35$. When 35 is divided by 3, the remainder is 2.

## LESSON 3-4  *(Multiple Choice)*

1. **E**  Multiplying both sides of the given equation, $2b = -3$, by $-2$ gives
$$-4b = -2(2b) = -2(-3) = 6$$
Since $-4b = 6$, then $1 - 4b = 1 + 6 = 7$.

2. **C**  Using the given facts that $a$ is a negative integer and $b$ is a positive integer,

*(Lesson 3-4   Multiple Choice continued)*

determine whether each Roman numeral statement is true or false.

- I. $a + b$ may be negative, 0, or positive, depending on the particular values of $a$ and $b$. For example, if $a = -2$ and $b = 2$, $a + b = 0$. If, however, $a = -3$, and $b = 4$, $a + b > 0$. Statement I is false.
- II. $\frac{b-a}{a} = \frac{b}{a} - \frac{a}{a} = \frac{b}{a} - 1$. Since a positive integer divided by a negative integer is negative, $\frac{b}{a}$ is negative, making $\frac{b}{a} - 1$ negative. Hence $\frac{b-a}{a} < 0$. Statement II is true.
- III. when $b$ is an even integer, $a^b > 0$. Statement III, $a^b < 0$, is not always true.

Hence, only statement II must be true.

3. **D**
$$\frac{(-2)^3 - (-4)^2}{(-1)(-2)(-3)} = \frac{-8 - (16)}{-6}$$
$$= \frac{-24}{-6}$$
$$= 4$$

4. **A**  If $k$ equals 0, 2, or $-1$, $k(k-2)(k+1)$ evaluates to 0. Hence, you can eliminate answer choices (B), (C), and (D). Plug each of the two remaining answer choices into $k(k-2)(k+1)$ to see which results in a negative value. For answer choice (A), $k = -2$:
$$k(k-2)(k+1) = -2(-2-2)(-2+1)$$
$$= -2 \ (-4) \ (-1)$$
$$= -8$$
The correct choice is (A).

5. **C**  $p^2(2-5) + (-p)^2 = -3p^2 + p^2 = -2p^2$

6. **C**  If $n + 5$ is an odd integer, then $n$ could be $-2$ since $-2 + 5 = 3$.

7. **D**  Determine whether each Roman numeral statement is true or false when $a < 0$ and $b > 0$.

- I. $a + b$ may be negative, 0, or positive, depending on the particular values of $a$ and $b$. For example, if $a = -2$ and $b = 2$, then $a + b = 0$. If $a = -3$ and $b = 2$, then $a + b < 0$. Hence, statement I is false.
- II. Pick numbers for $a$ and $b$. If $a = -3$ and $b = 2$, then $b - a = 2 - (-3) = 2 + 3 = 5$. Since $b - a > 0$, statement II is true.

- III. Since the square of a negative number is positive, $a\left[\frac{a}{b}\right] = \frac{a^2}{b} > 0$, so statement III is true.

Only Roman numeral statements II and III are true.

8. **D**  Since $(-2)^3 = (-2) \times (-2) \times (-2) = -8$ and $(-3)^2 = (-3) \times (-3) = 9$, then
$$(-2)^3 + (-3)^2 = -8 + 9 = 1$$

**Questions 9 and 10.** From the diagram, $a < -1$, $-1 < b < 0$, $0 < c < 1$, and $d > 1$.

9. **B**  Determine whether each Roman numeral statement is true or false.

- I. Pick a number for $c$. If $c = \frac{1}{2}$, then $c^2 = \frac{1}{2} \times \frac{1}{2} = \frac{1}{4}$. Since $\frac{1}{4} < \frac{1}{2}$, $c^2 < c$. Hence, statement I is true.
- II. Since $a < -1$, then $a^2 > 1$. For example, if $a = -3$, then $a^2 = 9$. Since $0 < c < 1$, $a^2 > c$. Hence, statement II is true.
- III. Pick a number for $b$. If $b = -\frac{1}{2}$, then $\frac{1}{b} = -\frac{2}{1} = -2$. Since $-\frac{1}{2} > -2$, then $b > \frac{1}{b}$. Hence, statement III is false.

Only Roman numeral statements I and II are true.

10. **C**  Determine whether each Roman numeral statement is true or false.

- I. On the basis of the diagram, pick numbers for $a$ and $d$. If $a = -2$ and $d = 2$, then $ad = -4 < b$, so statement I is false.
- II. Since $ab$ is the product of two negative numbers, $ab > 0$. Since $ad$ is the product of numbers with opposite signs, $ad < 0$. Since $ab > ad$, statement II is true.
- III. Since $a < b$, the reciprocals of $a$ and $b$ have the opposite size relationship. Since $\frac{1}{a} > \frac{1}{b}$, statement III is true.

Hence, only Roman numeral statements II and III are true.

11. **B**  If $(y-3)^2 = 16$, then the number inside the parentheses must be either 4 or $-4$. If $y - 3 = 4$, then $y = 7$ and $y^2 = 49$. If $y - 3 = -4$, then $y = 3 - 4 = -1$, so $y^2 = 1$. The smallest possible value of $y^2$ is 1.

12. **D**  You are told that a set of five integers contains at least one positive integer. If the product of these five integers is negative,

*(Lesson 3-4   Multiple Choice continued)*

there must be an odd number of negative integers. Hence, of the remaining four integers, at most three can be negative.

13. **A**   Rewrite $a^2b^3c$ as $(a^2b^2)bc > 0$. Then determine whether each Roman numeral statement is true or false.

- I. Since $(a^2b^2)bc > 0$, $a^2 > 0$, and $b^2 > 0$, it must be the case that $bc > 0$. Hence, statement I is true.
- II. Using $(a^2b^2)bc > 0$, you cannot tell whether $ac$ is positive or negative. Hence, statement II is false.
- III. In the product $(a^2b^2)bc > 0$, there is no restriction on the signs of $a$ and $b$, so their product can be positive or negative. Hence, statement III is false.

Only Roman numeral statement I must be true.

14. **A**   Since $a^2b^3c^5 = (a^2b^2c^4)bc < 0$ and $a^2b^2c^4$ is always positive, it must be the case that $bc$ is negative.

15. **B**   Determine whether each Roman numeral statement is true or false when $a \neq 0$.

- I. $(-a)^2 = (-a) \times (-a) = a^2$. Since $(-a)^2 \neq -a^2$, statement I is false.
- II. Since $-(b-a) = -b-(-a) = b+a = a-b$, statement II is true.
- III. If $a > 0$, then $a > -a$. However, if $a < 0$, then $a < -a$, so statement III is false.

Hence, only Roman numeral statement II is always true.

*(Quantitative Comparison)*

1. **B**   Since $(+3) \times (-9) = -27$ and $(+3) + (-9) = -6$, Column B > Column A.

2. **C**   Since $-2(x-4) = -2x - 2(-4) = -2x + 8 = 8 - 2x$, Column A = Column B.

3. **D**   If $(1 + k)^2 = 9$, then $1 + k$ can be either 3 or $-3$. If $1 + k = 3$, then $k = 2$, so Column A = Column B. If, however, $1 + k = -3$, then $k = -4$, so Column B > Column A.
   Since two different answers are possible, the correct choice is (D).

4. **B**   If $-3 \leq x \leq 2$, the largest possible value of $4 - x^2$ in Column A occurs when $x^2$ has its smallest value. Hence, the largest possible value of $4 - x^2$ is $4 - 0 = 4$. The largest possible value of $x^2 - 4$ in Column B

occurs when $x^2$ has its greatest value. Since $-3 \leq x \leq 2$, the largest possible value of $x^2 - 4$ is $(-3)^2 - 4 = 9 - 4 = 5$. Hence, Column B > Column A.

5. **B**   If $a < 0$ and $b > 0$, then $a - b < 0$. Since $-a > 0$, then $b - a > 0$. Hence, Column B > Column A.

6. **B**   In Column A, $(-1)^{99} = -1$ since $-1$ raised to an odd integer power always equals $-1$. In Column B,
$$\frac{1-100}{(-33)+(-33)+(-33)} = \frac{-99}{-99} = 1$$
Since $1 > -1$, Column B > Column A.

7. **D**   You can easily find numbers such that $a + b > 0$ or $a + b < 0$ when $-1 < ab < 0$. For example, if $a = 2$ and $b - \frac{1}{4}$, then
$$a + b = 1\frac{3}{4} > ab$$
so Column A > Column B. If, however, $a = -2$ and $b = \frac{1}{4}$, then
$$a + b = -1\frac{3}{4} < ab$$
so Column B > Column A.
Since two different answers are possible, the correct choice is (D).

8. **B**   Rewrite Column A as $(-a)^3 = -a^3$, and rewrite Column B as $(-b)^3 = -b^3$. Since $a > b > 1$, then $a^3 > b^3$, so $-b^3 > -a^3$. Hence, Column B > Column A.

9. **B**   *Solution 1*: You need to compare $a^2 - b$ in Column A to $a(a - b) = a^2 - ab$ in Column B. Since $0 < a < b < 1$, $b$ is a positive fraction less than 1, so $ab < b$. Since the quantity $ab$ that is being subtracted from $a^2$ in Column B is smaller than the quantity $b$ that is being subtracted from $a^2$ in Column A, the difference in Column B must be greater than the difference in Column A.
   *Solution 2*: Pick numbers for $a$ and $b$. Since you are told that $0 < a < b < 1$, let $a = 0.5$ and $b = 0.8$. In Column A,
   $a^2 - b = (0.5)^2 - 0.8 = 0.25 - 0.8 = -0.55$
   In Column B,
   $a(a-b) = 0.5(0.5-0.8) = 0.5(-0.3) = -0.15$
   Since $-0.15 > -0.55$, Column B > Column A.

10. **D**   Since $m^2 = n^2$ and $m \neq 0$, the factors of $n^2$ and $m^2$ do not affect the comparison. If $m^2 = n^2$, then $m = n$ or $m = -n$. If $m = n$,

*(Lesson 3-4 Quantitative Comparison continued)*

Column A = Column B. If, however, $m = -n$, the two columns are unequal.

Since more than one answer is possible, the correct choice is (D).

*(Grid In)*

1. **10** Since $p^2 = 16$, $p = 4$ or $-4$. Also, $q^2 = 36$, so $q = 6$ or $-6$. Hence, the largest possible value of $q - p$ is $6 - (-4) = 6 + 4 = 10$.

2. **16** Follow these steps:
   - Find the largest possible value of $y$. Since $x^2$ is always nonnegative, the largest possible value of $y = 1 - x^2$ occurs when $x^2$ has its smallest value. Since $-4 \le x \le 2$, the smallest value of $x^2$ is 0. The *largest* possible value of $y$ is $y = 1 - x^2 = 1 - 0 = 1$.
   - Find the smallest possible value of $y$. The smallest possible value of $y = 1 - x^2$ occurs when $x^2$ has its largest value. The largest value of $x^2$ is $(-4)^2 = 16$. The *smallest* possible value of $y$ is $y = 1 - x^2 = 1 - 16 = -15$.
   - Subtract. The number obtained when the *smallest* possible value of $y$ is subtracted from the *largest* possible value of $y$ is $1 - (-15) = 1 + 15 = 16$.

**LESSON 3-5** *(Multiple Choice)*

1. **D** To find the number of minutes that elapse from 4:56 P.M. to 5:32 P.M., subtract 4 hours and 56 minutes from 5 hours and 32 minutes:

   5 hours 32 minutes $\Rightarrow$ 4 hours 92 minutes
   − 4 hours 56 minutes    − 4 hours 56 minutes
                                                  36 minutes

   Since,
   $$\frac{36 \text{ minutes}}{1 \text{ hour}} = \frac{36}{60} = \frac{12 \times 3}{12 \times 5} = \frac{3}{5}$$
   the number of minutes that elapse from 4:56 P.M. to 5:32 P.M. is $\frac{3}{5}$ of an hour.

2. **B** If each of the fractions $\frac{3}{k}, \frac{4}{k}, \frac{5}{k}$ is in lowest terms, then k cannot be divisible by 3, 4, and 5. Of the answer choices, only 49 is not divisible by 3, 4, and 5.

3. **D** To find the number that has the greatest value, compare each digit, reading from left to right. The numbers in choices (A) and (C) can be eliminated since they are less than any of the other choices. Since the thousandths digit of 0.2938 is greater than the thousandths digit of each of the remaining answer choices, 0.2938 is the largest number.

4. **D** Four hours elapse from 11:00 A.M. to 3:00 P.M. on Wednesday of the same day, and $52 (= 24 + 24 + 4)$ hours elapse from 11:00 A.M. on Wednesday to 3:00 P.M. on Friday of the same week. To compare these elapsed times, form the fraction
   $$\frac{4}{52} = \frac{\cancel{4}}{\cancel{4} \times 13} = \frac{1}{13}$$

5. **D** In 0.030, the digit 3 is located two places to the *right* of the decimal point, so it is in the hundredths place.

6. **A** Since $\frac{9.5}{19} = \frac{1}{2}$, then $\frac{9}{19} < \frac{1}{2}$. Similarly, $\frac{7.5}{15} = \frac{1}{2}$, so $\frac{8}{15} > \frac{1}{2}$. Hence, you can eliminate answer choices (B), (C), (D), and (E). Choice (A) is the correct answer.

7. **C** Since $\left[\frac{1}{2}\right]^3 = \frac{1}{8}$, the new value of $V$ is $\frac{1}{8}$ of the original value of $V$.

8. **C** Since each inch on ruler $A$ is marked in equal $\frac{1}{8}$-inch units, a side that measures 12 of these $\frac{1}{8}$-inch units is
   $$12 \times \frac{1}{8} = \frac{3}{2} = 1\frac{1}{2} \text{ inches long}$$
   If the same side is measured with ruler $B$, which is marked in equal $\frac{1}{12}$-inch units, the side will measure 18 of these $\frac{1}{12}$-inch units since
   $$1\frac{1}{2} = \frac{12}{12} + \frac{6}{12}$$
   and $12 + 6 = 18$.

9. **C**
   $$60 + 2 + \frac{4}{8} + \frac{3}{500} = 62 + 0.5 + 0.006$$
   $$= 62.506.$$

10. **B** Since $N \times \frac{7}{12} = \frac{7}{12} \times \frac{3}{14}$, you can cancel $\frac{7}{12}$ on each side to get $N = \frac{3}{14}$, so $\frac{1}{N} = \frac{14}{3}$.

11. **E** If $n = 2.5 \times 10^{25}$, then
    $$\sqrt{n} = \sqrt{2.5 \times 10^{25}}$$
    $$= \sqrt{25 \times 10^{24}}$$
    $$= \sqrt{25} \times \sqrt{10^{24}}$$
    $$= 5 \times 10^{\frac{24}{2}}$$
    $$= 5 \times 10^{12}$$

12. **A** Since division by 0 is not allowed, a variable in the denominator of a fraction cannot be equal to a number that makes the denominator evaluate to 0. If $y = \frac{x-2}{x+3}$, then $x$ cannot be equal to $-3$ since $-3 + 3 = 0$.

*(Lesson 3-5  Multiple Choice continued)*

13. **E**  Eight pencils cost $0.42, and $2.10 ÷ 0.42 = 5. Hence, 8 × 5 = 40 pencils can be purchased.

14. **C**  For a total of 48 ounces of orange juice, six 8-ounce containers must be purchased since 6 × 8 ounces = 48 ounces. If each 8-ounce container costs $0.69, six of these containers will cost 6 × $0.69 = $4.14.

For a total of 48 ounces of orange juice, four 12-ounce containers must be purchased since 4 × 12 ounces = 48 ounces. If each 12-ounce container costs $0.95, four of these containers will cost 4 × $0.95 = $3.80.

The amount of money that will be saved by purchasing the 12-ounce containers is $4.14 − $3.80 = $0.34.

15. **E**  If $b \neq 0$, then
$$\frac{(-2b)^3}{-8b^3} = \frac{(-2)^3 b^3}{-8b^3} = \frac{-8b^3}{-8b^3} = 1$$

*(Quantitative Comparison)*

1. **C**  The thousandths digit of 0.4648 is less than 5, so 0.4648 rounded to the nearest hundredth is 0.46. The thousandths digit of 0.4551 is 5 or greater, so 0.4551 rounded to the nearest hundredth is also 0.46. Hence, Column A = Column B.

2. **A**  Since $4\frac{3}{7} = \frac{28+3}{7} = \frac{31}{7}$,

Column A > Column B.

3. **C**  Since $\frac{63}{18} = \frac{9 \times 7}{9 \times 2} = \frac{7}{2}$,

Column A = Column B.

4. **A**  Use a calculator:
$$\frac{11}{15} = 11 \div 15 = 0.7333\ldots$$
and
$$\frac{5}{7} = 5 \div 7 = 0.7142\ldots$$

Hence, Column A > Column B.

5. **A**  Factor the fractions in the two columns:
$$\left(\frac{1}{2}\right)^3 = \left(\frac{1}{2}\right) \times \left(\frac{1}{2}\right) \times \left(\frac{1}{2}\right) = \frac{1}{8}$$
and
$$\left(\frac{1}{3}\right)^2 = \frac{1}{3} \times \frac{1}{3} = \frac{1}{9}$$

If two fractions have the same numerator, then the larger fraction is the fraction with the smaller denominator. Since $\frac{1}{8} > \frac{1}{9}$, Column A > Column B.

6. **B**  In Column A,
$$\frac{23}{5} = 4\frac{3}{5} \quad \text{and} \quad \frac{99}{5} = 19\frac{4}{5}$$
The whole numbers between $\frac{23}{5}$ and $\frac{99}{5}$ are the whole numbers from 5 to 19, inclusive. If $A$ and $B$ are whole numbers, then the number of whole numbers from $A$ to $B$ is $B - A + 1$. Thus, the number of whole numbers from 5 to 19 is $19 - 5 + 1 = 15$. Similarly, in Column B,
$$\frac{11}{3} = 3\frac{2}{3} \quad \text{and} \quad \frac{59}{3} = 19\frac{2}{3}$$
The number of whole numbers between $\frac{11}{3}$ and $\frac{59}{3}$ is $19 - 4 + 1 = 15 + 1 = 16$. Hence, Column B > Column A.

7. **A**  Since $x$ is a positive integer, simplify the comparison by cross-multiplying:

| Column A | Column B |
|---|---|
| $\dfrac{x+1}{x+2}$ | $\dfrac{1}{2}$ |
| $2(x+1)$ | $1(x+2)$ |
| $2x+2$ | $x+2$ |
| $2x$ | $x$ |

Since $2x > x$ for all positive values of $x$, Column A > Column B.

8. **B**  Try to make the columns look alike by changing the quantity in Column B so that 10 has the same power as 10 in Column A. Thus:
$$0.00312k \times 10^7 = 0.00312k \times 10^4 \times 10^3$$
$$= 31.2k \times 10^3$$
Since $k$ is a positive integer, $31.2k > 3.12k$, so Column B > Column A.

*(Grid In)*

1. **8**  If $\frac{4\triangle}{6\triangle} + \frac{5}{17} = 1$, then $\frac{4\triangle}{6\triangle} = \frac{12}{17}$ since $\frac{12}{17} + \frac{5}{17} = 1$. Since $\frac{4\triangle}{6\triangle} = \frac{12}{17}$, $4\triangle$ must be a multiple of 12. Hence, $4\triangle = 48$, so $\triangle = 8$.

2. **24**  Since 1.5 inches represents 45 miles, 1 inch represents $\frac{45}{1.5}$ or 30 miles. Hence, 0.8 inch represents $0.8 \times 30$ or 24 miles.

3. **7**  If four lemons cost $0.68, one lemon costs $\frac{\$0.68}{4}$ or $0.17. Since $\frac{\$1.19}{\$0.17} = 7$, there are typically seven lemons in 1 pound of lemons.

4. **15/8** or **1.88**  Since $2.50 was charged for the first $\frac{1}{2}$ mile, $10.75 − $2.50 or $8.25

*(Lesson 3-5 Grid In continued)*

represents the charge for the total number of additional $\frac{1}{8}$ miles, at the rate of $0.75 per $\frac{1}{8}$ mile. Dividing $8.25 by $0.75 gives the number of $\frac{1}{8}$ miles traveled. Since $8.25 ÷ $0.75 = 11, the taxi trip was $\frac{1}{2} + \frac{11}{8}$ miles.

Write $\frac{1}{2} + \frac{11}{8}$ as either a fraction or a decimal, and then grid in the result:

- since $\frac{1}{2} + \frac{11}{8} = \frac{4}{8} + \frac{11}{8} = \frac{15}{8}$, grid 15/8; *or*
- since $\frac{1}{2} + \frac{11}{8} = 0.5 + 1.375 = 1.875$, grid 1.88.

5. **5.6** Since $\frac{288}{1728} = \frac{1}{6}$, 288 cubic inches are equivalent to $\frac{1}{6}$ of a cubic foot. If 1 cubic foot of the metal weighs 8 pounds, then $\frac{1}{6}$ of a cubic foot will weigh

$$\frac{1}{\cancel{6}} \times \cancel{8}^{\,4} \text{ or } \frac{4}{3} \text{ pounds}$$

Since each pound of the metal costs $4.20, the cost of $\frac{4}{3}$ pounds equals

$$\frac{4}{3} \times \$4.20 = 4 \times \$1.40 = \$5.60$$

Omit the dollar sign and grid 5.6

## LESSON 3-6 *(Multiple Choice)*

1. **B** To subtract $7\frac{3}{4}$ minutes from $9\frac{1}{3}$ minutes, first change $\frac{3}{4}$ to $\frac{9}{12}$ and $\frac{1}{3}$ to $\frac{9}{12}$:

$$9\frac{3}{12} - 7\frac{9}{12} = 8\frac{15}{12} - 7\frac{9}{12} = 1\frac{6}{12} = 1\frac{1}{2} \text{ minutes}$$

Since 1 minute = 60 seconds and $\frac{1}{2}$ minute = 30 seconds, $1\frac{1}{2}$ minutes = 90 seconds.

2. **A** Since $\frac{1}{2} ÷ \frac{3}{4} = \frac{1}{2} \times \frac{4}{3} = \frac{2}{3}$, multiplying a number by $\frac{1}{2}$ and then dividing the result by $\frac{3}{4}$ is equivalent to multiplying the original number by $\frac{2}{3}$.

3. **D**
$$\left(\frac{1}{10}\right)^3 + \left(\frac{2}{10}\right)^3 + \left(\frac{3}{10}\right)^3$$
$$= 0.001 + 0.008 + 0.027$$
$$= 0.036$$

4. **D**
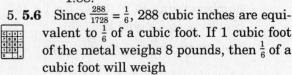
$$\frac{3}{4} \text{ of } \frac{2}{3} \text{ of } 12 = \frac{\cancel{3}}{\cancel{4}} \times \frac{2}{\cancel{3}} \times \frac{\cancel{12}}{1} = 6$$

5. **C** $\quad \frac{2}{21} \text{ of } 3 \times 5 \times 7 = \frac{2}{\cancel{21}} \times \cancel{3} \times 5 \times \cancel{7} = 10$

6. **C** You are told that $\frac{1}{4} < n < \frac{1}{3}$. Since each of the answer choices has a denominator of 24, change $\frac{1}{4}$ and $\frac{1}{3}$ into equivalent fractions that have 24 as their denominators.

Since
$$\frac{1}{4} = \frac{1}{4} \cdot \frac{6}{6} = \frac{6}{24}$$
and
$$\frac{1}{3} = \frac{1}{3} \cdot \frac{6}{6} = \frac{6}{18}$$
then $\frac{6}{24} < n < \frac{8}{24}$. A possible value of $n$ is $\frac{7}{24}$.

7. **A**
$$p = \frac{\frac{4}{\sqrt{3}} - \frac{1}{\sqrt{3}}}{\sqrt{3}} = \frac{\frac{3}{\sqrt{3}}}{\sqrt{3}}$$
$$= \frac{3}{\sqrt{3}} ÷ \sqrt{3}$$
$$= \frac{3}{\sqrt{3}} \times \left(\frac{1}{\sqrt{3}}\right)$$
$$= \frac{3}{3}$$
$$= 1$$
Since $p = 1$, then $p^2 = 1 \times 1 = 1$.

8. **C** If $k = \dfrac{c}{\frac{a}{b}}$, then
$$k = \frac{c}{\frac{a}{b}} = c ÷ \frac{a}{b} = c \times \frac{b}{a} = \frac{bc}{a}$$
so $\frac{1}{k} = \frac{a}{bc}$.

9. **A** If an item is on sale at $\frac{1}{3}$ off the list price of $d$ dollars, its sale price is $d - \frac{1}{3}d = \frac{2}{3}d$. If the sale price of $\frac{2}{3}d$ is reduced by $\frac{1}{4}$ of that price, then the final sale price is $\frac{2}{3}d - \frac{1}{4}\left(\frac{2}{3}d\right)$. Simplify:
$$\frac{2}{3}d - \frac{1}{4}\left(\frac{2}{3}d\right) = \frac{2}{3}d - \frac{1}{6}d$$
$$= \frac{4}{6}d - \frac{1}{6}d$$
$$= \frac{3}{6}d$$
$$= \frac{d}{2}$$

10. **E** *Solution 1*: If $\frac{a}{b}$ is a fraction less than 1, then its reciprocal is greater than 1. For example, $\frac{3}{4} < 1$ and $\frac{4}{3} > 1$. Since the reciprocal of $\frac{a}{b}$ may be written as
$$\frac{1}{\frac{a}{b}}, \quad \frac{1}{\frac{a}{b}} > 1$$
Two times this value, or
$$\frac{2}{\frac{a}{b}}$$
must also be greater than 1.

*(Lesson 3-6 Multiple Choice continued)*

*Solution 2*: Pick numbers for $a$ and $b$ that make $\frac{a}{b} < 1$. For example, let $a = 3$ and $b = 4$. Plug these numbers into each answer choice until you find the one (E) that produces a number greater than 1.

11. **B** After Juan gives $\frac{2}{3}$ of his $p$ pencils to Roger, Juan is left with $\frac{p}{3}$ pencils. When he gives $\frac{1}{4}$ of the pencils that he has left to Maria, Juan has $\frac{p}{3} - \frac{1}{4}\left[\frac{p}{3}\right]$ pencils left. Simplify:

$$\frac{p}{3} - \frac{1}{4}\left[\frac{p}{3}\right] = \frac{p}{3} - \frac{p}{12}$$
$$= \frac{4p}{12} - \frac{p}{12}$$
$$= \frac{3p}{12}$$
$$= \frac{1}{4}p$$

12. **C** Determine whether each Roman numeral inequality makes the statement true or false.

- I and II. If a number that is greater than 1 is raised to a power, then, as the exponent gets larger, the value of that expression also becomes larger. Thus $a^3 < a^2 < a$ is false when $a > 1$, and true when $0 < a < 1$. Hence, inequality I is false, and inequality II is true.

- III. If $-1 < a < 0$, then $a^2 > 0$ and $a < 0$, so $a^2 > a$. Therefore, $a^3 < a^2 < a$ is not true. Thus, inequality III is false.

Only Roman numeral inequality II is true.

*(Quantitative Comparison)*

1. **C** In Column A, $2x \div \frac{1}{4} = 2x \times 4 = 8x$. In Column B, $x$ multiplied by 8 is $8x$. Hence, Column A = Column B.

2. **A** In Column B, the number of minutes that elapse from 1:54 P.M. to 2:00 P.M. is 6, and the number of minutes that elapse from 2:00 P.M. to 2:28 P.M. is 28. Hence, 34 minutes elapse from 1:54 P.M. to 2:28 P.M.

In Column A, the number of minutes in $\frac{7}{12}$ of an hour is $\frac{7}{12} \times 60 = 35$. Hence, Column A > Column B.

3. **B** In Column A,

$$\frac{1}{4} = \frac{1}{4} \div 4 = \frac{1}{4} \times \frac{1}{4} = \frac{1}{16}$$

In Column B,

$$\frac{4}{\frac{1}{4}} = 4 \div \frac{1}{4} = 4 \times 4 = 16$$

Hence, Column B > Column A.

4. **A** In Column A, $\frac{3}{8}$ of $\frac{8}{3} = \frac{3}{8} \times \frac{8}{3} = 1$. Since $\frac{719}{720} < 1$, Column A > Column B.

5. **C** Add the fractions in Column A:

$$\frac{3}{9} + \frac{4}{12} + \frac{5}{15} = \frac{1}{3} + \frac{1}{3} + \frac{1}{3} = 1$$

Hence, Column A = Column B.

6. **C** Using $a = \frac{7}{10}$ and $b = 1\frac{3}{4} = \frac{7}{4}$, find the value of $\frac{a}{b}$:

$$\frac{a}{b} = a \div b = \frac{7}{10} \div \frac{7}{4}$$
$$= \frac{7}{10} \times \frac{4}{7}$$
$$= \frac{4}{10}$$
$$= \frac{2}{5}$$

Hence, Column A = Column B.

7. **B** Since $\frac{1}{n} > 1$, the denominator of $\frac{1}{n}$ must be less than the numerator; that is, $n < 1$. The square root of a positive number less than 1 is always greater than the number. Since $\sqrt{n} > n$, Column B > Column A.

8. **A** In Column A,

$$\frac{x}{0.2} = x \div 0.2$$
$$= x \div \frac{1}{5}$$
$$= x \times 5$$
$$= 5x$$

Since $x > 0$, then $5x > 4x$ and Column A > Column B.

9. **B** In Column A, $(a^3)^2 = a^{3 \times 2} = a^6$. As the exponent of a number that is less than 1 gets larger, the value of that expression becomes smaller. Since $0 < a < 1$, then $a^5 > a^6$ and Column B > Column A.

10. **C** In Column A,

$$\frac{1}{1 - \left(\frac{1}{2}\right)^2} = \frac{1}{1 - \frac{1}{4}}$$
$$= 1 \div \frac{3}{4}$$
$$= \frac{4}{3}$$

*(Lesson 3-6 Quantitative Comparison continued)*

Hence, Column A = Column B.

11. **A** Since $0 < x < 1$, then $\sqrt{x} < 1$, so $\frac{1}{\sqrt{x}}$ must be greater than 1. Hence, Column A > Column B.

12. **A** Since $-1 < b < 0$, $\frac{1}{b}$ must be a negative number less than $-1$. Then, since $b$ is a negative number greater than $-1$, Column A > Column B.

13. **B** Since $a + \frac{1}{5} = b$, then $b > a$. Subtracting the same number from both sides of an inequality does not change the inequality, so $b - 1 > a - 1$. Hence, Column B > Column A.

14. **B** As the exponent of a number between 0 and 1 increases, the value of that expression decreases. Since $0 < x < 1$, then $x^3 < x^2$. In Column B, the number that is being subtracted from 1 is smaller than the number that is being subtracted from 1 in Column A. Therefore the difference in Column B is greater than the difference in Column A.

**Questions 15 and 16.** $A$ is $\frac{2}{3}$ of $B$, $B$ is $\frac{3}{4}$ of $C$, and $A, B, C > 0$.

15. **B** Since $A = \frac{2}{3}B$ and $B = \frac{3}{4}C$,

$$A = \frac{2}{3}\left[\frac{3}{4}\right]C = \frac{2}{4}C = \frac{1}{2}C$$

Since $A = \frac{1}{2}C$, $\frac{A}{C} = \frac{1}{2}$, or, equivalently, $\frac{C}{A} = 2$. Since $2 > \frac{1}{2}$, Column B > Column A.

16. **C** In Column A, $8B = 8\left[\frac{3}{4}C\right] = 6C$. In Column B, $6A + 3C = 6\left[\frac{1}{2}C\right] + 3C = 6C$. Hence, Column A = Column B.

*(Grid In)*

1. **3/2** When $x = \frac{1}{2}$,

$$y = \sqrt{\frac{x+4}{2}} = \sqrt{\frac{1}{2}\left[\frac{1}{2}+4\right]} = \sqrt{\frac{1}{2}\left[\frac{9}{2}\right]} = \sqrt{\frac{9}{4}} = \frac{3}{2}$$

Grid in as 3/2.

2. **3/4** To find what fraction of $a$ is $b$, divide $b$ by $a$, as in

$$\frac{5}{6} \div \frac{10}{9} = \frac{5}{6} \times \frac{9}{10} = \frac{3}{4}$$

Grid in as 3/4.

3. **625** If $25\left(\frac{x}{y}\right) = 4$, $\frac{x}{y} = \frac{4}{25}$ so $\frac{y}{x} = \frac{25}{4}$. Hence, $100\left(\frac{y}{x}\right) = 100\left(\frac{25}{4}\right) = 25(25) = 625$.

**LESSON 3-7** *(Multiple Choice)*

1. **C** If $\frac{3}{8}$ of a number is 6, then $\frac{1}{8}$ of the same number is $\frac{1}{3}$ of 6, which is 2. Since $\frac{1}{8}$ of the

number is 2, $\frac{7}{8}$ of the same number is $7 \times 2$ or 14.

2. **B** If Claire has read $\frac{5}{8}$ of a book, $\frac{3}{8}$ of the book remains to be read. Since there are 120 pages left to read, $\frac{3}{8}$ of the total number of pages in the book equals 120. Hence, $\frac{1}{8}$ of the total number of pages in the book is $\frac{1}{3}$ of 120 or 40. Since there are 40 pages in $\frac{1}{8}$ of the book, there are $5 \times 40$ or 200 pages in the first $\frac{5}{8}$ of the book that Claire has read.

3. **B**

$$\frac{4}{9} \text{ of } 27 = \frac{4}{\cancel{9}} \times \frac{\overset{3}{\cancel{27}}}{1} = 12$$

If $\frac{6}{5}$ of a number is 12, then $\frac{1}{5}$ of that number is $\frac{1}{6}$ of 12 or 2. Since $\frac{1}{5}$ of a number is 2, the number must be $5 \times 2$ or 10.

4. **D** If $\frac{3}{5}$ of a class are boys, then $\frac{2}{5}$ are girls. Since there are 10 girls in the class, $\frac{2}{5}$ of the class represents 10 students, so in $\frac{1}{5}$ of the class there are $\frac{1}{2}$ of 10 or 5 students. Since there are 5 students in $\frac{1}{5}$ of the class, there is a total of $5 \times 5$ or 25 students in the whole class.

5. **E** The two-digit whole numbers in which either digit is $\frac{3}{4}$ of the other digit are 34, 43, 68, and 86. Use a calculator to find the sum of these four numbers: $34 + 43 + 68 + 86 = 231$.

6. **A** After Boris lends $\frac{1}{4}$ of $D$ dollars, he has $\frac{3}{4}D$ dollars left. After he spends $\frac{1}{3}$ of the money that he has left, the number of dollars he now has is $\frac{3}{4}D - \frac{1}{3}\left[\frac{3}{4}D\right]$. Simplify:

$$\frac{3}{4}D - \frac{1}{3}\left[\frac{3}{4}D\right] = \frac{3}{4}D - \frac{1}{4}D$$

$$= \frac{2}{4}D$$

$$= \frac{1}{2}D \text{ or } \frac{D}{2}$$

7. **A** The value obtained by increasing $a$ by $\frac{1}{5}$ of its value is

$$a + \frac{1}{5}a = a + 0.2a = 1.2a$$

The value obtained by decreasing $b$ by $\frac{1}{2}$ of its value is

$$b - \frac{1}{2}b = b - 0.5b = 0.5b$$

Since it is given that these two values are equal, $1.2a = 0.5b$.

*(Lesson 3-7 Multiple Choice continued)*

**8. C**

$$3\frac{1}{2} \div \frac{1}{4} = \frac{7}{2} \times \frac{4}{1} = 14$$

**9. B** If a chocolate bar that weighs $\frac{9}{16}$ of a pound is cut into seven equal parts, three parts weigh $\frac{3}{7} \times \frac{9}{16} = \frac{27}{112}$ pound.

**10. C** At the basketball game, $\frac{3}{5}$ of the students who attended were seniors, so $\frac{2}{5}$ were not seniors. Of the students who were not seniors, $\frac{1}{3}$ were juniors.

- The number of juniors who attended the game comprised $\frac{1}{3}$ of $\frac{2}{5}$ or $\frac{2}{15}$ of the total number of students.
- The number of seniors and juniors who attended represented
$$\frac{3}{5} + \frac{2}{15} = \frac{9}{15} + \frac{2}{15} = \frac{11}{15}$$
of the total number of students.
- The remaining 80 students were all sophomores. Since $1 - \frac{11}{15} = \frac{4}{15}$, $\frac{4}{15}$ of the students in attendance were sophomores. Thus,
$$\frac{4}{15} \to 80, \text{ so } \frac{1}{15} \to \frac{1}{4} \times 80 = 20$$
- Since $\frac{1}{15}$ of the total number of students is 20 and $15 \times 20 = 300$, a total of 300 students attended the game.
- Since the number of seniors comprised $\frac{3}{5}$ of the total number of students, the number of seniors who attended the game was
$$\frac{3}{5} \text{ of } 300 = \frac{3}{5} \times 300 = 180$$
Hence, 180 seniors attended the basketball game.

**11. C** If, of the 75 people in a room, $\frac{2}{5}$ are college graduates, then $\frac{3}{5}$ are *not* college graduates. If $\frac{4}{9}$ of the students who are *not* college graduates are seniors in high school, then $\frac{5}{9}$ of the students who are *not* college graduates are *not* seniors in high school. Thus, $\frac{5}{9} \times \frac{3}{5}$ or $\frac{1}{3}$ of the number of people in the room are neither college graduates nor high school seniors.
Since there are 75 people in the room and $\frac{1}{3} \times 75 = 25$, 25 people are neither college graduates nor high school seniors.

**12. B** You are given that $\frac{2}{3}$ of $\frac{3}{4}$ of a number is 24. Since

$$\frac{\overset{1}{\cancel{2}}}{\cancel{3}} \times \frac{\overset{1}{\cancel{3}}}{\cancel{4}} = \frac{1}{2}$$

$\frac{1}{2}$ of the number is 24, so the number is $2 \times 24$ or 48. Hence, $\frac{1}{4}$ of the number is $\frac{1}{4} \times 48$ or 12.

**13. C** If a man paints $\frac{3}{4}$ of a house in 2 days, then he painted $\frac{1}{4}$ of the house in $\frac{1}{3} \times 2$ or $\frac{2}{3}$ of a day. Since $\frac{1}{4}$ of the house remains to be painted, it will take him $\frac{2}{3}$ of a day to complete the job.

**14. E** If a water tank drops from being $\frac{3}{5}$ full to $\frac{1}{3}$ full, then the level of the water tank drops $\frac{4}{15}$ of a full tank since
$$\frac{3}{5} - \frac{1}{3} = \frac{9}{15} - \frac{5}{15} = \frac{4}{15}$$
The water tank drops $\frac{4}{15}$ of a full tank as a result of 12 gallons being poured out. Thus,
$$\frac{4}{15} \to 12, \text{ so } \frac{1}{15} \to \frac{1}{4} \times 12 = 3$$
Since $\frac{1}{15}$ of a full tank holds 3 gallons, $\frac{15}{15}$ or a full tank holds $15 \times 3$ or 45 gallons.

**15. C** Since Susan received $\frac{2}{3}$ of the ballots cast, $\frac{1}{3}$ of the ballots were cast for the other candidates. Bill received $\frac{5}{5} - \frac{1}{5} = \frac{4}{5}$ of the remaining ballots, which amounted to 48 votes. Bill's votes represent $\frac{4}{5} \times \frac{1}{3}$ or $\frac{4}{15}$ of the total number of ballots. Hence,
$$\frac{4}{15} \to 48, \text{ so } \frac{1}{15} \to \frac{1}{4} \times 48 = 12$$
Since 12 votes are $\frac{1}{15}$ of the total number cast, the total number of votes cast was $12 \times 15 = 180$. Since Susan received $\frac{2}{3}$ of the 180 ballots cast, she received $\frac{2}{3} \times 180 = 120$ votes.

**16. D** You can find the capacity of the gas tank by reasoning as follows:
- If a gas gauge moves from exactly $\frac{1}{8}$ full to exactly $\frac{7}{8}$ full, then the amount of gas put into the tank represents $\frac{3}{4}\left(= \frac{7}{8} - \frac{1}{8}\right)$ of the capacity of the tank.
- If the gas costs $1.50 per gallon and Arlene is charged $18.00 for the gas, then the number of gallons put into the tank equals $18 \div 1.5 = 12$.
- Since $\frac{3}{4} \to 12$, $\frac{1}{4} \to \frac{1}{3} \times 12 = 4$. Hence,
$$\frac{4}{4} \text{ (full tank)} \to 4 \times 4 = 16 \text{ gallons}$$

*(Lesson 3-7 Multiple Choice continued)*

The capacity of the gas tank is 16 gallons.

17. **E** Suppose the prices of stocks $A$ and $B$ were each $100 at the beginning of the day. Then:

- If the price of stock $A$ increases by $\frac{1}{10}$ of its original value, the new price of stock $A$ is

$$\$100 + \left[\frac{1}{10} \times \$100\right] = \$100 + \$10 = \$110$$

- If the price of stock $B$ decreases by $\frac{1}{10}$ of its original value, the new price of stock $B$ is

$$\$100 - \left[\frac{1}{10} \times \$100\right] = \$100 - \$10 = \$90$$

- The relationship between the new price of stock $A$ and the new price of stock $B$ can be found by forming a fraction in which the numerator is the new price of stock $A$ and the denominator is the new price of stock $B$: $\frac{110}{90} = \frac{11}{9}$.

The new price of stock $A$ is $\frac{11}{9}$ of the new price of stock $B$.

18. **D** Suppose $n$ represents the original number of people in the room. When $\frac{3}{4}$ of these $n$ people leave the room, $\frac{1}{4}n$ people remain. After three people enter the same room, there are $\frac{1}{4}n + 3$ people in the room. Since the number of people now in the room is $\frac{1}{3}$ of the original number of people in the room, $\frac{1}{4}n + 3 = \frac{1}{3}n$. Hence:

$$3 = \frac{1}{3}n - \frac{1}{4}n$$

Simplify the right side of this equation by subtracting fractions after changing them to their lowest common denominator:

$$3 = \frac{4}{12}n - \frac{3}{12}n$$
$$= \frac{1}{12}n$$
$$3 \times 12 = n$$
$$36 = n$$

You now know that the original number of people in the room was 36. The question asks for the number of people who left the room: $\frac{3}{4} \times 36 = 27$.

*(Grid In)*

1. **18** After a number is increased by $\frac{1}{3}$ of its value, the new number is $\frac{4}{3}$ of its original value. Since the result of the increase is 24,

$$\frac{4}{3} \to 24, \text{ so } \frac{1}{3} \to \frac{1}{4} \times 24 = 6$$

Since $\frac{1}{3}$ of the original number is 6, multiplying by 3:

$$\frac{3}{3} \to 3 \times 6 = 18$$

gives 18 as the original number.

2. **3/5** Pick a convenient number, say 100, for the number of male voters. You are told that the number of female voters was $1\frac{1}{2}$ times the number of male voters. Then there were $100 \times 1\frac{1}{2}$ or 150 female voters.

- Since $\frac{1}{2}$ of the male voters cast their ballots for candidate $A$, this candidate received $\frac{1}{2} \times 100$ or 50 votes from the males.
- Since $\frac{2}{3}$ of the female voters cast their ballots for candidate $A$, this candidate received $\frac{2}{3} \times 150$ or 100 votes from the females.
- A total of $100 + 150$ or 250 ballots were cast. Candidate $A$ received $50 + 100$ or 150 votes. Since

$$\frac{150}{250} = \frac{3 \times 50}{5 \times 50} = \frac{3}{5},$$

candidate $A$ received $\frac{3}{5}$ of the total number of votes cast.

Grid in as 3/5.

**LESSON 3-8** *(Multiple Choice)*

1. **B** $20\%$ of $\frac{2}{3}$ of $15 = \frac{1}{5} \times \frac{2}{3} \times 15 = 2$

2. **C** To change from a percent to a decimal, divide the percent by 100 and drop the percent sign:

$$\frac{2}{5}\% = \frac{\frac{2}{5}}{100} = \frac{0.4}{100} = 0.004$$

3. **E** $\frac{0.25 + 0.15}{0.50} = \frac{0.40}{0.50} = \frac{4}{5} = 0.80 = 80\%$

4. **C** If 480 of 500 seats were occupied, then 20 seats were *not* occupied. Hence, the percent of the seats that were *not* occupied is

$$\frac{20}{500} \times 100\% = \frac{1}{25} \times 100\% = 4\%$$

*(Lesson 3-8   Multiple Choice continued)*

**5. A**

$$\text{Percent decrease} = \frac{\text{Amount of decrease}}{\text{Original amount}} \times 100\%$$

$$= \frac{168-147}{168} \times 100\%$$

$$= \frac{21}{168} \times 100\%$$

$$= 0.125 \times 100\%$$

$$= 12.5\% \text{ or } 12\frac{1}{2}\%$$

**6. E**   $0.01\% = \dfrac{0.01}{100} = \dfrac{1}{10,000} = \dfrac{1}{10^4}$

**7. B**   If the result of increasing $a$ by 300% of $a$ is $b$, then $a + 3a = 4a = b$. Dividing both sides of $4a = b$ by $b$ gives $4\frac{a}{b} = 1$, so $\frac{a}{b} = \frac{1}{4}$. Since $\frac{1}{4} = 25\%$, $a$ is 25% of $b$.

**8. D**   After a 20% price increase, the new price of a radio is $78.00. Hence,

$$\text{original price} = \frac{\$78.00}{1+20\%}$$

$$= \frac{\$78.00}{1.2}$$

$$= \$78.00 \div 1.2$$

$$= \$65.00$$

**9. C**   After a discount of 15%, the price of a shirt is $51. Hence,

$$\text{original price} = \frac{\$51}{1-15\%}$$

$$= \frac{\$51}{1-0.15}$$

$$= \$51 \div 0.85$$

$$= \$60.00$$

**10. C**   Two students use the computer a total of 28% + 52% or 80% of the total time of 3 hours or 180 minutes. The third student uses the computer for 100% − 80% or 20% of 180 minutes, which equals $0.2 \times 180 = 36$ minutes.

**11. D**

$$20\% \text{ of } 25\% \text{ of } \frac{4}{5} = \frac{1}{5} \times \frac{1}{4} \times \frac{4}{5} = \frac{1}{25} = 0.04$$

**12. C**   According to the given table, a person can obtain a driver's license at 14 years of age in 7 states and at 15 years of age in 12 states. Hence, there are 7 + 12 or 19 states in which a person under the age of 16 can obtain a driver's license. Since the total number of states is 50, the percent of the states in which a person under the age of 16 can obtain a driver's license is $\frac{19}{50} \times 100\%$ or 38%.

**13. A**   The sum of all of the sectors of a circle graph is 100% Hence,

$$11\%+35\%+23\%+x\%+7\%+3x\% = 100\%$$

$$76\% + x\% + 3x\% = 100\%$$

$$4x\% = 100\% - 76\%$$

$$= 24\%$$

$$x = \frac{24\%}{4\%} = 6$$

Since $3x\% = 3(6)\% = 18\%$, municipal bonds make up 18% of the investment portfolio. To find the amount of money invested in municipal bonds, multiply the total value of the portfolio by 18%: 18% of \$250,000 = 0.18 × \$250,000 = \$45,000 Hence, \$45,000 is invested in municipal bonds.

**14. E**   The original amount invested in health stocks is 11% of the total investment. You are told that 20% of the 23% that is invested in technology stocks is reinvested in health stocks. Since 20% of 23% = 0.2 × 23% = 4.6%, a total of 11% plus 4.6% or 15.6% is now invested in health stocks. Since

$$15.6\% \text{ of } \$250,000 = 0.156 \times \$250,000$$

$$= \$39,000$$

the amount of money now invested in health stocks is \$39,000.

**15. E**   According to the height of the bars in the given graph, in 1991 the number of households with computers was 40 million and in 1995 the number was 100 million. Hence,

$$\text{Percent increase} = \frac{\text{Amount of increase}}{\text{Original amount}} \times 100\%$$

$$= \frac{100-40}{40} \times 100\%$$

$$= \frac{60}{40} \times 100\%$$

$$= 1.5 \times 100\%$$

$$= 150\%$$

**16. A**   Calculate the percent of increase for each 2 consecutive years:

➤ From 1991 to 1992:

$$\text{Percent increase} = \frac{\text{Amount of increase}}{\text{Original amount}} \times 100\%$$

$$= \frac{55-40}{40} \times 100\%$$

$$= \frac{15}{40} \times 100\%$$

$$= 37.5\%$$

*(Lesson 3-8  Multiple Choice continued)*

• From 1992 to 1993:

$$\text{Percent increase} = \frac{\text{Amount of increase}}{\text{Original amount}} \times 100\%$$

$$= \frac{65-55}{55} \times 100\%$$

$$= \frac{10}{55} \times 100\%$$

$$= 18.18\%$$

• From 1993 to 1994:

$$\text{Percent increase} = \frac{\text{Amount of increase}}{\text{Original amount}} \times 100\%$$

$$= \frac{80-65}{65} \times 100\%$$

$$= \frac{15}{65} \times 100\%$$

$$= 23.08\%$$

• From 1994 to 1995:

$$\text{Percent increase} = \frac{\text{Amount of increase}}{\text{Original amount}} \times 100\%$$

$$= \frac{100-80}{80} \times 100\%$$

$$= \frac{20}{80} \times 100\%$$

$$= 25\%$$

Hence, the greatest percent of increase occurred from 1991 to 1992.

17. **A**  Since

$$0.04\% = \frac{0.04}{100} = \frac{4}{10,000} = \frac{1}{2500}$$

it can be expected that one of every 2500 light bulbs manufactured will be defective.

18. **C**  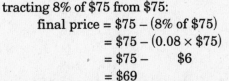 Suppose the pair of shoes cost $100. After a discount of 25%, the price of the shoes is $75. If this new price is then discounted 8%, the final price is calculated by subtracting 8% of $75 from $75:

$$\begin{aligned}\text{final price} &= \$75 - (8\% \text{ of } \$75) \\ &= \$75 - (0.08 \times \$75) \\ &= \$75 - \qquad \$6 \\ &= \$69\end{aligned}$$

Since the final selling price of $69 is $31 less than the original selling price of $100 and $\frac{\$31}{\$100} = 31\%$, successive discounts of 25% and 8% are equivalent to a single discount of 31%.

19. **C**

$$\text{Percent of} \atop \text{alcohol} = \frac{\text{Amount of alcohol by weight}}{\text{Total weight of solution}} \times 100\%$$

$$= \frac{8}{8+17} \times 100\%$$

$$= \frac{8}{25} \times 100\%$$

$$= 8 \times 4\%$$

$$= 32\%$$

20. **D**  Since 40% of $\frac{2}{5}$ and $a$ is 40% of $b$, then $a = \frac{2}{5}b$. Thus:

$$b = \frac{5}{2}a = a + \frac{3}{2}a$$

Since $b$ exceeds $a$ by $\frac{3}{2}a$ and $\frac{3}{2} = 1.5 \times 100\% = 150\%$, then $b$ exceeds $a$ by 150% of $a$.

*(Quantitative Comparison)*

1. **C**  In Column A,
$$20\% \text{ of } 40 = 0.20 \times 40 = 8$$
In Column B,
$$\frac{1}{4} \text{ of } 32 = \frac{1}{4} \times 32 = 8$$
Hence, Column A = Column B.

2. **C**  In Column A,
$$25\% \text{ of } 0.16 = \frac{1}{4} \times 0.16 = 0.04$$
In Column B,
$$\frac{4}{100} = 0.04$$
Hence, Column A = Column B.

3. **A**  In Column A,
$$30\% \text{ of } \frac{10}{9} = \frac{30}{100} \times \frac{10}{9} = \frac{3}{9} = \frac{1}{3}$$
In Column B,
$$75\% \text{ of } \frac{1}{3} = \frac{3}{4} \times \frac{1}{3} = \frac{1}{4}$$
Since $\frac{1}{3} > \frac{1}{4}$, Column A > Column B.

4. **D**  In Column A
$$1.5\% \text{ of } 10x = \frac{1.5}{100} \times 10x = 0.15x$$
If $x > 0$, then Column A > Column B. If, however, $x = 0$, Column A = Column B. Since more than one answer is possible, the correct choice is (D).

5. **C**  In Column A,
$$\frac{1}{4\%} = 1 \div \frac{4}{100} = 1 \times \frac{100}{4} = 1 \times 25 = 25$$
In Column B,
$$25\% \text{ of } 100 = 0.25 \times 100 = 25$$
Hence, Column A = Column B.

6. **C**  If the selling price of a computer is $d$ dollars, then after a 20% discount the selling price equals $d - 20\%$ of $d = d - 0.2d = 0.8d$ dollars. Hence, Column A = Column B.

*(Lesson 3-8 Quantitative Comparison continued)*

7. **C** In Column A,
$$0.09\% = \frac{0.09}{100} = 0.0009$$
In Column B,
$$\left(\frac{3}{100}\right)^2 = (0.03)^2 = 0.0009$$
Hence, Column A = Column B.

8. **B** In Column A,
$$40\% \text{ of } \frac{x}{2} = \frac{2}{5} \times \frac{x}{2} = \frac{x}{5}$$
In Column B,
$$75\% \text{ of } \frac{x}{3} = \frac{3}{4} \times \frac{x}{3} = \frac{x}{4}$$
If two positive fractions have the same numerator, then the larger fraction is the fraction with the smaller denominator. Since $x > 0$, then $\frac{x}{4} > \frac{x}{5}$. Hence, Column B > Column A.

9. **C** In Column A,
$$250\% \text{ of } 5 = 2.5 \times 5 = 12.5$$
In Column B,
$$\frac{1}{2} \text{ of } 25 = \frac{1}{2} \times 25 = 0.5 \times 25 = 12.5$$
Hence, Column A = Column B.

10. **B** You are told that $x$ is 75% of $y$, that is, $x = 75\%$ of $y$. Thus, $x = \frac{3}{4}y$. In Column A, 50% of $y$ is $\frac{1}{2}y$. In Column B, 75% of $x$ is $\frac{3}{4}x$. Since $x = \frac{3}{4}y$, the quantities to be compared are as follows:

| Column A | Column B |
|---|---|
| $\frac{1}{2}y = \frac{8}{16}y$ | $\frac{3}{4}x = \frac{3}{4}\left(\frac{3}{4}y\right) = \frac{9}{16}y$ |

Since $y > 0$, then $\frac{9}{16}y > \frac{8}{16}y$, so Column B > Column A.

11. **C** In Column B, 150% of $a$ is $1.5 \times a = \frac{3}{2}a$. Since $a = \frac{2}{3}b$, the quantities to be compared are as follows:

| Column A | Column B |
|---|---|
| $b$ | $\frac{3}{2}a = \frac{3}{2}\left(\frac{2}{3}b\right) = b$ |

Hence, Column A = Column B.

*(Grid In)*

1. **87.5** The cash price of $84.00 reflects the new amount after a discount of 4%. Hence,

$$\text{Original amount} = \frac{\text{New amount after decrease of } P\%}{1 - P\%}$$
$$= \frac{\$84}{1 - 4\%}$$
$$= \frac{\$84}{1 - 0.04}$$
$$= \frac{\$84}{0.96}$$
$$= \$87.5$$

The credit-card purchase price of the same item is $87.5.

2. **50** After three boys are dropped from the class, 25 students remain. Of the 25 students, 44% are boys, so 56% are girls. Since 56% of 25 = $0.56 \times 25 = 14$, 14 girls are enrolled in the class. Hence, 14 of the 28 students in the original class were girls. Thus, the number of girls in the original class comprised $\frac{1}{2}$ or 50% of that class.

CHAPTER

# 4

# Algebraic Methods

This chapter reviews equations and inequalities, as well as some related algebraic methods that you need to know for the SAT I.

---

## LESSONS IN THIS CHAPTER

---

# Solving Equations

## OVERVIEW

A **linear** or **first-degree equation** is an equation such as $2y + 1 = 13$ in which the greatest exponent of the variable is 1. To solve a linear equation do the same things to both sides of the equation until the resulting equation has the form

$$variable = \text{number}$$

## SOLVING LINEAR EQUATIONS BY USING INVERSE OPERATIONS

To isolate the variable in a linear equation, you must perform the same arithmetic operation on each side of the equation.

- In the equation $x - 7 = -3$, variable $x$ can be isolated by adding 7 on each side of the equation:

$$
\begin{array}{r}
x - 7 = -3 \\
\underline{+7 = +7} \\
x + 0 = 4 \\
x = 4
\end{array}
$$

- In the equation $x + 5 = 3$, variable $x$ can be isolated by subtracting 5 on each side of the equation:

$$
\begin{array}{r}
x + 5 = 3 \\
\underline{-5 = -5} \\
x + 0 = -2 \\
x = -2
\end{array}
$$

- In the equation $3x = 21$, variable $x$ can be isolated by dividing each side of the equation by 3:

$$3x = 21$$

$$\frac{3x}{3} = \frac{21}{3}$$

$$x = 7$$

- In the equation $\frac{2}{3}x = -8$, variable $x$ can be isolated by multiplying each side of the equation by the reciprocal of the fractional coefficient of the variable:

$$\frac{2}{3}x = -8$$

$$\frac{3}{2}\left(\frac{2}{3}x\right) = \frac{3}{\cancel{2}}\overset{-4}{(\cancel{-8})}$$

$$x = -12$$

## SOLVING LINEAR EQUATIONS BY USING TWO OPERATIONS

To isolate the variable in a linear equation, it may be necessary to add or subtract and then to multiply or divide. For example, to solve $2y + 1 = 13$, begin by subtracting 1 on each side of the equation to get $2y = 12$. Dividing each side of this equation by 2 gives $y = 6$.

## SOLVING EQUATIONS WITH PARENTHESES

If an equation contains parentheses, remove them by applying the distributive law. Then solve the resulting equation.

**EXAMPLE:**    Solve for $b$:  $3(b + 2) + 2b = 21$

**SOLUTION:**    $3(b + 2) + 2b = 21$

- Remove the parentheses by multiplying each term inside the parentheses by 3:    $3b + 6 + 2b = 21$

- Combine like terms:    $5b + 6 = 21$

- Subtract 6 from each side of the equation:    $5b + 6 - 6 = 21 - 6$

- Simplify:    $5b = 15$

- Divide each side of the equation by 5:    $\frac{5b}{5} = \frac{\overset{3}{\cancel{15}}}{\cancel{5}}$

    $b = 5$

## SOLVING EQUATIONS THAT CONTAIN FRACTIONS

If an equation contains fractions, eliminate them by multiplying each member of *both* sides of the equation by the LCD of all the denominators. Then solve the resulting equation.

**EXAMPLE:**  Solve for $m$: $\dfrac{m}{3} - \dfrac{m}{4} = 2$

**SOLUTION:**  Since the LCD of 3 and 4 is 12, multiply each member of the given equation by 12:

$$12\left[\frac{m}{3}\right] - 12\left[\frac{m}{4}\right] = 12\,(2)$$

$$4m \quad - \quad 3m \quad = 24$$

$$m = 24$$

## SOLVING EQUATIONS BY CROSS-MULTIPLYING

Since $\frac{4}{8} = \frac{1}{2}$, $4 \times 2 = 8 \times 1$. This operation is sometimes referred to as **cross-multiplying**. If an equation sets one fraction equal to another fraction, then cross-multiply and solve the equation that results.

**EXAMPLE:**  If $\dfrac{x}{3} = \dfrac{x+4}{5}$, what is the value of $x$?

**SOLUTION:**  Cross-multiply by setting the products of opposite pairs of terms equal:

$$\frac{x}{3} \diagdown\hspace{-0.9em}\diagup \frac{(x+4)}{5}$$

$$5x = 3\,(x+4)$$

$$5x = 3x + 12$$

$$2x = 12$$

$$x = \frac{12}{2} = 6$$

## SOLVING EQUATIONS WITH VARIABLE EXPONENTS

The SAT I may include an equation in which the variable is in an exponent. To solve this type of equation, write each side of the equation as a power of the same base. Then set the exponents equal to each other. The result will be another equation that you already know how to solve.

**EXAMPLE:**  If $3^{2x-5} = 27$, what is the value of $x$?

**SOLUTION:**

- Write 27 as a power of 3:  $3^{2x-5} = 27 = 3^3$
- Since the bases are the same, equate the exponents:  $2x - 5 = 3$
- Solve for $x$:  $2x = 8$
  $$x = \frac{8}{2} = 4$$

## USING A ROOT OF AN EQUATION TO ANSWER A QUESTION

To find the value of an algebraic expression that involves a particular variable, you may need to first find the value of that variable by solving an equation.

**EXAMPLE:**    If $3y - 2 = 4$, what is the value of $y - 2$?

**SOLUTION:**

- If $3y - 2 = 4$, then $3y = 6$, so $y = \frac{6}{3} = 2$.
- Since $y = 2$, then $y - 2 = 2 - 2 = 0$.
- The value of $y - 2$ is 0.

## TIPS FOR SCORING HIGH

1. If time permits, plug your solution of an equation back into the original equation to make sure that it works.
2. If you get stuck on a multiple choice question that asks for the value of a variable in an equation, you may be able to find the solution by testing each of the numerical answer choices in the equation until you find the one that works.
3. Answer the question asked. For example, if the question is "If $3y - 2$ = 4, what is the value of $y - 2$?" make sure your answer is the value of $y - 2$, and not the value of $y$.

# LESSON 4-1  TUNE-UP EXERCISES

| Multiple Choice |
|---|

**1** If $x + 4x - 3x + 8 = 0$, then $x =$

(A) −4
(B) −2
(C) 0
(D) 2
(E) 6

**2** If $(9 - 4)(x + 4) = 30$, then $x =$

(A) 2
(B) 4
(C) 6
(D) 8
(E) 12

**3** If $2(1 + 5) = 3(w - 4)$, then $w =$

(A) 0
(B) 2
(C) 4
(D) 6
(E) 8

**4** If $\frac{x}{3} - 1 = 1 - 3$, then $x =$

(A) −6
(B) −3
(C) −1
(D) 3
(E) 6

**5** If $3(x - 2) - 2x = 7$, then $x =$

(A) 1
(B) 4
(C) 8
(D) 13
(E) 14

**6** If $3x - 6 = 18$, then $x - 2 =$

(A) 2
(B) 4
(C) 6
(D) 8
(E) 10

**7** If $2x + y = y + 14$, then $x =$

(A) $\frac{7}{2}$
(B) 7
(C) 14
(D) 21
(E) It cannot be determined.

**8** If $\frac{4}{7}k = 36$, then $\frac{3}{7}k =$

(A) 21
(B) 27
(C) 32
(D) 35
(E) 42

**9** If $\frac{1}{2}x + \frac{1}{4}x + \frac{1}{8}x = 14$, then $x =$

(A) 4
(B) 8
(C) 12
(D) 16
(E) 24

**10** If $2s + 3t = 12$ and $4s = 36$, then $t =$

(A) −3
(B) −2
(C) 2
(D) 3
(E) 9

**11** If $2^{x+1} = 32$, then $(x+1)^2 =$

(A) 4
(B) 9
(C) 16
(D) 25
(E) 36

$2^{x+1} = 2^5$
$x+1 = 5 \quad (5)^2 = 25$
$x = 4$

**12** If $\frac{2}{x} = 2$, then $x + 2 =$

(A) $\frac{3}{2}$
(B) $\frac{5}{2}$
(C) 3
(D) 4
(E) 6

**13** If $\frac{y-2}{2} = y + 2$, then $y =$

(A) −6
(B) −4
(C) −2
(D) 4
(E) 6

**14** If $\frac{2y}{7} = \frac{y+3}{4}$, then $y =$

(A) 5
(B) 9
(C) 13
(D) 17
(E) 21

$-7y + 21 = 8y - 7y$
$y = 21$

**15** If $\frac{y}{3} = 4$, then $3y =$

(A) 4
(B) 12
(C) 24
(D) 30
(E) 36

**16** If $\frac{1}{q} + \frac{3}{q} = 12$, then $q =$

(A) $\frac{1}{4}$
(B) $\frac{1}{3}$
(C) $\frac{1}{2}$
(D) 3
(E) 4

**17** If $8 - 3x = 2x + 13$, then $x =$

(A) −4
(B) −3
(C) −2
(D) −1
(E) 1

**18** If $(1-2-3-4)\,w = (-1)(-2)(-3)(-4)$, then $w =$

(A) −3
(B) −2
(C) −1
(D) 2
(E) 3

**19** If $\frac{1}{3} + \frac{3}{p} = 1$, then $p =$

(A) $\frac{3}{2}$
(B) 2
(C) $\frac{5}{2}$
(D) $\frac{9}{2}$
(E) 6

$\frac{3}{p} = \frac{2}{3}$

$\frac{2p = 9}{2 \quad 2}$

**20** If $\frac{3}{4} = \frac{6}{x} = \frac{9}{y}$, then $x + y =$

(A) 4
(B) 8
(C) 12
(D) 16
(E) 20

---

### Quantitative Comparison

Each question consists of two quantities in boxes, one in Column A and one in Column B. You are to compare the two quantities and on the answer sheet fill in

    A  if the quantity in Column A is greater;
    B  if the quantity in Column B is greater;
    C  if the two quantities are equal;
    D  if the relationship cannot be determined from the information given

AN E RESPONSE WILL NOT BE SCORED

---

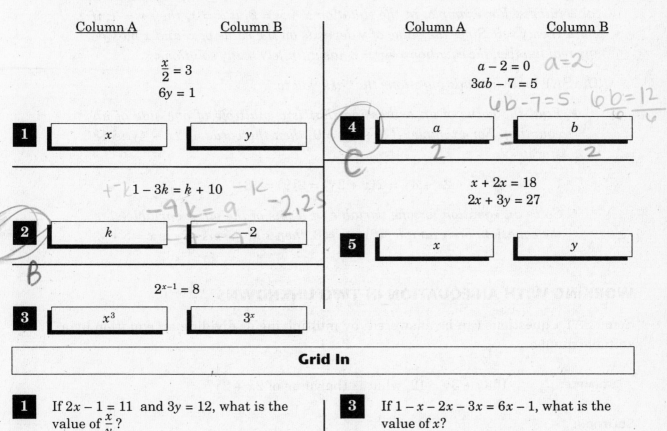

| Column A | Column B | | Column A | Column B |
|---|---|---|---|---|

$$\frac{x}{2} = 3$$
$$6y = 1$$

**1**    $x$      $y$

$$a - 2 = 0 \quad a = 2$$
$$3ab - 7 = 5$$
$$6b - 7 = 5 \quad 6b = 12$$

**4**    $a$  =  $b$

$$1 - 3k = k + 10$$
$$-9k = 9 \quad -2.25$$

**2**    $k$      $-2$

$$x + 2x = 18$$
$$2x + 3y = 27$$

**5**    $x$      $y$

$$2^{x-1} = 8$$

**3**    $x^3$      $3^x$

---

### Grid In

**1**   If $2x - 1 = 11$ and $3y = 12$, what is the value of $\frac{x}{y}$?

**3**   If $1 - x - 2x - 3x = 6x - 1$, what is the value of $x$?

**2**   If $7(a + b) - 4(a + b) = 24$, what is the value of $a + b$?

**4**   In the equation $p = \frac{5b}{c^2}$, what is a value of $c$ when $p = 9$ and $b = 20$?

$$7a + 7b - 4a - 4b = 24$$
$$3a + 3b = 24$$
$$a + b = 8$$

$$\frac{c^2 9}{9} = \frac{5(20)}{c^2} \cdot \frac{c^2}{9}$$
$$c^2 = 11.\overline{1}$$
$$c = 3.\overline{3}$$

# Equations with More Than One Variable

## OVERVIEW

*An equation that has more than one variable can have many different numerical solutions. For example, in the equation $x + y = 8$, if $x = 1$, then $y = 7$; if $x = 2$, then $y = 6$. Since the value of $y$ depends on the value of $x$, and $x$ may be any real number, the equation $x + y = 8$ has infinitely many solutions.*

*The SAT I may include questions that ask you to:*

- *Find the value of an expression that is a multiple of one side of an equation. For example, if $x + 2y = 9$, then the value of $2x + 4y$ is 18 since*

$$2x + 4y = 2(x + 2y) = 2(9) = 18$$

- *Solve an equation for one variable in terms of the other variable(s) of the equation. For example, if $x + y = 8$, then $x$ in terms of $y$ is $x = 8 - y$.*

## WORKING WITH AN EQUATION IN TWO UNKNOWNS

Some SAT I questions can be answered by multiplying or dividing an equation by a suitable number.

**EXAMPLE:**    If $3x + 3y = 12$, what is the value of $2x + 2y$?

**SOLUTION:**

- Divide each member of $3x + 3y = 12$ by 3:

$$\frac{\overset{1}{\cancel{3}}x}{\cancel{3}} + \frac{\overset{1}{\cancel{3}}y}{\cancel{3}} = \frac{12}{3}$$
$$x + y = 4$$

- Multiply each member of $x + y = 4$ by 2:

$$2x + 2y = 2(4) = 8$$

## SOLVING FOR ONE VARIABLE IN TERMS OF ANOTHER

To solve an equation that includes more than one variable, isolate in the usual way the variable that is being solved for.

**EXAMPLE:**    If $2x - z = y$, what is $x$ in terms of $y$ and $z$?

**SOLUTION:**    Treat $y$ and $z$ as constants, and isolate $x$ on one side of the equation.

- Add $z$ on both sides of $2x - z = y$ to get $2x = y + z$.
- Divide both sides of $2x = y + z$ by 2:

$$x = \frac{y+z}{2}$$

# LESSON 4-2 TUNE-UP EXERCISES

**Multiple Choice**

**1** If $6 = 2x + 4y$, what is the value of $x + 2y$?

(A) 2
(B) 3
(C) 6
(D) 8
(E) 12

**2** If $\frac{a}{2} + \frac{b}{2} = 3$, what is the value of $2a + 2b$?

(A) 6
(B) 8
(C) 12
(D) 16
(E) 24

**3** If $2s - 3t = 3t - s$, what is $s$ in terms of $t$?

(A) $\frac{t}{2}$
(B) $2t$
(C) $t + 2$
(D) $\frac{t}{2} + 1$
(E) $3t$

**4** If $a + b = 5$ and $\frac{c}{2} = 3$, what is the value of $2a + 2b + 2c$?

(A) 12
(B) 14
(C) 16
(D) 20
(E) 22

**5** If $xy + z = y$, what is $x$ in terms of $y$ and $z$?

(A) $\frac{y+z}{y}$
(B) $\frac{y-z}{z}$
(C) $\frac{y-z}{y}$
(D) $1 - z$
(E) $\frac{z-y}{y}$

**6** If $4^{x+y} = 8$, what is the value of $2x + 2y$?

(A) $\frac{2}{3}$
(B) $\frac{3}{2}$
(C) 3
(D) 6
(E) 8

**7** If $2x^2 + 3y^2 = 0$, what is the value of $3x + 2y$?

(A) $-1$
(B) 0
(C) 1
(D) 3
(E) 5

**8** If $\frac{a-b}{b} = \frac{2}{3}$, what is the value of $\frac{a}{b}$?

(A) $\frac{1}{2}$
(B) $\frac{3}{5}$
(C) $\frac{3}{2}$
(D) $\frac{5}{3}$
(E) 2

**9** If $\dfrac{10^x}{10^y} = 100$, then $x =$

(A) $y + 2$
(B) $y - 2$
(C) $2 - y$
(D) $2y$
(E) $\dfrac{y}{2}$

**10** If $\dfrac{1}{p+q} = r$ and $p \neq -q$, what is $p$ in terms of $r$ and $q$?

(A) $\dfrac{rq-1}{q}$
(B) $\dfrac{1+rq}{q}$
(C) $\dfrac{r}{1+rq}$
(D) $\dfrac{1-rq}{r}$
(E) $\dfrac{1-q}{rq}$

**11** If $a = 2b = 5c$, then $4a$ is equal to which of the following expressions?

  I. $8c$
 II. $4b + 10c$
III. $2b + 15c$

(A) I, II, and III
(B) II and III only
(C) II only
(D) III only
(E) None

**12** If $\dfrac{a+b+c}{3} = \dfrac{a+b}{2}$, then $c =$

(A) $\dfrac{a-b}{2}$
(B) $\dfrac{a+b}{2}$
(C) $5a + 5b$
(D) $\dfrac{a+b}{5}$
(E) $-a - b$

**13** If $wx = z$, which of the following expressions is equal to $xz$?

(A) $\dfrac{w}{z^2}$
(B) $\dfrac{w^2}{z}$
(C) $wz^2$
(D) $w^2z$
(E) $\dfrac{z^2}{w}$

**14** If $q = \dfrac{p}{\sqrt{r}}$ and $p \neq 0$, then $\dfrac{1}{p^2} =$

(A) $\dfrac{r}{q^2}$
(B) $\dfrac{1}{q^2r}$
(C) $\dfrac{1}{qr}$
(D) $\dfrac{q}{r^2}$
(E) $\dfrac{1}{qr^2}$

**15** If $\dfrac{c}{d} - \dfrac{a}{b} = x$, $a = 2c$, and $b = 5d$, what is the value of $\dfrac{c}{d}$ in terms of $x$?

(A) $\dfrac{2}{3}x$
(B) $\dfrac{3}{4}x$
(C) $\dfrac{4}{3}x$
(D) $\dfrac{5}{3}x$
(E) $\dfrac{7}{2}x$

## Quantitative Comparison

Each question consists of two quantities in boxes, one in Column A and one in Column B. You are to compare the two quantities and on the answer sheet fill in

    A  if the quantity in Column A is greater;
    B  if the quantity in Column B is greater;
    C  if the two quantities are equal;
    D  if the relationship cannot be determined from the information given

AN E RESPONSE WILL NOT BE SCORED

| Column A | Column B |
| --- | --- |

$$2^x \cdot 2^y = 8$$

**1**    $2x + 2y$      $6$

$$x - y = 2$$

**2**    $(-1)^{3x-3y}$      $(-1)^7$

$$\frac{1}{a} = b, \ \frac{1}{b} = c, \ b \neq 0$$

**3**    $a$      $c$

| Column A | Column B |
| --- | --- |

$$8^x = 4^y$$

**4**    $\dfrac{y}{x}$      $2$

$$\frac{a}{b} = \frac{7}{8} \ \text{and} \ \frac{c}{a} = \frac{4}{7}$$
$$a, b, c > 0$$

**5**    $b$      $3c$

## Grid In

**1**   If $16 \times a^2 \times 64 = (4 \times b)^2$ and $a$ and $b$ are positive integers, then $b$ is how many times greater than $a$?

**2**   If $3^{x-1} + 3^{x-1} + 3^{x-1} = 81^y$, what is the value of $\dfrac{x}{y}$?

# Polynomials and Algebraic Fractions

## OVERVIEW

A **polynomial** *is a single term or the sum or difference of two or more unlike terms. For example, the polynomial* $a + 2b + 3c$ *represents the sum of the three unlike terms* $a$, $2b$, *and* $3c$. *Since polynomials represent real numbers, they can be added, subtracted, multiplied, and divided using the laws of arithmetic.*

*Whenever a variable appears in the denominator of a fraction, you may assume that it cannot have a value that makes the denominator of the fraction equal to 0.*

## CLASSIFYING POLYNOMIALS

A polynomial can be classified according to the number of terms it contains.

- A polynomial with one term, as in $3x^2$, is called a **monomial**.
- A polynomial with two unlike terms, as in $2x + 3y$, is called a **binomial**.
- A polynomial with three unlike terms, as in $x^2 + 3x - 5$, is called a **trinomial**.

## OPERATIONS WITH POLYNOMIALS

Polynomials may be added, subtracted, multiplied, and divided.

- To add polynomials, write one polynomial on top of the other one so that like terms are aligned in the same vertical columns. Then combine like terms. For example:

$$\begin{array}{r} 2x^2 - 3x + 7 \\ + \quad x^2 + 5x - 9 \\ \hline 3x^2 + 2x - 2 \end{array}$$

- To subtract polynomials take the opposite of each term of the polynomial that is being subtracted. Then add the two polynomials. For instance, the difference

$$(7x - 3y - 9z) - (5x + y - 4z)$$

can be changed into a sum by adding the opposite of each term of the second polynomial to the first polynomial:

$$7x - 3y - 9z$$
$$+ \quad -5x - \ y + 4z$$
$$\overline{2x - 4y - 5z}$$

- To multiply monomials, multiply their numerical coefficients and multiply *like* variable factors by *adding* their exponents. For example:

$$(-2a^2b)(4a^3b^2) = (-2)(4)(a^2 \cdot a^3)(b \cdot b^2)$$

$$= -8(a^{2+3})(b^{1+2})$$

$$= -8a^5b^3$$

- To divide monomials, divide their numerical coefficients and divide *like* variable factors by *subtracting* their exponents. For example:

$$\frac{14x^5y^2}{21x^2y^2} = \left(\frac{14}{21}\right)\left(\frac{x^5}{x^2}\right)\left(\frac{y^2}{y^2}\right)$$

$$= \left(\frac{2}{3}\right)(x^{5-2})(1)$$

$$= \frac{2}{3}x^3$$

- To multiply a polynomial by a monomial, multiply each term of the polynomial by the monomial and add the resulting products. For example:

$$5x(3x - y + 2) = 5x(3x) + 5x(-y) + 5x(2)$$

$$= 15x^2 \ - \ 5xy \ + \ 10$$

The expression $-(a - b)$ can be interpreted as "take the opposite of whatever is inside the parentheses." The result is $b - a$ since

$$-(a - b) = -1(a - b)$$

$$= (-1)a + (-1)(-b)$$

$$= -a \quad + \ b \ \text{ or } \ b - a$$

- To divide a polynomial by a monomial, divide each term of the polynomial by the monomial and add the resulting quotients. For example:

$$\frac{6x + 15}{3} = \frac{6x}{3} + \frac{15}{3} = 2x + 5$$

## MULTIPLYING BINOMIALS USING FOIL

To find the product of two binomials, write them next to each other and then add the products of their *First*, *Inner*, *Outer*, and *Last* pairs of terms.

**EXAMPLE:** Multiply: $(2x + 1)(x - 5)$.

**SOLUTION:**   For the product $(2x + 1)(x - 5)$, $2x$ and $x$ are the *F*irst pair of terms; $2x$ and $-5$ are the *O*utermost pairs of terms; 1 and $x$ are the *I*nnermost terms; 1 and $-5$ are the *L*ast terms of the binomials. Thus:

$$\begin{array}{cccc} \overbrace{F} & \overbrace{O} & \overbrace{I} & \overbrace{L} \end{array}$$

$$(2x + 1)(x - 5) = (2x)(x) + (2x)(-5) + (1)(x) + (1)(-5)$$

$$= 2x^2 \quad + \quad [-10x + x] \quad - 5$$

$$= 2x^2 \quad - \quad 9x \quad - 5$$

## PRODUCTS OF SPECIAL PAIRS OF BINOMIALS

You can save some time if you memorize and learn to recognize when these special multiplication rules can be applied:

---
### Special Multiplication Rules

- $(a - b)(a + b) = a^2 - b^2$
- $(a + b)^2 = (a + b)(a + b) = a^2 + 2ab + b^2$
- $(a - b)^2 = (a - b)(a - b) = a^2 - 2ab + b^2$
---

**EXAMPLE:**     Express $(x + 3)(x - 3)$ as a binomial.

**SOLUTION:**   $(x + 3)(x - 3) = (x)^2 - (3)^2$
$$= x^2 - 9$$

**EXAMPLE:**     Express $(2y - 1)(2y + 1)$ as a binomial.

**SOLUTION:**   $(2y - 1)(2y + 1) = (2y)^2 - (1)^2$
$$= 4y^2 - 1$$

**EXAMPLE:**     If $(x + y)^2 - (x - y)^2 = 28$, what is the value of $xy$?

**SOLUTION:**

- Square each binomial:

$$(x + y)^2 \quad - \quad (x - y)^2 \quad = 28$$

$$(x^2 + 2xy + y^2) - (x^2 - 2xy + y^2) = 28$$

- Write the first squared binomial without the parentheses. Then remove the second set of parentheses by changing the sign of each term inside the parentheses to its opposite.

$$x^2 + 2xy + y^2 - x^2 + 2xy - y^2 = 28$$

- Combine like terms. Adding $x^2$ and $-x^2$ gives 0, as does adding $y^2$ and $-y^2$. The sum of $2xy$ and $2xy$ is $4xy$. The result is

$$4xy = 28$$

$$xy = \frac{28}{4} = 7$$

## COMBINING ALGEBRAIC FRACTIONS

Algebraic fractions are combined in much the same way as fractions in arithmetic.

**EXAMPLE:**   Write $\frac{w}{2} - \frac{w}{3}$ as a single fraction.

**SOLUTION:**   The LCD of 2 and 3 is 6. Change each fraction into an equivalent fraction that has 6 as its denominator.

$$\frac{w}{2} - \frac{w}{3} = \frac{3}{3}\left(\frac{w}{2}\right) - \frac{2}{2}\left(\frac{w}{3}\right)$$

$$= \frac{3w}{6} - \frac{2w}{6}$$

$$= \frac{3w - 2w}{6}$$

$$= \frac{w}{6}$$

**EXAMPLE:**   If $h = \frac{y}{x - y}$, what is $h + 1$ in terms of $x$ and $y$?

**SOLUTION:**

- Add 1 on both sides of the given equation:

$$h + 1 = \frac{y}{x - y} + 1$$

- On the right side of the equation, replace 1 with $\frac{x - y}{x - y}$:

$$h + 1 = \frac{y}{x - y} + \frac{x - y}{x - y}$$

- Write the sum of the numerators over the common denominator:

$$h + 1 = \frac{y + x - y}{x - y} = \frac{x}{x - y}$$

## THE SUM AND DIFFERENCE OF TWO RECIPROCALS

The formulas for the sum and the difference of the reciprocals of two nonzero numbers, $x$ and $y$, are worth remembering:

If $x$ and $y$ are not 0, then

$$\frac{1}{x} + \frac{1}{y} = \frac{x+y}{xy} \quad \text{and} \quad \frac{1}{x} - \frac{1}{y} = \frac{y-x}{xy}$$

**EXAMPLE:** If $t = \dfrac{1}{r} + \dfrac{1}{s}$, then what is $\dfrac{1}{t}$ in terms of $r$ and $s$?

**SOLUTION:** Since $t = \dfrac{1}{r} + \dfrac{1}{s} = \dfrac{r+s}{rs}$, then $\dfrac{1}{t} = \dfrac{rs}{r+s}$ .

## LESSON 4-3  TUNE-UP EXERCISES

---

**Multiple Choice**

---

**1** $\dfrac{20b^3 - 8b}{4b} =$

(A) $5b^2 - 2b$
(B) $5b^3 - 2$
(C) $5b^2 - 8b$
(D) $5b^2 - 2$
(E) $5b - 2$

**2** $(39)^2 + 2(39)(61) + (61)^2 =$

(A) 9,099
(B) 9,909
(C) 9,990
(D) 10,000
(E) 10,001

**3** If $p = 5$, then $\dfrac{3p^2 + 12p}{3p} =$

(A) 5
(B) 9
(C) 12
(D) 15
(E) 18

**4** Which expression is equal to $\left[\left(\dfrac{a^5b^4c^2}{ab^2c}\right)^2\right]^3$?

(A) $a^{20}b^{10}c^5$
(B) $a^{24}b^{12}c^6$
(C) $a^{25}b^{12}c^6$
(D) $a^{24}b^{12}c^5$
(E) $a^{24}b^{12}$

**5** If $(x - y)^2 = 50$ and $xy = 7$, what is the value of $x^2 + y^2$?

(A) 8
(B) 36
(C) 43
(D) 57
(E) 64

**6** If $p = \dfrac{a}{a - b}$ and $a \neq b$, then, in terms of $a$ and $b$, $1 - p =$

(A) $\dfrac{a}{b - a}$

(B) $\dfrac{b}{b - a}$

(C) $\dfrac{a}{a - b}$

(D) $\dfrac{b}{a - b}$

(E) $\dfrac{a + b}{a - b}$

**7** If $(p - q)^2 = 25$ and $pq = 14$, what is the value of $(p + q)^2$?

(A) 25
(B) 36
(C) 53
(D) 64
(E) 81

**8** If $a - b = p$ and $a + b = k$, then $a^2 - b^2 =$

(A) $pk$
(B) $p^2 - k^2$
(C) $p + k$
(D) $\dfrac{p^2}{k^2}$
(E) $k^2 - p^2$

**9** If $(2y + k)^2 = 4y^2 - 12y + k^2$, what is the value of $k$?

(A) 3
(B) 1
(C) −1
(D) −2
(E) −3

**10** Which statement is true for all real values of $x$ and $y$?

(A) $(x + y)^2 = x^2 + y^2$
(B) $x^2 + x^2 = x^4$
(C) $\dfrac{2^{x+2}}{2^x} = 4$
(D) $(3x)^2 = 6x^2$
(E) $x^5 - x^3 = x^2$

**11** If $(px + q)^2 = 9x^2 + kx + 16$, what is the value of $k$?

(A) 3
(B) 4
(C) 12
(D) 24
(E) 30

**12** If $\left(\dfrac{x}{100} - 1\right)\left(\dfrac{x}{100} + 1\right) = kx^2 - 1$, then $k =$

(A) 0.1
(B) 0.01
(C) 0.001
(D) 0.0001
(E) 0.000001

**13** If $(x + 5)(x + p) = x^2 + 2x + k$, then

(A) $p = 3$ and $k = 5$
(B) $p = -3$ and $k = 15$
(C) $p = 3$ and $k = -15$
(D) $p = 3$ and $k = 15$
(E) $p = -3$ and $k = -15$

**14** For what value of $p$ is $(x - 2)(x + 2) = x(x - p)$?

(A) $-4$
(B) 0
(C) $\dfrac{2}{x}$
(D) $\dfrac{4}{x}$
(E) $-\dfrac{4}{x}$

**15** For $x, y > 0$, which expression is equal to $\dfrac{1}{\dfrac{1}{x} + \dfrac{1}{y}}$ ?

(A) $x + y$
(B) $\dfrac{x + y}{xy}$
(C) $\dfrac{xy}{x + y}$
(D) $\dfrac{x}{y} + \dfrac{y}{x}$
(E) $\dfrac{1}{xy}$

**16** If $\left(k + \dfrac{1}{k}\right)^2 = 16$, then $k^2 + \dfrac{1}{k^2} =$

(A) 4
(B) 8
(C) 12
(D) 14
(E) 18

**17** If $\dfrac{a}{b} = 1 - \dfrac{x}{y}$, then $\dfrac{b}{a} =$

(A) $\dfrac{x}{y - x}$
(B) $\dfrac{y}{x} - 1$
(C) $\dfrac{y}{x - y}$
(D) $\dfrac{x}{y} + 1$
(E) $\dfrac{y}{y - x}$

**18** If $\dfrac{a}{b} = \dfrac{1}{x} - \dfrac{1}{y}$, then $\dfrac{b}{a} =$

(A) $\dfrac{xy}{y - x}$
(B) $\dfrac{xy}{x + y}$
(C) $\dfrac{xy}{x - y}$
(D) $\dfrac{x + y}{y - x}$
(E) $\dfrac{y - x}{y + x}$

## Quantitative Comparison

Each question consists of two quantities in boxes, one in Column A and one in Column B. You are to compare the two quantities and on the answer sheet fill in

A if the quantity in Column A is greater;
B if the quantity in Column B is greater;
C if the two quantities are equal;
D if the relationship cannot be determined from the information given

AN E RESPONSE WILL NOT BE SCORED

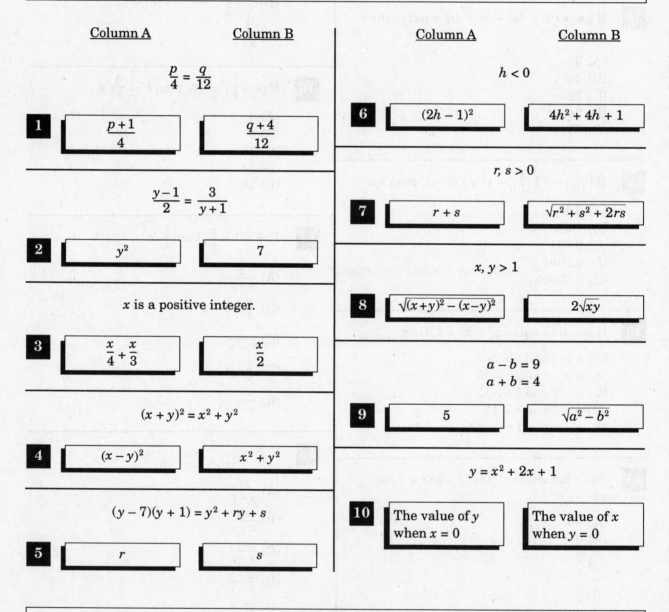

| Column A | Column B |
| --- | --- |

$$\frac{p}{4} = \frac{q}{12}$$

**1** $\dfrac{p+1}{4}$     $\dfrac{q+4}{12}$

$$\frac{y-1}{2} = \frac{3}{y+1}$$

**2** $y^2$     $7$

$x$ is a positive integer.

**3** $\dfrac{x}{4} + \dfrac{x}{3}$     $\dfrac{x}{2}$

$(x+y)^2 = x^2 + y^2$

**4** $(x-y)^2$     $x^2 + y^2$

$(y-7)(y+1) = y^2 + ry + s$

**5** $r$     $s$

| Column A | Column B |
| --- | --- |

$h < 0$

**6** $(2h-1)^2$     $4h^2 + 4h + 1$

$r, s > 0$

**7** $r + s$     $\sqrt{r^2 + s^2 + 2rs}$

$x, y > 1$

**8** $\sqrt{(x+y)^2 - (x-y)^2}$     $2\sqrt{xy}$

$a - b = 9$
$a + b = 4$

**9** $5$     $\sqrt{a^2 - b^2}$

$y = x^2 + 2x + 1$

**10** The value of $y$ when $x = 0$     The value of $x$ when $y = 0$

## Grid In

**1** If $(3y - 1)(2y + k) = ay^2 + by - 5$ for all values of $y$, what is the value of $a + b$?

**2** If $4x^2 + 20x + r = (2x + s)^2$ for all values of $x$, what is the value of $r - s$?

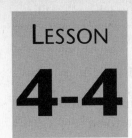

# LESSON 4-4

# Factoring

## OVERVIEW

*Factoring* reverses multiplication.

| Operation | Example |
|-----------|---------|
| Multiplication | $2(x + 3y) = 2x + 6y$ |
| Factoring | $2x + 6y = 2(x + 3y)$ |

*There are three basic types of factoring that you need to know for the* SAT I:

- *Factoring the difference of two squares using the rule*

$$a^2 - b^2 = (a + b)(a - b)$$

- *Factoring out a common monomial factor, as in*

$$4x^2 - 6x = 2x(x - 3) \quad \text{and} \quad ay - by = y(a - b)$$

- *Factoring a quadratic trinomial using the reverse of FOIL, as in:*

$$x^2 - x - 6 = (x - 3)(x + 2) \quad \text{and} \quad x^2 - 2x + 1 = (x - 1)(x - 1) = (x - 1)^2$$

## FACTORING A POLYNOMIAL BY REMOVING A COMMON FACTOR

If all of the terms of a polynomial have factors in common, the polynomial can be factored by using the reverse of the distributive property to remove these common factors. For example, in

$$24x^3 + 16x = 8x(3x^2 + 2)$$

$8x$ is the *Greatest Common Factor* (GCF) of $24x^3$ and $16x$ since 8 is the GCF of 24 and 16, and $x$ is the greatest power of that variable that is contained in both $24x^3$ and $16x$. The factor that corresponds to $8x$ can be obtained by dividing $24x^3 + 16x$ by $8x$:

$$\frac{24x^3 + 16x}{8x} = \frac{24x^3}{8x} + \frac{16x}{8x} = 3x^2 + 2$$

You can check that the factorization is correct by multiplying $3x^2 + 2$ by $8x$ and verifying that the product is $24x^3 + 16x$.

**145**

## USING FACTORING TO ISOLATE VARIABLES IN EQUATIONS

It may be necessary to use factoring to help isolate a variable in an equation in which terms involving the variable cannot be combined into a single term.

**EXAMPLE:** If $ax - c = bx + d$, what is $x$ in terms of $a$, $b$, $c$, and $d$?

**SOLUTION:** Isolate terms involving $x$ on the same side of the equation.

- On each side of the equation add $c$ and subtract $bx$:

$$ax - bx = c + d$$

- Factor out $x$ from the left side of the equation:

$$x(a - b) = c + d$$

- Divide both sides of the new equation by the coefficient of $x$:

$$\frac{\overset{1}{\cancel{x(a-b)}}}{\cancel{a-b}} = \frac{c+d}{a-b}$$

$$x = \frac{c+d}{a-b}$$

## FACTORING A QUADRATIC TRINOMIAL

The quadratic trinomial $x^2 - 7x + 12$ can be factored as the product of two binomials by reversing the FOIL multiplication process.

- Think: "What two integers when multiplied together give $+12$ and when added together give $-7$?"
- Recall that, since the product of these integers is *positive* 12, the two integers must have the same sign. Hence, the integers are limited to the following pairs of factors of $+12$:

$$1 \text{ and } 12; \quad -1 \text{ and } -12$$

$$2 \text{ and } 6; \quad -2 \text{ and } -6$$

$$3 \text{ and } 4; \quad -3 \text{ and } -4$$

- Choose $-3$ and $-4$ as the factors of 12 since they add up to $-7$. Thus,

$$x^2 - 7x + 12 = (x - 3)(x - 4)$$

- Use FOIL to check that the product $(x - 3)(x - 4)$ is $x^2 - 7x + 12$.

**EXAMPLE:** Factor: $n^2 - 5n - 14$.

**SOLUTION:**  Find two integers that when multiplied together give −14 and when added together give −5. The two factors of −14 must have different signs since their product is negative. Since $(+2)(-7) = -14$ and $2 + (-7) = -5$, the factors of −14 you are looking for are $+2$ and $-7$. Thus:

$$n^2 - 5n - 14 = (n + 2)(n - 7)$$

## FACTORING THE DIFFERENCE OF TWO SQUARES

Since $(a + b)(a - b) = a^2 - b^2$, any binomial of the form $a^2 - b^2$ can be rewritten as $(a + b)$ times $(a - b)$.

$$a^2 - b^2 = (a + b)(a - b)$$

Here are some examples in which this factoring rule is used:

- $y^2 - 16 = (y + 4)(y - 4)$
- $x^2 - 0.25 = (x + 0.5)(x - 0.5)$
- $100 - x^4 = (10 - x^2)(10 + x^2)$

## FACTORING COMPLETELY

To factor a polynomial into factors that cannot be further factored, it may be necessary to use more than one factoring technique. For example:

- $3t^2 - 75 = 3(t^2 - 25) = 3(t + 5)(t - 5)$
- $t^3 - 6t^2 + 9t = t(t^2 - 6t + 9) = t(t - 3)(t - 3)$ or $t(t - 3)^2$

## USING FACTORING TO SIMPLIFY ALGEBRAIC FRACTIONS

To simplify an algebraic fraction, factor the numerator and the denominator. Then cancel any factor that is in both the numerator and the denominator of the fraction since any nonzero quantity divided by itself is 1.

**EXAMPLE:**  Simplify: $\dfrac{2b - 2a}{a^2 - b^2}$ .

**SOLUTION:**

- Factor the numerator and the denominator.

$$\frac{2b - 2a}{a^2 - b^2} = \frac{2(b - a)}{(a + b)(a - b)}$$

- Rewrite $b - a$ as $-(a - b)$:

$$= \frac{2[-(a - b)]}{(a + b)(a - b)}$$

- Cancel $a - b$ since it is a factor of the numerator and a factor of the denominator:

$$= \frac{-2\overset{1}{\cancel{(a - b)}}}{(a + b)\cancel{(a - b)}}$$

$$= \frac{-2}{a + b}$$

# LESSON 4-4   TUNE-UP EXERCISES

| **Multiple Choice** |
|---|

**1** $\dfrac{4x+4y}{2x^2-2y^2} =$

(A) $\dfrac{2}{x+y}$

(B) $2(x-y)$

(C) $\dfrac{xy}{2(x-y)}$

(D) $2\left(\dfrac{1}{x+y}\right)$

(E) $\dfrac{2}{x-y}$

**2** The sum of $\dfrac{a}{a^2-b^2}$ and $\dfrac{b}{a^2-b^2}$ is

(A) $\dfrac{1}{a-b}$

(B) $\dfrac{a}{a-b}$

(C) $\dfrac{b}{a-b}$

(D) $\dfrac{a+b}{a-b}$

(E) $\dfrac{ab}{a-b}$

**3** If $ax + x^2 = y^2 - ay$, what is $a$ in terms of $x$ and $y$?

(A) $y-x$

(B) $x-y$

(C) $x+y$

(D) $\dfrac{x^2+y^2}{x+y}$

(E) $\dfrac{x^2+y^2}{x-y}$

**4** If $\dfrac{xy}{x+y} = 1$ and $x \neq -y$, what is $x$ in terms of $y$?

(A) $\dfrac{y+1}{y-1}$

(B) $\dfrac{y+1}{y}$

(C) $\dfrac{y}{y-1}$

(D) $\dfrac{y}{y+1}$

(E) $1 - \dfrac{1}{y}$

**5** If $x^2 = r^2 + 2rs + s^2$, $y^2 = r^2 - s^2$, $x > 0$, and $y > 0$, then $\dfrac{x}{y} =$

(A) $\dfrac{r+s}{r-s}$

(B) $\sqrt{\dfrac{r+s}{r-s}}$

(C) $\dfrac{r-s}{rs}$

(D) $\sqrt{\dfrac{r^2+s^2}{r^2-s^2}}$

(E) $\sqrt{r+s}$

**6** If $h = \dfrac{x^2-1}{x+1} + \dfrac{x^2-1}{x-1}$, what is $x$ in terms of $h$?

(A) $\dfrac{h}{2}$

(B) $2h + 1$

(C) $2h - 1$

(D) $\sqrt{\dfrac{h}{2}}$

(E) $\sqrt{2h}$

**7** If $ax^2 - bx = ay^2 + by$, then $\frac{a}{b} =$

(A) $\frac{1}{x-y}$

(B) $\frac{1}{x+y}$

(C) $\frac{x-y}{x+y}$

(D) $\frac{x+y}{x-y}$

(E) $\frac{x}{y}$

**8** If $a \neq b$ and $\frac{a^2 - b^2}{9} = a + b$, then what is the value of $a - b$?

(A) $\frac{1}{3}$

(B) 3

(C) 9

(D) 12

(E) It cannot be determined from the information given.

**9** $\dfrac{x+y}{\frac{x^2}{y} - y} =$

(A) $\frac{y}{x+y}$

(B) $\frac{y}{x-y}$

(C) $\frac{x}{x^2+y^2}$

(D) $\frac{x-y}{x+y}$

(E) $\frac{x+y}{x-y}$

**10** If $k > 1$ and $\frac{k-1}{\sqrt{k}-1} = p$, what is the value of $k$ in terms of $p$?

(A) $\sqrt{p-1}$

(B) $\sqrt{1+p}$

(C) $(p-1)^2$

(D) $(1+p)^2$

(E) $\left(1 - \frac{1}{p}\right)^2$

---

## Quantitative Comparison

Each question consists of two quantities in boxes, one in Column A and one in Column B. You are to compare the two quantities and on the answer sheet fill in

A  if the quantity in Column A is greater;
B  if the quantity in Column B is greater;
C  if the two quantities are equal;
D  if the relationship cannot be determined from the information given

AN E RESPONSE WILL NOT BE SCORED

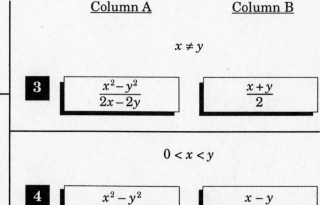

| Column A | Column B |
|---|---|
| **1** $\dfrac{9^8 - 9^7}{8}$ | $9^6$ |

$x \neq 1$

| Column A | Column B |
|---|---|
| **2** $\dfrac{1-x^2}{1-x}$ | $x + 1$ |

$x \neq y$

| Column A | Column B |
|---|---|
| **3** $\dfrac{x^2 - y^2}{2x - 2y}$ | $\dfrac{x+y}{2}$ |

$0 < x < y$

| Column A | Column B |
|---|---|
| **4** $x^2 - y^2$ | $x - y$ |

| Column A | Column B |
|---|---|

$$0 < s < r < 1$$

**5** | $r(r-s) - s(r-s)$ | $r - s$ |

---

$$a > b > c > 0$$

**6** | $a^2 - b^2$ | $ac + bc$ |

---

$$\frac{x^2 - y^2}{x + y} = 7,\ x \neq -y$$

**7** | $x$ | $y + \sqrt{7}$ |

| Column A | Column B |
|---|---|

$$r \neq s$$

**8** | $(r^2 - s^2)(r + s)$ | $(r + s)^2(r - s)$ |

---

$$x > 2$$

**9** | $x^2 - 2x$ | $x^2 - x - 2$ |

---

$$y^2 + ay + b = (y - 3)(y + 1)$$

**10** | $a$ | $b$ |

# LESSON 4-5

# Quadratic Equations

## OVERVIEW

A **quadratic equation** is an equation in which the greatest exponent of the variable is 2, as in $x^2 + 3x - 10 = 0$. A quadratic equation has two roots, which can be found by breaking down the quadratic equation into two first-degree equations.

## ZERO-PRODUCT RULE

If the product of two or more real numbers is 0, at least one of these numbers is 0.

> **EXAMPLE:** For what values of $x$ is $(x - 1)(x + 3) = 0$?

**SOLUTION:** Since $(x - 1)(x + 3) = 0$, either $x - 1 = 0$ or $x + 3 = 0$.

- If $x - 1 = 0$, then $x = 1$.
- If $x + 3 = 0$, then $x = -3$.

The possible values of $x$ are 1 and $-3$.

## SOLVING A QUADRATIC EQUATION BY FACTORING

The two roots of a quadratic equation may or may not be equal. For example, the equation $(x - 3)^2 = 0$ has a double root of $x = 3$. The quadratic equation $x^2 = 9$, however, has two unequal roots, $x = 3$ and $x = -3$. More complicated quadratic equations on the SAT I can be solved by factoring the quadratic expression.

> **EXAMPLE:** Solve: $x^2 + 2x = 0$.

**SOLUTION:**

- Factor the left side of the quadratic equation: 
$$x^2 + 2x = 0$$
$$x(x + 2) = 0$$

- Form two first-degree equations by setting each factor equal to 0: 
$$x = 0 \quad \text{or} \quad x + 2 = 0$$

- Solve each first-degree equation: 
$$x = 0 \quad \text{or} \quad x = -2$$

The two roots are 0 and $-2$.

**151**

If a quadratic equation does not have all of its nonzero terms on the same side of the equation, you must put the equation into this form before factoring.

    **EXAMPLE:**        Solve: $x^2 + 3x = 10$.

**SOLUTION:**

- To rewrite the quadratic equation so that all the nonzero terms are on the same side, subtract 10 from both sides of $x^2 + 3x = 10$:          $x^2 + 3x - 10 = 0$

- Factor the quadratic polynomial:          $(x + 5)(x - 2) = 0$

- Set each factor equal to 0:          $x + 5 = 0$    or    $x - 2 = 0$

- Solve each equation:          $x = -5$    or    $x = 2$

The two roots are $-5$ and 2.

You can check that $x = -5$ and $x = 2$ are the roots by plugging each value into $x^2 + 3x = 10$ and verifying that the left side then equals 10, the right side.

# LESSON 4-5   TUNE-UP EXERCISES

**1** If $(x + 7)(x - 3) = 0$, then $x =$

(A) 7 or 3
(B) 7 or −3
(C) −7 or 3
(D) −7 or −3
(E) −4 or −3

**2** If $\frac{a^2}{2} = 2a$, then $a$ equals

(A) 0 or −2
(B) 0 or 2
(C) 0 or −4
(D) 0 or 4
(E) 2 or −2

**3** If $x$ is a positive integer and $(x - 6)^2 = 9$, then $x^2 =$

(A) 3
(B) 9
(C) 16
(D) 36
(E) 49

**4** If $a(x - y) = 0$, then which of the following statements is (are) always true?

I. $a = 0$
II. $y = 0$
III. $x = y$

(A) I and II only
(B) I and III only
(C) II and III only
(D) I, II, and III
(E) None

**5** If $(x - 1)^2 - (x - 1) = 0$, then,

(A) $x = 0$ or $x = 2$
(B) $x = 0$ or $x = 1$
(C) $x = -1$ or $x = 2$
(D) $x = 1$ or $x = 2$
(E) $x = 0$ or $x = -1$

**6** If $w$ and $y$ are positive integers, $w^2 - 2w = 0$, and $2y^2 - y = 0$, then $\frac{w}{y} =$

(A) $\frac{1}{2}$
(B) $\frac{2}{3}$
(C) 1
(D) 2
(E) 4

**7** For which of the following equations is $x = -2$ a root?

I. $\frac{2}{x} - x = 0$
II. $x^2 + 4 = 0$
III. $x^2 + 4x + 4 = 0$

(A) I only
(B) II only
(C) III only
(D) I and III
(E) II and III

**8** If $\frac{x^2}{3} = x$, then $x =$

(A) 0 or −3
(B) 3 or −3
(C) 3 only
(D) 0 only
(E) 0 or 3

**9** By how much does the sum of the roots of the equation $(x + 1)(x - 3) = 0$ exceed the product of its roots?

(A) 1
(B) 2
(C) 3
(D) 4
(E) 5

**10** If $x^2 - 63x - 64 = 0$ and $p$ and $n$ are integers such that $p^n = x$, which of the following CANNOT be a value for $p$?

(A) $-8$
(B) $-4$
(C) $-1$
(D) 4
(E) 64

**11** If $r > 0$ and $r^t = 6.25r^{t+2}$, then $r =$

(A) $\frac{2}{5}$
(B) $\frac{4}{9}$
(C) $\frac{5}{8}$
(D) $\frac{3}{4}$
(E) $\frac{5}{4}$

## Quantitative Comparison

Each question consists of two quantities in boxes, one in Column A and one in Column B. You are to compare the two quantities and on the answer sheet fill in

    A  if the quantity in Column A is greater;
    B  if the quantity in Column B is greater;
    C  if the two quantities are equal;
    D  if the relationship cannot be determined from the information given

AN E RESPONSE WILL NOT BE SCORED

| Column A | Column B |
|----------|----------|

$$x^2 + 8x + 16 = 0$$

**1**    $x$      $-3$

$$x(x + p) = 0$$

**2**    $x$      $p$

$a$ and $b$ are the solutions of $x^2 + 2x - 3 = 0$.

**3**    $a + b$      $ab$

| Column A | Column B |
|----------|----------|

$$\frac{x-1}{5} = \frac{7}{x+1}$$

**4**    $x$      6

$$y^2 + ky - 9 = 0$$
$y$ and $k$ are nonzero integers.

**5**    $k$      8

## Grid In

**1** If $(4p + 1)^2 = 81$ and $p > 0$, what is a possible value of $p$?

**2** If $(x - 1)(x - 3) = -1$, what is a possible value of $x$?

**3** By what amount does the sum of the roots exceed the product of the roots of the equation $(x - 5)(x + 2) = 0$?

# Systems of Equations

## OVERVIEW

A **system of equations** *is a set of equations whose solution makes each of the equations true at the same time.* SAT I *questions involving systems of two equations in two variables can usually be solved by:*

- *substituting the solution of one equation into the other equation to eliminate one of the variables in the second equation; or*
- *adding or subtracting corresponding sides of the two equations.*

## SOLVING A SYSTEM OF TWO EQUATIONS BY SUBSTITUTION

If one of the equations in a system of two equations has the form

$$variable = \text{math expression},$$

the same variable in the other equation can be replaced with that math expression. For example, if

$$3x = 2 \quad \text{and} \quad 5y + 3x = 7,$$

we can find the value of $y$ by replacing $3x$ with 2 in the second equation to obtain $5y + 2 = 7$. Hence, $5y = 5$, so $y = 1$.

**EXAMPLE:** If $y = 2x - 3$ and $x + y = 18$, what is the value of $y$?

**SOLUTION:**

- Substitute $2x - 3$ for $y$ in the equation $x + y = 18$. Then solve for $x$.

$$x + y = 18$$

$$x + \overbrace{2x - 3}^{y} = 18$$

$$3x - 3 = 18$$

$$3x = 21$$

$$x = \frac{21}{3} = 7$$

- Find the corresponding value of $y$ by substituting 7 for $x$ in either of the original equations.

$$x + y = 18$$

$$7 + y = 18$$

$$y = 11$$

**EXAMPLE:** If $2x - 3y = 6$ and $y - 5 = 3 - y$, what is the value of $x$?

**SOLUTION:**

- Since the second equation does not depend on $x$, solve it for $y$:

$$y - 5 = 3 - y$$

$$2y = 8$$

$$y = \frac{8}{2} = 4$$

- Substitute 4 for $y$ in the first equation:

$$2x - 3y = 6$$

$$2x - 3(4) = 6$$

$$2x - 12 = 6$$

$$2x = 18$$

$$x = \frac{18}{2} = 9$$

## SOLVING A SYSTEM OF EQUATIONS BY COMBINING CORRESPONDING SIDES

It may be possible to solve a system of two equations by writing one equation above the other equation and then adding or subtracting the like terms in each column so that one variable is eliminated.

**EXAMPLE:** If $x - 2y = 5$ and $x + 2y = 11$, what is the value of $x$?

**SOLUTION:** Since the numerical coefficients of $y$ in the two equations are opposites, adding the corresponding sides of the equations will eliminate $y$.

$$x - 2y = 5$$

$$+ \quad x + 2y = 11$$

$$2x + 0 = 16$$

$$x = \frac{16}{2} = 8$$

The value of $x$ is 8.

Before combining the equations in a system of equations, it may be necessary to multiply one or both equations by a number that will eliminate one of the two variables when the two equations are added together.

**EXAMPLE:**  If $2a = b + 7$ and $5a = 2b + 15$, what is the value of $a$?

**SOLUTION:**

- Rewrite each equation so that its variable terms are on the same side:

$$2a = b + 7 \qquad \rightarrow \qquad 2a - b = 7$$

$$5a = 2b + 15 \qquad \rightarrow \qquad 5a - 2b = 15$$

- Multiply the first equation by $-2$, so that the coefficient of $b$ becomes $+2$. Then add the two equations to eliminate $b$:

$$2a - b = 7 \qquad \rightarrow \qquad -4a + 2b = -14$$

$$5a - 2b = 15 \qquad \rightarrow \qquad 5a - 2b = 15$$

$$a = 1$$

The value of $a$ is 1.

## SOLVING OTHER TYPES OF SYSTEMS OF EQUATIONS

If a system of equations has more variables than equations, you may be asked to solve for some combination of variables.

**EXAMPLE:**  If $2r = s$ and $24t = 3s$, what is $r$ in terms of $t$?

**SOLUTION:**  Since the question asks for $r$ in terms of $t$, work towards eliminating variable $s$.

- Substitute $2r$ for $s$ in the second equation:

$$24t = 3s = 3(2r) = 6r$$

- Solve for $r$ in the equation:     $24t = 6r$:

$$\frac{24t}{6} = \frac{6r}{6}$$

$$4t = r$$

Hence, $r = 4t$.

**EXAMPLE:**     If $ab - 3 = 12$ and $2bc = 5$, what is the value of $\frac{a}{c}$?

**SOLUTION:**    Since the question asks for $\frac{a}{c}$, you must eliminate variable $b$.

- Find the value of $ab$ in the first equation. Since $ab - 3 = 12$, then $ab = 15$.
- To eliminate $b$, divide corresponding sides of $ab = 15$ and $2bc = 5$:

$$\frac{ab}{2bc} = \frac{15}{5}$$

$$\frac{ab}{2bc} = 3$$

$$\frac{a}{2c} = 3$$

Solve the resulting equation
for $\frac{a}{c}$:                     $2\left(\frac{a}{2c}\right) = 2\,(3)$

$$\frac{a}{c} = 6$$

The value of $\frac{a}{c}$ is 6.

# LESSON 4-6   TUNE-UP EXERCISES

| **Multiple Choice** |
| --- |

**1**  If $2x - 3y = 11$ and $3x + 15 = 0$, what is the value of $y$?

(A) $-7$
(B) $-5$
(C) $\frac{1}{3}$
(D) $3$
(E) $10$

**2**  If $2a = 3b$ and $4a + b = 21$, then $b =$

(A) $1$
(B) $3$
(C) $4$
(D) $7$
(E) $8$

**3**  If $2p + q = 11$ and $p + 2q = 13$, then $p + q =$

(A) $6$
(B) $8$
(C) $9$
(D) $12$
(E) $18$

**4**  If $m + p + k = 70$, $p = 2m$, and $k = 2p$, then $m =$

(A) $2$
(B) $5$
(C) $7$
(D) $10$
(E) $14$

**5**  If $x - y = 3$ and $x + y = 5$, what is the value of $y$?
(A) $-4$
(B) $-2$
(C) $-1$
(D) $1$
(E) $2$

**6**  If $5x + y = 19$ and $x - 3y = 7$, then $x + y =$

(A) $-4$
(B) $-1$
(C) $3$
(D) $4$
(E) It cannot be determined from the information given.

**7**  If $x - 9 = 2y$ and $x + 3 = 5y$, what is the value of $x$?

(A) $-2$
(B) $4$
(C) $11$
(D) $15$
(E) $17$

**8**  If $\frac{1}{x} + \frac{1}{y} = \frac{1}{4}$ and $\frac{1}{x} - \frac{1}{y} = \frac{3}{4}$, then $x =$

(A) $\frac{1}{4}$
(B) $\frac{1}{2}$
(C) $1$
(D) $2$
(E) $4$

**9**  If $5a + 3b = 35$ and $\frac{a}{b} = \frac{2}{5}$, what is the value of $a$?

(A) $\frac{14}{5}$
(B) $\frac{7}{2}$
(C) $5$
(D) $7$
(E) $9$

**10** If $\frac{x}{y} = 6, \frac{y}{w} = 4$, and $x = 36$, what is the value of $w$?

(A) $\frac{1}{2}$
(B) 2
(C) $\frac{3}{2}$
(D) 4
(E) 6

**11** If $4r + 7s = 23$ and $r - 2s = 17$ then $3r + 3s =$

(A) 8
(B) 24
(C) 32
(D) 40
(E) 48

**12** If $\frac{p-q}{2} = 3$ and $rp - rq = 12$, then $r =$

(A) $-1$
(B) 1
(C) 2
(D) 4
(E) It cannot be determined from the information given.

**13** If $(a + b)^2 = 9$ and $(a - b)^2 = 49$, what is the value of $a^2 + b^2$?

(A) 17
(B) 20
(C) 29
(D) 58
(E) 116

**14** If $\frac{r+s}{r} = 3$ and $\frac{t+r}{t} = 5$, what is the value of $\frac{s}{t}$?

(A) $\frac{1}{2}$
(B) $\frac{3}{5}$
(C) 4
(D) 8
(E) 16

**15** If $3x + y = c$ and $x + y = b$, what is the value of $x$ in terms of $c$ and $b$?

(A) $\frac{c-b}{3}$
(B) $\frac{c-b}{2}$
(C) $\frac{b-c}{3}$
(D) $\frac{b-c}{2}$
(E) $\frac{c-b}{4}$

**16** If $a + b = 11$ and $a - b = 7$, then $ab =$

(A) 6
(B) 8
(C) 10
(D) 12
(E) 18

$$x - z = 7$$
$$x + y = 3$$
$$z - y = 6$$

**17** For the above system of three equations, $x =$

(A) 5
(B) 6
(C) 7
(D) 8
(E) 9

**18** In a drama club, $x$ students each contributed $y$ dollars to buy a $60 gift for their advisor. If four more students had contributed, each pupil could have contributed 3 dollars less to buy the same gift. Which of the following pairs of equations expresses this relationship?

(A) $xy = 60$
   $(x + 4)(y - 3) = 60$
(B) $xy = 60$
   $(x - 4)(y + 3) = 60$
(C) $xy = 60$
   $(x + 3)(y - 4) = 60$
(D) $xy = 60$
   $xy = 60 - (4 \times 3)$
(E) $x + y = 60$
   $(4x)(3y) = 60$

**19** If $r^8 = 5$ and $r^7 = \dfrac{3}{t}$, what is the value of $r$ in terms of $t$?

(A) $\dfrac{5}{3}t$

(B) $\dfrac{3}{5}t$

(C) $5 - \dfrac{3}{t}$

(D) $3 + \dfrac{5}{t}$

(E) $\dfrac{t}{15}$

**20** If $\dfrac{a}{b} = \dfrac{6}{7}$ and $\dfrac{a}{c} = \dfrac{2}{5}$, what is the value of $3b + c$ in terms of $a$?

(A) $12a$

(B) $9a$

(C) $8a$

(D) $6a$

(E) $4a$

---

## Quantitative Comparison

Each question consists of two quantities in boxes, one in Column A and one in Column B. You are to compare the two quantities and on the answer sheet fill in

A if the quantity in Column A is greater;
B if the quantity in Column B is greater;
C if the two quantities are equal;
D if the relationship cannot be determined from the information given

AN E RESPONSE WILL NOT BE SCORED

---

| Column A | Column B |
|---|---|

$3p + 2q = 17$
$2q - 1 = 7$

**1** | $p$ | $q$

$xy - 5 = 7$
$y + 5 = 2$

**2** | $x$ | $y$

$4a + 4b = 9$
$3a - 3b = 7$

**3** | $a - b$ | $a + b$

$x + y = 9$
$2x - y = 3$

**4** | $x$ | $y$

| Column A | Column B |
|---|---|

$a - b + c = 3$
$a + b + c = 7$

**5** | $a + c$ | $4$

$x + y = 9$
$y + z = 7$
$x + z = 6$

**6** | $x + y + z$ | $11$

$\dfrac{2}{x} = \dfrac{1}{y}$
$2x - y = 3$

**7** | $x$ | $y$

## Grid In

**1** If 5 sips + 4 gulps = 1 glass and 13 sips + 7 gulps = 2 glasses, how many sips equal a gulp?

**2** If $2a = 9 - b$ and $4a = 3b - 12$, what is the value of $a$?

**3** John and Sara each bought the same type of pen and notebook in the school bookstore, which does not charge sales tax. John paid $5.55 for two pens and three notebooks, and Sara paid $3.50 for one pen and two notebooks. How much does the school bookstore charge for one notebook?

# Algebraic Inequalities

## OVERVIEW

*Linear inequalities* such as $2x - 3 \le 7$ are solved by isolating the variable in much the same way that linear (first-degree) equations are solved. Multiplying or dividing both sides of an inequality by a negative number reverses the direction of the inequality sign. For example:

$$5 > 3 \text{ so } 5 \times (-2) < 3 \times (-2)$$

*Any general property of inequalities that is stated in this lesson for the "is less than" $(<)$ relation is also true for each of the other $(\le, >, \text{ and } \ge)$ inequality relations.*

## SOME PROPERTIES OF INEQUALITIES

An equivalent inequality results when

- The same number, positive or negative, is added or subtracted on both sides of the inequality. For example:

$$6 > 4, \quad \text{so} \quad 6 + 2 > 4 + 2$$
$$6 > 4, \quad \text{so} \quad 6 - 2 > 4 - 2$$

- Both sides of the inequality are multiplied or divided by the same *positive* number. For example:

$$6 > 4, \quad \text{so} \quad 6 \times 2 > 4 \times 2$$
$$6 > 4, \quad \text{so} \quad \frac{6}{2} > \frac{4}{2}$$

- Both sides of the inequality are multiplied or divided by the same *negative* number and the direction of the inequality is reversed. For example:

$$6 > 4, \quad \text{so} \quad 6 \times (-2) < 4 \times (-2)$$
$$6 > 4, \quad \text{so} \quad \frac{6}{-2} < \frac{4}{-2}$$

## SOLVING LINEAR INEQUALITIES

To solve a linear inequality, isolate the variable by performing the same arithmetic operation on both sides of the inequality. Remember to reverse the direction of the inequality sign whenever multiplying or dividing the inequality by a negative number.

**EXAMPLE:**    If $3x - 2 < 10$, find $x$.

**SOLUTION:**
$$3x - 2 < 10$$
$$3x < 12$$
$$x < \frac{12}{3}$$
$$x < 4$$

The solution consists of all real numbers less than 4.

**EXAMPLE:**    What is the *greatest* integer value of $x$ such that $1 - 2x > 6$?

**SOLUTION:**
$$1 - 2x > 6$$
$$-2x > 5$$

Reverse the inequality:    $x < \dfrac{5}{-2}$

Since $x$ is less than $\frac{-5}{2}$, the greatest integer value of $x$ is $-3$.

## SOLVING COMBINED INEQUALITIES

To solve an inequality that has the form $c \leq ax + b \leq d$, where $a$, $b$, $c$, and $d$ stand for numbers, isolate the variable by performing the same operation on each member of the inequality.

**EXAMPLE:**    If $-2 \leq 3x - 7 \leq 8$, find $x$.

**SOLUTION:**

- Add 7 to each member of the inequality:

$$-2 \leq 3x - 7 \leq 8$$
$$-2 + 7 \leq 3x - 7 + 7 \leq 8 + 7$$
$$5 \leq 3x \leq 15$$

- Divide each member of the inequality by 3:

$$\frac{5}{3} \leq \frac{3x}{3} \leq \frac{15}{3}$$
$$\frac{5}{3} \leq x \leq 5$$

The solution consists of all real numbers greater than or equal to $\frac{5}{3}$ and less than or equal to 5.

## ORDERING PROPERTIES OF INEQUALITIES

- If $a < b$ and $b < c$, then $a < c$. For example:

$$2 < 3 \text{ and } 3 < 4, \quad \text{so} \quad 2 < 4$$

- If $a < b$ and $x < y$, then $a + x < b + y$. For example:

$$2 < 3 \text{ and } 4 < 5, \quad \text{so} \quad 2 + 4 < 3 + 5$$

- If $a < b$ and $x > y$, then the relationship between $a + x$ and $b + y$ cannot be determined until each of the letters is replaced by a specific number.

# LESSON 4-7   TUNE-UP EXERCISES

| **Multiple Choice** |
|---|

**1** What is the largest integer value of $p$ that satisfies the inequality $4 + 3p < p + 1$?

(A) $-2$
(B) $-1$
(C) $0$
(D) $1$
(E) $2$

**2** If $-3 < 2x + 5 < 9$, which CANNOT be a possible value of $x$?

(A) $-2$
(B) $-1$
(C) $0$
(D) $1$
(E) $2$

**3** If the sum of a number and the original number increased by 5 is greater than 11, which could be a possible value of the number?

(A) $-5$
(B) $-1$
(C) $1$
(D) $3$
(E) $4$

**4** If $0 < a^2 < b$, which of the following statements is (are) always true?

I. $a < \dfrac{b}{a}$
II. $a^4 < a^2 b$
III. $\dfrac{a^2}{b} < 1$

(A) I only
(B) I and II
(C) II and III
(D) I and III
(E) I, II, and III

**5** What is the smallest integer value of $x$ that satisfies the inequality $4 - 3x < 11$?

(A) $-3$
(B) $-2$
(C) $-1$
(D) $0$
(E) $1$

**6** If $a > b > c > 0$, which of the following statements must be true?

I. $\dfrac{a-c}{b-a} > \dfrac{b-c}{b-a}$
II. $ab > ac$
III. $\dfrac{b}{a} > \dfrac{b}{c}$

(A) I only
(B) II only
(C) III only
(D) I and II
(E) II and III

**7** If $n$ is an integer, how many different values of $n$ satisfy the inequality $-4 \le 3n \le 87$?

(A) 32
(B) 31
(C) 30
(D) 29
(E) 28

**8** Which of the following statements must be true when $a^2 < b^2$ and $a$ and $b$ are not 0?

I. $\dfrac{a^2}{a} < \dfrac{b^2}{a}$
II. $\dfrac{1}{a^2} > \dfrac{1}{b^2}$
III. $(a + b)(a - b) < 0$

(A) I only
(B) II only
(C) III only
(D) I and II
(E) II and III

**9** For how many integer values of $b$ is $b + 3 > 0$ and $1 > 2b - 9$?

(A) Four
(B) Five
(C) Six
(D) Seven
(E) Eight

**10** If $xy > 1$ and $z < 0$, which of the following statements must be true?

  I. $x > z$
 II. $xyz < -1$
III. $\frac{xy}{z} < \frac{1}{z}$

(A) I only
(B) II only
(C) III only
(D) II and III
(E) None

## Quantitative Comparison

Each question consists of two quantities in boxes, one in Column A and one in Column B. You are to compare the two quantities and on the answer sheet fill in

A if the quantity in Column A is greater;
B if the quantity in Column B is greater;
C if the two quantities are equal;
D if the relationship cannot be determined from the information given

AN E RESPONSE WILL NOT BE SCORED

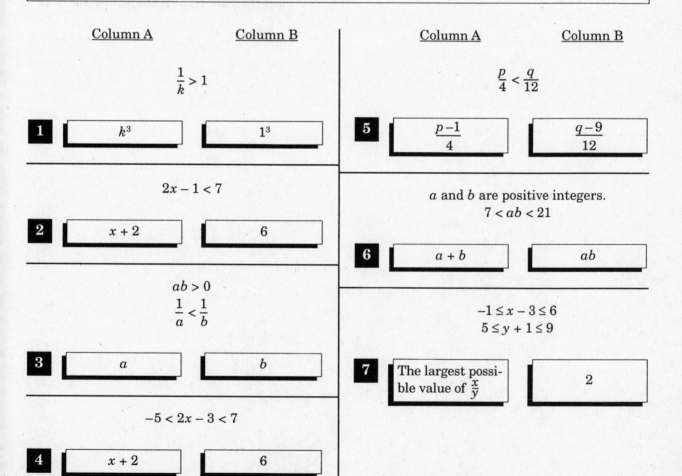

Column A    Column B

$$\frac{1}{k} > 1$$

**1**   $k^3$      $1^3$

$$2x - 1 < 7$$

**2**   $x + 2$      $6$

$$ab > 0$$
$$\frac{1}{a} < \frac{1}{b}$$

**3**   $a$      $b$

$$-5 < 2x - 3 < 7$$

**4**   $x + 2$      $6$

Column A    Column B

$$\frac{p}{4} < \frac{q}{12}$$

**5**   $\frac{p-1}{4}$      $\frac{q-9}{12}$

$a$ and $b$ are positive integers.
$$7 < ab < 21$$

**6**   $a + b$      $ab$

$$-1 \le x - 3 \le 6$$
$$5 \le y + 1 \le 9$$

**7**   The largest possible value of $\frac{x}{y}$      $2$

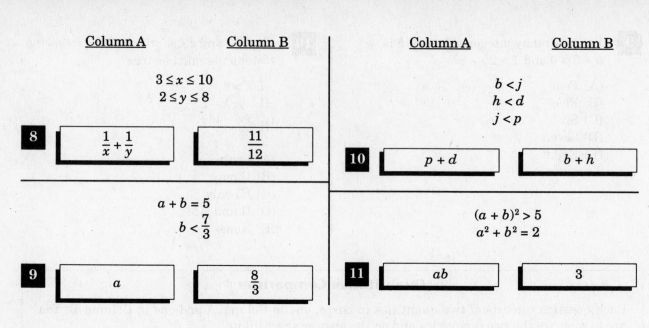

| Column A | Column B | | Column A | Column B |
|---|---|---|---|---|

$3 \le x \le 10$
$2 \le y \le 8$

**8** | $\frac{1}{x} + \frac{1}{y}$ | $\frac{11}{12}$

$b < j$
$h < d$
$j < p$

**10** | $p + d$ | $b + h$

$a + b = 5$
$b < \frac{7}{3}$

**9** | $a$ | $\frac{8}{3}$

$(a + b)^2 > 5$
$a^2 + b^2 = 2$

**11** | $ab$ | $3$

---

## Grid In

**1** For what integer value of $y$ is $y + 5 > 8$ and $2y - 3 < 7$?

**2** If 2 times an integer $x$ is increased by 5, the result is always greater than 16 and less than 29. What is the least value of $x$?

# ANSWERS TO CHAPTER 4 TUNE-UP EXERCISES

**LESSON 4-1** *(Multiple Choice)*

1. **A**
$$x + 4x - 3x + 8 = 0$$
$$5x - 3x + 8 = 0$$
$$2x + 8 = 0$$
$$2x = -8$$
$$x = \frac{-8}{2} = -4$$

2. **A**
$$(9-4)(x+4) = 30$$
$$5(x+4) = 30$$
$$x + 4 = \frac{30}{5} = 6$$
$$x = 6 - 4$$
$$= 2$$

3. **E**
$$2(1+5) = 3(w-4)$$
$$2(6) = 3(w) + 3(-4)$$
$$12 = 3w - 12$$
$$12 + 12 = 3w$$
$$24 = 3w$$
$$\frac{24}{3} = w$$
$$8 = w$$

4. **B**
$$\frac{x}{3} - 1 = 1 - 3$$
$$\frac{x}{3} - 1 = -2$$
$$\frac{x}{3} = -2 + 1$$
$$\frac{x}{3} = -1$$
$$x = -1(3) = -3$$

5. **D**
$$3(x-2) - 2x = 7$$
$$3x + 3(-2) - 2x = 7$$
$$3x - 6 - 2x = 7 + 6$$
$$x = 13$$

6. **C** *Solution 1*: Solve $3x - 6 = 18$ for $x - 2$ by dividing each member of the equation by 2:
$$\frac{3x}{3} - \frac{6}{3} = \frac{18}{3}$$
$$x - 2 = 6$$

*Solution 2*: The answer choices represent possible values for $x - 2$. Add 2 to each answer choice to obtain the possible values for $x$. Then substitute each of these values into the given equation until you find one that works:

$$3x - 6 = 18$$
$$3(6 + 2) - 6 = 18$$
$$24 - 6 = 18$$

7. **B** If $2x + y = y + 14$, subtracting $y$ from each side of the equation gives $2x = 14$, so
$$x = \frac{14}{2} = 7$$

8. **B** If $\frac{4}{7}k = 36$, then
$$\frac{1}{7}k = \frac{1}{4}(36) = 9$$
Since $\frac{1}{7}k = 9$, then
$$\frac{3}{7}k = 3(9) = 27$$

9. **D**
$$\frac{1}{2}x + \frac{1}{4}x + \frac{1}{8}x = 14$$
$$\frac{4}{8}x + \frac{2}{8}x + \frac{1}{8}x = 14$$
$$\frac{7}{8}x = 14$$
$$\frac{8}{7}\left(\frac{7}{8}x\right) = \frac{8}{7}(14)$$
$$x = 16$$

10. **B** If $2s + 3t = 12$ and $4s = 36$, the second equation can be used to eliminate variable $s$ in the first equation. Since $2s = \frac{36}{2} = 18$, replace $2s$ with 18 in the first equation. Then solve for $t$:
$$18 + 3t = 12$$
$$3t = 12 - 18$$
$$t = -6$$
$$t = \frac{-6}{3} = -2$$

11. **D** Since $2^{x+1} = 32$ and $32 = 2^5$, then $2^{x+1} = 2^5$. Hence, $x + 1 = 5$, so
$$(x+1)^2 = (5)^2 = 25$$

12. **C** If $\frac{2}{x} = 2$, then $x = 1$ since $\frac{2}{1} = 2$. Hence,
$$x + 2 = 1 + 2 = 3$$

13. **A** If $\frac{y-2}{2} = y + 2$, then $y - 2 = 2(y + 2)$. Eliminate the parentheses, and then collect all the terms involving $y$ on the same side of the equation.
$$y - 2 = 2(y + 2)$$
$$= 2y + 4$$
$$y - 2y = 4 + 2$$
$$-y = 6, \text{ so } y = -6$$

14. **E** If $\frac{2y}{7} = \frac{y+3}{4}$, set the cross-products equal and then solve the resulting equation.
$$\frac{2y}{7} = \frac{y+3}{4}$$

*(Lesson 4-1 Multiple Choice continued)*

$$4(2y) = 7(y + 3)$$
$$8y = 7y + 21$$
$$8y - 7y = 21$$
$$y = 21$$

15. **E**  If $\frac{y}{3} = 4$, then $y = 3(4) = 12$, so
$$3y = 3(12) = 36$$

16. **B**  To add fractions that have the same denominator, write the sum of the numerators over the common denominator. Since $\frac{1}{q} + \frac{3}{q} = 12$ , then $\frac{4}{q} = 12$, so $4 = 12q$. Dividing each side of the equation $4 = 12q$ by 12 gives
$$q = \frac{4}{12} = \frac{1}{3}$$

17. **D**  To solve $8 - 3x = 2x + 13$, collect the constants on the left side of the equation and the variable terms on the right side.
$$8 - 3x = 2x + 13$$
$$8 - 13 = 2x + 3x$$
$$-5 = 5x$$
$$\frac{-5}{5} = \frac{5x}{5}$$
$$-1 = x$$

18. **A**  To solve $(1-2-3-4)w = (-1)(-2)(-3)(-4)$, first simplify the left side of the equation. Since $1 - 2 - 3 - 4 = 1 - 9 = -8$, then $-8w = (-1)(-2)(-3)(-4)$. The product of an even number of negative numbers is positive, so $(-1)(-2)(-3)(-4) = 24$. The original equation simplifies to
$$-8w = 24$$
$$w = \frac{24}{-8} = -3$$

19. **D**  Since $\frac{1}{3} + \frac{3}{p} = 1$, isolate the variable term by subtracting $\frac{1}{3}$ from each side of the equation:
$$\frac{3}{p} = 1 - \frac{1}{3} = \frac{2}{3}$$
Eliminate the fractions in this equation by cross-multiplying:
$$\frac{3}{p} = \frac{2}{3}$$
$$2p = 9$$
$$p = \frac{9}{2}$$

20. **E**  Write the equation $\frac{3}{4} = \frac{6}{x} = \frac{9}{y}$ as two equations:
$$\frac{3}{4} = \frac{6}{x} \quad \text{and} \quad \frac{3}{4} = \frac{9}{y}$$

Solve the first equation for $x$:
$$\frac{3}{4} = \frac{6}{x}$$
$$3x = 24$$
$$x = \frac{24}{3} = 8$$
Solve the second equation for $y$:
$$\frac{3}{4} = \frac{9}{y}$$
$$3y = 36$$
$$y = \frac{36}{3} = 12$$
Hence,
$$x + y = 8 + 12 = 20$$

*(Quantitative Comparison)*

1. **A**  It is given that $\frac{x}{2} = 3$ and $6y = 1$. In Column A, since $\frac{x}{2} = 3$, then $x = 2(3) = 6$. In Column B, since $6y = 1$, then $y = \frac{1}{6}$. Hence, Column A > Column B.

2. **B**  Solve the given equation. If $1 - 3k = k + 10$, then $1 - 10 = k + 3k$, so $-9 = 4k$. Thus, $k = -\frac{9}{4}$, which is the quantity in Column A. In Column B, $-2 = -\frac{8}{4}$. Since $-\frac{8}{4} > -\frac{9}{4}$, Column B > Column A.

3. **B**  It is given that $2^{x-1} = 8$. Since $8 = 2^3$, then $2^{x-1} = 2^3$, so $x - 1 = 3$ and $x = 4$. In Column A,
$$x^3 = 4^3 = 4 \times 4 \times 4 = 64$$
In Column B,
$$3^x = 3^4 = 3 \times 3 \times 3 \times 3 = 81$$
Hence, Column B > Column A.

4. **C**  From the given information, $a - 2 = 0$, so in Column A $a = 2$. You are also told that $3ab - 7 = 5$, so $3(2)b = 5 + 7$. Since $6b = 12$, in Column B $b = \frac{12}{6} = 2$. Hence, Column A = Column B.

5. **A**  From the given information, $x + 2x = 18$, so $3x = 18$ and in Column A $x = \frac{18}{3} = 6$. You are also told that $2x + 3y = 27$, so $2(6) + 3y = 27$ and $12 + 3y = 27$. Since $3y = 27 - 12 = 15$, in Column B $y = \frac{15}{3} = 5$. Hence, Column A > Column B.

*(Grid In)*

1. **6/4**  If $2x - 1 = 11$, then $2x = 12$, so $x = \frac{12}{2} = 6$. Since $3y = 12$, then $y = \frac{12}{3} = 4$. Hence, $\frac{x}{y} = \frac{6}{4}$. Grid is as 6/4 or 1.5.

2. **8**  If $7(a + b) - 4(a + b) = 24$, then $3(a + b) = 24$, so
$$a + b = \frac{24}{3} = 8$$

*(Lesson 4-1   Grid In continued)*

3. **2/12**  If $1 - x - 2x - 3x = 6x - 1$, then $1 - 6x = 6x - 1$, so $1 + 1 = 6x + 6x$ and $2 = 12x$. Thus, $x = \frac{2}{12}$. Grid is as 2/12.

4. **10/3**  In the equation $p = \frac{5b}{c^2}$, if $p = 9$ and $b = 20$, then $9 = \frac{5(20)}{c^2}$ or $9c^2 = 100$, so $c^2 = \frac{100}{9}$. Taking the positive square root of both sides of the equation gives

$$c = \frac{\sqrt{100}}{\sqrt{9}} = \frac{10}{3}$$

Grid is as 10/3.

## LESSON 4-2  *(Multiple Choice)*

1. **B**  Dividing each member of the given equation, $6 = 2x + 4y$, by 2 will leave $x + 2y$ on the right side of the equation:

$$\frac{6}{2} = \frac{2x}{2} + \frac{4y}{2}$$
$$3 = x + 2y$$

The value of $x + 2y$ is 3.

2. **C**  Multiplying each member of the given equation, $\frac{a}{2} + \frac{b}{2} = 3$, by 4 will make the left side of the equation equal to $2a + 2b$:

$$4\left(\frac{a}{2}\right) + 4\left(\frac{b}{2}\right) = 4(3)$$
$$2a + 2b = 12$$

The value of $2a + 2b$ is 12.

3. **A**  For the given equation, $2s - 3t = 3t - s$, finding $s$ in terms of $t$ means solving the equation for $s$ by treating $t$ as a constant. Work toward isolating variable $s$ by first adding $3t$ on each side of the equation:

$$2s = 3t + 3t - s$$
$$= 6t - s$$

Next, add $s$ on each side of the equation:

$$2s + s = 6t$$
$$3s = 6t$$
$$s = \frac{6t}{3} = \frac{t}{2}$$

4. **E**  • If $a + b = 5$, then $2(a + b) = 2(5)$, so $2a + 2b = 10$.
   • If $\frac{c}{2} = 3$, then $4(\frac{c}{2}) = 4(3)$, so $2c = 12$.
   Hence, $2a + 2b + 2c = 10 + 12 = 22$.

5. **C**  If $xy + z = y$, then $xy = y - z$, so $x = \frac{y-z}{y}$.

6. **B**  In the given equation $4^{x+y} = 8$, 4 and 8 are each powers of 2. Since $8 = 2^3$ and $4^{x+y} = (2^2)^{x+y} = 2^{2(x+y)}$, then $2^{2(x+y)} = 2^3$. If powers of the same base are equal, then their exponents must be equal. Hence, $2^{(x+y)} = 3$, so $x + y = \frac{3}{2}$.

7. **B**  Since $x^2$ and $y^2$ are both greater than or equal to 0, the only values of $x^2$ and $y^2$ for which $2x^2 + 3y^2 = 0$ are $x^2 = y^2 = 0$. If $x^2 = y^2 = 0$, then $x = y = 0$, so

$$3x + 2y = 3(0) + 2(0) = 0$$

8. **D**  If $\frac{a-b}{b} = \frac{2}{3}$, then

$$\frac{a}{b} - \frac{b}{b} = \frac{2}{3}$$
$$\frac{a}{b} - 1 = \frac{2}{3}$$
$$\frac{a}{b} = 1 + \frac{2}{3}$$
$$= \frac{5}{3}$$

The value of $\frac{a}{b}$ is $\frac{5}{3}$.

9. **A**  To divide powers that have the same base, *subtract* their exponents. If $\frac{10^x}{10^y} = 100$, then $100^{x-y} = 100 = 10^2$, so $x - y = 2$. Hence, $x = y + 2$.

10. **D**  Since $\frac{1}{p+q} = r = \frac{r}{1}$, eliminate the fractions by cross-multiplying:

$$r(p + q) = 1(1)$$
$$rp + rq = 1$$
$$rp = 1 - rq$$
$$p = \frac{1 - rq}{r}$$

11. **B**  Determine whether each Roman numeral expression is equal to $4a$.
   • I. Since $a = 2b = 5c$, then $a = 5c$, so
   $$4a = 4(5c) = 20c$$
   Expression I is not equal to $4a$.
   • II. Rewrite $4a$ as $2a + 2a$. Since $a = 2b = 5c$, then $2a = 2(2b) = 4b$ and $2a = 2(5c) = 10c$, so
   $$4a = 2a + 2a = 4b + 10c$$
   Expression II is equal to $4a$.
   • III. Rewrite $4a$ as $a + 3a$. Since $a = 2b$ and $a = 5c$, then $3a = 3(5c) = 15c$. Hence,
   $$4a = a + 3a = 2b + 15c$$
   Expression III is equal to $4a$.
   Only Roman numeral expressions II and III are equal to $4a$.

12. **B**  Since $\frac{a+b+c}{3} = \frac{a+b}{2}$, eliminate the fractions by cross-multiplying:

$$2(a + b + c) = 3(a + b)$$
$$2a + 2b + 2c = 3a + 3b$$
$$2c = (3a - 2a) + (3b - 2b)$$
$$= a + b$$
$$c = \frac{a+b}{2}$$

*(Lesson 4-2 Multiple Choice continued)*

**13. E**   If $wx = z$, then $x = \frac{z}{w}$, so

$$xz = z\left(\frac{z}{w}\right) = \frac{z^2}{w}$$

**14. B**   Multiplying both sides of the equation $q = \frac{p}{\sqrt{r}}$ by $\sqrt{r}$ gives $p = q\sqrt{r}$. Squaring each side of this equation gives

$$p^2 = (q\sqrt{r})^2$$
$$= (q)^2 (\sqrt{r})^2$$
$$= q^2 r$$

Since $p^2 = q^2 r$, then $\frac{1}{p^2} = \frac{1}{q^2 r}$.

**15. D**   Since $a = 2c$ and $b = 5d$, replace $a$ in the equation $\frac{c}{d} - \frac{a}{b} = x$, with $2c$ and replace $b$ with $5d$:

$$\frac{c}{d} - \frac{2c}{5d} = x$$
$$\frac{5c}{5d} - \frac{2c}{5d} = x$$
$$\frac{5c - 2c}{5d} = x$$
$$\frac{3c}{5d} = x$$

To find the value of $\frac{c}{d}$ in terms of $x$, multiply both sides of the equation $\frac{3c}{5d} = x$ by the reciprocal of $\frac{3}{5}$:

$$\frac{5}{3}\left(\frac{3c}{5d}\right) = \frac{5}{3}x$$
$$\frac{c}{d} = \frac{5}{3}x$$

*(Quantitative Comparison)*

**1. C**   To multiply powers of the same base, *add* their exponents. Since $2^x \cdot 2^y = 8$, then $2^{x+y} = 8 = 2^3$. Hence, $x + y = 3$, so in Column A

$$2x + 2y = 2(3) = 6$$

Since Column B is also 6, Column A = Column B.

**2. A**   Since $x - y = 2$, then $3x - 3y = 3(2) = 6$. Negative 1 raised to any even power is 1, and $-1$ raised to any odd power is $-1$. In Column A,

$$(-1)^{3x-3y} = (-1)^6 = 1$$

In Column B, $(-1)^7 = -1$. Hence, Column A > Column B.

**3. C**   Since $\frac{1}{a} = b$, then $a = \frac{1}{b}$. It is also given that $\frac{1}{b} = c$. Since $a$ and $c$ are each equal to the same quantity, they are equal to each other. Hence, Column A = Column B.

**4. B**   Express each side of the given equation, $8^x = 4^y$, as a power of 2 and then set the exponents equal. Thus:

$$8^x = 4^y$$
$$2^{3x} = 2^{2y}$$
$$3x = 2y$$

Dividing each side of $3x = 2y$ by 2 gives $\frac{3}{2}x = y$, so $\frac{3}{2} = \frac{y}{x}$. Since $2 > \frac{3}{2}$, Column B > Column A.

**5. B**   Eliminate variable $a$ by multiplying corresponding sides of $\frac{a}{b} = \frac{7}{8}$ and $\frac{c}{a} = \frac{4}{7}$:

$$\frac{a}{b} \times \frac{c}{a} = \frac{7}{8} \times \frac{4}{7}$$
$$\frac{c}{b} = \frac{4}{8}$$
$$\frac{c}{b} = \frac{1}{2}, \text{ so } b = 2c$$

Since $b = 2c$, rewrite Column A as $2c$. You are told that $c > 0$, so Column B > Column A.

*(Grid In)*

**1. 8**   Since $(4 \times b)^2 = 4^2 \times b^2 = 16 \times b^2$,

$$16 \times a^2 \times 64 = 16 \times b^2$$
$$\cancel{16} \times a^2 \times 64 = \cancel{16} \times b^2$$
$$a^2 \times 64 = b^2$$
$$\sqrt{a^2 \times 64} = \sqrt{b^2}$$
$$a \times 8 = b$$

Hence, $b$ is 8 times as great as $a$.

**2. 4**   Express each side of the given equation, $3^{x-1} + 3^{x-1} + 3^{x-1} = 81^y$, as a power of three. Since the left side of the equation contains three terms that are the same and $81 = 3^4$, the given equation can be written as $3 \cdot 3^{x-1} = 3^{4y}$. Hence, $3^{1+x-1} = 3^{4y}$ or $3^x = 3^{4y}$. If powers of the same base are equal, the exponents must be equal. Thus, $x = 4y$, so $\frac{x}{y} = 4$.

**LESSON 4-3**   *(Multiple Choice)*

**1. D**   Write each term of the polynomial numerator separately over the monomial denominator. Then divide powers of the same base by subtracting their exponents.

$$\frac{20b^3 - 8b}{4b} = \frac{20b^3}{4b} - \frac{8b}{4b}$$
$$= 5b^{3-1} - 2$$
$$= 5b^2 - 2$$

**2. D**   The given expression $(39)^2 + 2(39)(61) + (61)^2$, has the form

$$x^2 + 2xy + y^2 = (x + y)^2$$

*(Lesson 4-3   Multiple Choice continued)*

where $x = 39$ and $y = 61$. Hence,

$$(39)^2 + 2(39)(61) + (61)^2 = (39 + 61)^2$$
$$= (100)^2$$
$$= 10,000$$

3. **B**   Simplify $\frac{3p^2 + 12p}{3p}$; then evaluate the resulting expression for $p = 5$:

$$\frac{3p^2 + 12p}{3p} = \frac{3p^2}{3p} + \frac{12p}{3p} = p + 4$$

If $p = 5$, then

$$\frac{3p^2 + 12p}{3p} = p + 4 = 5 + 4 = 9.$$

4. **B**   To simplify the expression inside the parentheses, divide powers of the same base by subtracting their exponents. Then raise the powers to higher powers by multiplying their exponents. Thus,

$$\left[\left(\frac{a^5 b^4 c^2}{ab^2 c}\right)^2\right]^3 = \left[(a^{5-1}b^{4-2}c^{2-1})^2\right]^3$$
$$= \left[(a^4 b^2 c)^2\right]^3$$
$$= \left[a^{4\times 2}b^{2\times 2}c^2\right]^3$$
$$= \left[a^8 b^4 c^2\right]^3$$
$$= a^{24}b^{12}c^6$$

5. **E**   If $(x - y)^2 = 50$ and $xy = 7$, then

$$(x - y) = x^2 - 2xy + y^2 = 50$$
$$x^2 - 2(7) + y^2 = 50$$
$$x^2 - 14 + y^2 = 50$$
$$x^2 + y^2 = 50 + 14$$
$$= 64$$

6. **B**   *Solution 1:* Do the algebra. If $p = \frac{a}{a-b}$, then

$$1 - p = 1 - \frac{a}{a-b}$$
$$= \frac{a-b}{a-b} - \frac{a}{a-b}$$
$$= \frac{a-b-a}{a-b}$$
$$= \frac{-b}{a-b}$$

Since $\frac{-b}{a-b}$ is not one of the answer choices, eliminate the negative sign in the numerator by multiplying the numerator and the denominator by $-1$:

$$1 - p = \left(\frac{-1}{-1}\right)\frac{-b}{a-b}$$
$$= \frac{(-1)(-b)}{(-1)(a-b)} = \frac{b}{-a+b} = \frac{b}{b-a}$$

*Solution 2:* Substitute numbers for the letters. Let $a = 3$ and $b = 2$; then

$$p = \frac{a}{a-b} = \frac{3}{3-2} = 3$$

so $1 - p = 1 - 3 = -2$. When you plug in 3 for $a$ and 2 for $b$ in each of the answer choices, you find that only choice (B) produces $-2$.

7. **E**   You are given that $(p - q)^2 = 25$ and $pq = 14$. Use the formula for the square of a binomial to expand $(p - q)^2$:

$$(p - q)^2 = p^2 - 2pq + q^2 = 25$$
$$p^2 - 2(14) + q^2 = 25$$
$$p^2 + q^2 = 25 + 28 = 53$$

Now use the formula for $(p + q)^2$:

$$(p + q)^2 = p^2 + 2pq + q^2$$
$$= 53 + 2(14)$$
$$= 53 + 28 = 81$$

8. **A**   If $a - b = p$ and $a + b = k$, then

$$a^2 - b^2 = (a - b)(a + b)$$
$$= (p)(k)$$
$$= pk$$

9. **E**   You are told that $(2y + k)^2 = 4y^2 - 12y + k^2$. Use the formula for expanding the square of a binomial:

$$(2y + k)^2 = (2y)^2 + 2(2yk) + k^2$$
$$= 4y^2 \quad + 4yk \quad + k^2$$

Compare $4y^2 - 12y + k^2$ with $4y^2 + 4yk + k^2$. Since the two trinomials must be the same, then $-12y = 4yk$, so

$$k = \frac{-12y}{4y} = -3$$

10. **C**   Examine each choice in turn:
- (A) $(x + y)^2 = x^2 + y^2$ is false since $(x + y)^2 = x^2 + 2xy + y^2$.
- (B) $x^2 + x^2 = x^4$ is false since $x^2 + x^2 = 1x^2 + 1x^2 = 2x^2$.
- (C) $\frac{2^{x+2}}{2^x} = 4$ is true since
$$\frac{2^{x+2}}{2^x} = 2^{(x+2)-x}$$
$$= 2^2$$
$$= 4$$
- (D) $(3x)^2 = 6x^2$ is false since $(3x)^2 = 3^2 \cdot x^2 = 9x^2$.
- (E) $x^5 - x^3 = x^2$ is false since only like terms can be subtracted.

11. **D**   Since

$$(px + q)^2 = p^2 x^2 + 2pq + q^2$$

and

$$(px + q)^2 = 9x^2 + kx + 16,$$

then $p^2 = 9$ and $q^2 = 16$, so $p = \pm 3$ and $q = \pm 4$. Since the answer choices contain only positive numbers, use $p = 3$ and $q = 4$. The middle terms of $p^2 x^2 + 2pq + q^2$ and

*(Lesson 4-3   Multiple Choice continued)*

$9x^2 + kx + 16$ must be equal, so
$$k = 2pq = 2(3)(4) = 24.$$

**12. C**
$$\left(\frac{x}{100} - 1\right)\left(\frac{x}{100} + 1\right) = kx^2 - 1$$
$$(0.01x - 1)(0.01x + 1) = kx^2 - 1$$
$$0.0001x - 1 = kx^2 - 1$$

Since the coefficients of $x$ on both sides of the equation must be the same, $k = 0.0001$.

**13. E**  Use FOIL to multiply the left side of the given equation:
$$(x + 5)(x + p) = x^2 + 2x + k$$
$$x^2 + 5x + px + 5p = x^2 + 2x + k$$
Since $5x + px$ must be equal to $2x$, $p = -3$. Hence,
$$k = 5p = 5(-3) = -15$$

**14. D**  Multiply on each side of the given equation, $(x - 2)(x + 2) = x(x - p)$. The result is
$$x^2 - 4 = x^2 - xp$$
so $4 = xp$ and $p = \frac{4}{x}$.

**15. C**  Use the formula $\frac{1}{x} + \frac{1}{y} = \frac{x+y}{xy}$. Since
$$\cfrac{1}{\cfrac{1}{x} + \cfrac{1}{y}}$$
is the reciprocal of $\frac{x+y}{xy}$, then
$$\cfrac{1}{\cfrac{1}{x} + \cfrac{1}{y}} = \frac{xy}{x+y}$$

**16. D**  Use the formula for the square of a binomial to expand the left side of the given equation:
$$\left(k + \frac{1}{k}\right)^2 = 16$$
$$k^2 + 2(k)\left(\frac{1}{k}\right) + \left(\frac{1}{k}\right)^2 = 16$$
$$k^2 + 2 + \frac{1}{k^2} = 16$$
$$k^2 + \frac{1}{k^2} = 16 - 2 = 14$$

**17. E**  Change the right side of the given equation, $\frac{a}{b} = 1 - \frac{x}{y}$, into a single fraction:
$$\frac{a}{b} = 1 - \frac{x}{y}$$
$$= \frac{y}{y} - \frac{x}{y}$$
$$= \frac{y - x}{y}$$
Since $\frac{b}{a}$ is the reciprocal of $\frac{a}{b}$,
$$\frac{b}{a} = \frac{y}{y-x}$$

**18. A**  Change the right side of the given equation, $\frac{a}{b} = \frac{1}{x} - \frac{1}{y}$, into a single fraction:
$$\frac{a}{b} = \frac{1}{x} - \frac{1}{y}$$
$$= \frac{1}{x} \cdot \frac{y}{y} - \frac{1}{y} \cdot \frac{x}{x}$$
$$= \frac{y - x}{xy}$$
Since $\frac{b}{a}$ is the reciprocal of $\frac{a}{b}$,
$$\frac{b}{a} = \frac{xy}{y-x}$$

*(Quantitative Comparison)*

**1. B**  For Column A, write each term of the numerator separately over the denominator. Then do the same for Column B:

| Column A | Column B |
| --- | --- |
| $\dfrac{p}{4} + \dfrac{1}{4}$ | $\dfrac{q}{12} + \dfrac{4}{12}$ |
| | $\dfrac{q}{12} + \dfrac{1}{3}$ |

From the centered information, you know that $\frac{p}{4} = \frac{q}{12}$, so $\frac{p}{4}$ in Column A and $\frac{q}{12}$ in Column B can be ignored. Since $\frac{1}{3} > \frac{1}{4}$, Column B > Column A.

**2. C**  Using the centered information, cross-multiply:
$$\frac{y-1}{2} = \frac{3}{y+1}$$
$$(y - 1)(y + 1) = 2(3)$$
$$y^2 - 1 = 6$$
$$y^2 = 7$$
Hence, Column A = Column B.

**3. A**  Try to make the columns look alike by combining fractions in Column A:
$$\frac{x}{4} + \frac{x}{3} = \frac{3}{3} \cdot \frac{x}{4} + \frac{4}{4} \cdot \frac{x}{3}$$
$$= \frac{3x}{12} + \frac{4x}{12}$$
$$= \frac{3x + 4x}{12}$$
$$= \frac{7x}{12}$$
In Column B, $\frac{x}{2} = \frac{6x}{12}$. Since $\frac{7}{12} > \frac{6}{12}$ and $x$ is a positive integer, Column A > Column B.

**4. C**  You are given that $(x + y)^2 = x^2 + y^2$. In general,
$$(x + y)^2 = x^2 + 2xy + y^2$$
It must be the case that $2xy = 0$. Since $2xy = 0$, then $x$ or $y$, or both $x$ and $y$, are equal to 0.

*(Lesson 4-3 Quantitative Comparison continued)*

- If $x = 0$, then, in Column A, $(x - y)^2 = (-y)^2 = 0$. In Column B, $x^2 + y^2 = y^2$, so Column A = Column B.
- If $y = 0$, then, in Column A, $(x + y)^2 = x^2$. In Column B, $x^2 + y^2 = x^2$, so Column A = Column B.
- If $x = y = 0$, then Columns A and B are both 0, so Column A = Column B.

5. **A** Work with the centered information:
$$(y - 7)(y + 1) = y^2 + ry + s$$
$$y^2 + y - 7y - 7 = y^2 + ry + s$$
$$y^2 - 6y - 7 = y^2 + ry + s$$
Comparison of the left and right sides of this equation shows that $r = -6$ and $s = -7$. Since $-6 > -7$, Column A > Column B.

6. **A** Try to make Column A look like Column B by squaring $2h - 1$. Use the formula
$$(a - b)^2 = a^2 - 2ab + b^2$$
where $a = 2h$ and $b = 1$. Then
$$(2h - 1)^2 = (2h)^2 - 2(2h)(1) + (1)^2$$
$$= 4h^2 - 4h + 1$$
Now compare the two columns:

| Column A | Column B |
|----------|----------|
| $4h^2 - 4h + 1$ | $4h^2 + 4h + 1$ |

Since you are given that $h < 0$, then $-4h > 0$ and $4h < 0$. Hence, Column A > Column B.

7. **C** Simplify the radical in Column B:
$$\sqrt{r^2 + s^2 + 2rs} = \sqrt{(r + s)^2} = r + s$$
since $r, s > 0$. Hence, Column A = Column B.

8. **C** Simplify the radical in Column A:
$$\sqrt{(x + y)^2 - (x - y)^2} = \sqrt{(x^2 + 2xy + y^2) - (x^2 - 2xy + y^2)}$$
$$= \sqrt{x^2 + 2xy + y^2 - x^2 + 2xy - y^2}$$
$$= \sqrt{4xy}$$
$$= \sqrt{4}\sqrt{xy}$$
$$= 2\sqrt{xy}$$
Hence, Column A = Column B.

9. **B** Evaluate the radical in Column B using the given equations:
$$\sqrt{a^2 - b^2} = \sqrt{(a - b)(a + b)}$$
$$= \sqrt{(9)(4)}$$
$$= \sqrt{36}$$
$$= 6$$
Since Column A = 5, Column B > Column A.

10. **A** If $y = x^2 + 2x + 1$, the value of $y$ when $x = 0$ is
$$y = 0^2 + 2(0) + 1 = 1$$
Place 1 under Column A.

The value of $x$ when $y = 0$ is the solution of the equation $0 = x^2 + 2x + 1$. Since
$$0 = x^2 + 2x + 1 = (x + 1)^2$$
Then $x + 1 = 0$, so $x = -1$.
Place $-1$ under Column B. Hence, Column A > Column B.

*(Grid In)*

1. **19** Since
$$(3y - 1)(2y + k) = ay^2 + by - 5$$
and the product of the last terms of the two binomial factors is equal to the constant term, $(-1)(k) = -5$, so $k = 5$. Now multiply the two binomials together:
$$(3y-1)(2y+5) = (3y)(2y) + (3y)(5) + (-1)(2y) + (-1)(5)$$
$$= 6y^2 + 15y - 2y - 5$$
$$= 6y^2 + 13y - 5$$
Since
$$(3y - 1)(2y + 5) = 6y^2 + 13y - 5 = ay^2 + by - 5$$
equating the coefficients makes $a = 6$ and $b = 13$, so
$$a + b = 6 + 13 = 19$$

2. **20**
$$4x^2 + 20x + r = (2x + s)^2$$
$$= (2x + s)(2x + s)$$
$$= (2x)(2x) + (2x)(s) + (s)(2x) + (s)(s)$$
$$= 4x^2 + 2sx + 2sx + s^2$$
$$= 4x^2 + 4sx + s^2$$
Since the coeffients of $x$ on each side of the equation must be the same, $20 = 4s$, so $s = 5$. Comparing the last terms of the polynomials on the two sides of the equation makes $r = s^2 = 5^2 = 25$. Hence,
$$r - s = 25 - 5 = 20$$

## LESSON 4-4 *(Multiple Choice)*

1. **E** Factor the numerator and the denominator. Then divide out any factor that is common to both the numerator and the denominator:
$$\frac{4x + 4y}{2x^2 - 2y^2} = \frac{4(x + y)}{2(x^2 - y^2)}$$
$$= \frac{\overset{1}{\cancel{4(x+y)}}}{\underset{}{2\cancel{(x+y)}(x - y)}}$$
$$= \frac{2}{x - y}$$

2. **A** Combine the fractions and then simplify:
$$\frac{a}{a^2 - b^2} + \frac{b}{a^2 - b^2} = \frac{a + b}{a^2 - b^2}$$

*(Lesson 4-4   Multiple Choice continued)*

$$= \frac{1}{\dfrac{(\cancel{a+b})}{(\cancel{a+b})(a-b)}}$$

$$= \frac{1}{a-b}$$

**3. A**  To solve $ax + x^2 = y^2 - ay$ for $a$ in terms of $x$ and $y$, isolate variable $a$ on the left side of the equation:

$$ax + x^2 = y^2 - ay$$
$$ax + ay = y^2 - x^2$$
$$a(x+y) = (y-x)(y+x)$$
$$\frac{a(x+y)}{(x+y)} = \frac{(y-x)\cancel{(y+x)}^{\,1}}{\cancel{(x+y)}}$$
$$a = y - x$$

**4. C**  If $\frac{xy}{x+y} = 1$, then multiplying both sides of the equation by $x + y$ gives $xy = x + y$, so $xy - x = y$. Hence, $x(y-1) = y$, so

$$x = \frac{y}{y-1}$$

**5. B**  Since

$$x^2 = r^2 + 2rs + s^2 = (r+s)^2$$

and

$$y^2 = r^2 - s^2 = (r+s)(r-s),$$

then

$$\frac{x^2}{y^2} = \frac{(r+s)(r+s)}{(r+s)(r-s)} = \frac{r+s}{r-s}$$

You are told that $x$ and $y > 0$, so

$$\frac{x}{y} = \sqrt{\frac{r+s}{r-s}}$$

**6. A**  Simplify each fraction, then add:

$$h = \frac{x^2-1}{x+1} + \frac{x^2-1}{x-1}$$
$$= \frac{(x+1)(x-1)}{x+1} + \frac{(x+1)(x-1)}{x-1}$$
$$= (x-1) + (x+1)$$
$$= 2x$$
$$\frac{h}{2} = x$$

**7. A**  Collect the terms involving $a$ on one side of the given equation and the terms involving $b$ on the opposite side of the equation:

$$ax^2 - bx = ay^2 + by$$
$$ax^2 - ay^2 = bx + by$$

Factor each side of the equation:

$$ax^2 - ay^2 = bx + by$$
$$a(x^2 - y^2) = b(x+y)$$

Divide each side of the equation by $b$ and $x^2 - y^2$:

$$\frac{a}{b} = \frac{x+y}{x^2-y^2}$$
$$= \frac{x+y}{(x+y)(x-y)}$$
$$= \frac{1}{x-y}$$

**8. C**  Solve the given equation, $\frac{a^2-b^2}{9} = a + b$, for $a^2 - b^2$. Then solve the equation that results for $a - b$.

$$\frac{a^2-b^2}{9} = a + b$$
$$a^2 - b^2 = 9(a+b)$$
$$(a+b)(a-b) = 9(a+b)$$
$$a - b = \frac{9(a+b)}{a+b} = 9$$

**9. B**  Simplify the complex fraction by multiplying its numerator and denominator by the LCD of the denominator of the fractions it contains, which is $y$:

$$\frac{x+y}{\dfrac{x^2}{y}-y} = \frac{y}{y}\left(\frac{x+y}{\dfrac{x^2}{y}-y}\right)$$
$$= \frac{y(x+y)}{y\left(\dfrac{x^2}{y}-y\right)}$$
$$= \frac{y(x+y)}{y\left(\dfrac{x^2}{y}\right)+y(-y)}$$
$$= \frac{y(x+y)}{x^2-y^2}$$
$$= \frac{y\cancel{(x+y)}^{\,1}}{(x+y)(x-y)}$$
$$= \frac{y}{x-y}$$

**10. C**  *Solution 1*: Factor $k - 1$ as the difference of two squares. After simplifying, solve for $k$.

$$\frac{k-1}{\sqrt{k}-1} = \frac{(\sqrt{k}-1)(\sqrt{k}+1)}{\sqrt{k}-1} = p$$
$$\sqrt{k} + 1 = p$$
$$\sqrt{k} = p - 1$$
$$k = (\sqrt{k})^2 = (p-1)^2$$

*Solution 2*: Plug in a perfect square for $k$, such as 4. Then find the value of $p$. If $k = 4$, then

$$\frac{k-1}{\sqrt{k}-1} = \frac{4-1}{\sqrt{4}-1} = \frac{3}{2-1} = 3 = p$$

Using $p = 3$, evaluate each answer choice. Pick (C), which gives $(3-1)^2$ or 4, the value of $k$.

*(Lesson 4-4 Quantitative Comparison)*

1. **A** In Column A, factor the numerator:
$$\frac{9^8 - 9^7}{8} = \frac{9^7(9-1)}{8}$$
$$= \frac{9^7(8)}{8} = 9^7$$
Since $9^7 > 9^6$, Column A > Column B.

2. **C** Simplify the fraction in Column A:
$$\frac{1-x^2}{1-x} = \frac{(1-x)(1+x)}{1-x}$$
$$= 1+x$$
Since $1 + x = x + 1$, Column A = Column B.

3. **C** Simplify the fraction in Column A:
$$\frac{x^2 - y^2}{2x - 2y} = \frac{(x+y)(x-y)}{2(x-y)}$$
$$= \frac{x+y}{2}$$
Hence, Column A = Column B.

4. **D** In column A,
$$x^2 - y^2 = (x-y)(x+y)$$
If $x + y = 1$, Column A = Column B. If, however, $x + y \neq 1$, Column A $\neq$ Column B. Since more than one answer is possible, the correct choice is (D).

5. **B** In Column A, factoring out the common *binomial* factor of $(r-s)$ from $r(r-s) - s(r-s)$ gives $(r-s)(r-s)$ or $(r-s)^2$. Since $0 < s < r < 1$, $r-s$ in Column B is a number between 0 and 1. Since any number between 0 and 1 is greater than its square, Column B > Column A.

6. **D** Factor the quantities in the two columns:

| Column A | Column B |
|---|---|
| $(a-b)(a+b)$ | $c(a+b)$ |
| $a-b$ | $c$ |

Since $a > b > c > 0$, we can ignore the common factor $(a + b)$. The size relationship between Column A and Column B depends on the values chosen for $a$, $b$, and $c$. For example,
- If $a = 5$, $b = 4$, and $c = 3$, then $a - b = 1$, so Column B > Column A.
- If $a = 8$, $b = 4$, and $c = 3$, $a - b = 4$, so Column A > Column B.

Since two different answers are possible, the correct choice is (D).

7. **A** Since you are given that $\frac{x^2 - y^2}{x+y} = 7$,
$$\frac{(x+y)(x-y)}{x+y} = 7$$

$$x - y = 7$$
$$x = y + 7$$
Hence, Column A is equivalent to $y + 7$, and Column B is $y + \sqrt{7}$. Since $7 > \sqrt{7}$, Column A > Column B.

8. **C** Rewrite each column in factored form.

| Column A | Column B |
|---|---|
| $(r^2 - s^2)(r+s)$ | $(r+s)^2(r-s)$ |
| $(r-s)(r+s)(r+s)$ | $(r+s)(r+s)(r-s)$ |

Since the order in which the same three terms are multiplied does not matter, Column A = Column B.

9. **B** Factor each column. In Column A,
$$x^2 - 2x = x(x-2)$$
In column B,
$$x^2 - x - 2 = (x+1)(x-2)$$
You are given that $x > 2$, so $(x - 2)$ is a positive number. Since $(x - 2)$ appears as a factor in both columns, it does not affect the comparison and so can be ignored. Since $(x + 1) > x$, Column B > Column A.

10. **A** Using FOIL, multiply the two factors in the given product:
$$y^2 + ay + b = (y-3)(y+1)$$
$$= y^2 + y - 3y - 3$$
$$= y^2 - 2y - 3$$
Since $y^2 + ay + b = y^2 - 2y - 3$, then $a = -2$ and $b = -3$. Since $-2 > -3$, Column A > Column B.

## LESSON 4-5 *(Multiple Choice)*

1. **C** If $(x + 7)(x - 3) = 0$, then either or both factors may be equal to 0. If $x + 7 = 0$, then $x = -7$. Also, if $x - 3 = 0$, then $x = 3$. Hence, $x$ may be equal to $-7$ or 3.

2. **D** Multiplying both sides of the given equation, $\frac{a^2}{2} = 2a$, by 2 gives $a^2 = 4a$. To apply the zero-product rule, one side of the equation must be 0. After $4a$ is subtracted from both sides of the equation, $a^2 - 4a = 0$, which can be factored as $a(a - 4) = 0$. Thus, either $a = 0$ or $a - 4 = 0$. Hence, $a$ equals 0 or 4.

3. **B** If $x$ is a positive integer and $(x - 6)^2 = 9$, the expression inside the parentheses must be equal to 3 or $-3$.
- If $x - 6 = 3$, then $x = 9$ and $x^2 = 81$. But 81 is not one of the choices.
- If $x - 6 = -3$, then $x = -3 + 6 = 3$, so $x^2 = 9$.

*(Lesson 4-5   Multiple Choice continued)*

4. **E** If $a(x - y) = 0$, then either $a = 0$ *or* $x - y = 0$ *or* both factors are 0. Determine whether each Roman numeral statement is always true.

- I. It may be the case that $x - y = 0$ and $a \neq 0$, so statement I is not always true.
- II. Since $y$ can be any number, provided that $a = 0$ or $x = y$, statement II is not necessarily true.
- III. It may be the case that $a = 0$ and $x - y \neq 0$, so statement III is not always true.

None of the Roman numeral statements is always true.

5. **B** *Solution 1*: If $(x - 1)^2 - (x - 1) = 0$, then factoring the left side of the equation gives
$$(x - 1)\,[(x - 1) - 1] = 0,$$
which simplifies to $(x - 1)x = 0$. Hence, $x$ may be equal to 0 or 1.

*Solution 2*: Substitute the pair of values in each of the answer choices into the given equation, $(x - 1)^2 - (x - 1) = 0$, until you find a pair, $x = 0$ or $x = 1$, that works.

6. **E** If $w^2 - 2w = 0$, then $w(w - 2) = 0$, so $w = 0$ or $w = 2$.

- If $2y^2 - y = 0$, then $y(2y - 1) = 0$, so $y = 0$ or $2y - 1 = 0$.
- If $2y - 1 = 0$, then $y = \frac{1}{2}$.

Since you are told that $w$ and $y$ are positive integers, $w = 2$ and $y = \frac{1}{2}$. Hence:
$$\frac{w}{y} = 2 \div \frac{1}{2} = 2 \times 2 = 4$$

7. **C** For each Roman numeral equation, check whether $-2$ is a root.

- I. If $\frac{2}{x} - x = 0$, then $\frac{2}{x} = x$. If $x = -2$, then
$$\frac{2}{x} = \frac{2}{-2} = -1 \neq x$$
Hence, $-2$ is not a root of equation I.
- II. If $x^2 + 4 = 0$, then
$$(-2)^2 + 4 = 4 + 4 \neq 0$$
Hence, $-2$ is not a root of equation II.
- III. If $x^2 + 4x + 4 = 0$, then
$$(-2)^2 + 4(-2) + 4 = 4 - 8 + 2 = 0$$
Hence, $-2$ is a root of equation III.

Only equation III has $-2$ as one of its roots.

8. **E** If $\frac{x^2}{3} = x$, then $x^2 = 3x$, so $x^2 - 3x = 0$. Factoring $x^2 - 3x = 0$ gives $x(x - 3) = 0$. Thus, $x = 0$ or $x = 3$.

9. **E** If $(x + 1)(x - 3) = 0$, then $x + 1 = 0$ or $x - 3 = 0$. Hence, $x = -1$ or $x = 3$. The sum of these roots is $-1 + 3$ or 2, and their product is $(-1) \times (3) = -3$. Since
$$2 - (-3) = 2 + 3 = 5$$
the sum of the roots of the equation exceeds the product of its roots by 5.

10. **B** If $x^2 - 63x - 64 = 0$, then
$$(x - 64)(x + 1) = 0$$
so $x = 64$ or $x = -1$ If $p$ and $n$ are integers such that $p^n = x$, then either $p^n = 64$ or $p^n = -1$. Examine each answer choice in turn until you find a number that cannot be the value of $p$ in either $p^n = 64$ or $p^n = -1$.

- (A) If $p = -8$, then $(-8)^n = 64$, so $n = 2$.
- (B) If $p = -4$, then there is no integer value of $n$ for which $(-4)^n = 64$ or $(-4)^n = -1$.

11. **A** Isolate variable $r$ by dividing both sides of the equation by $r^t$:
$$r^t = 6.25r^{t+2}$$
$$1 = \frac{6.25r^{t+2}}{r^t}$$
$$= 6.25r^{(t+2)-t}$$
$$= 6.25r^2$$

Using a calculator, divide both sides of the equation by 6.25. Since $\frac{1}{6.25} = 0.16 = r^2$,
$$r = \sqrt{0.16} = 0.4 \text{ or } \frac{2}{5}$$

*(Quantitative Comparison)*

1. **B** If $x^2 + 8x + 16 = 0$, then
$$(x + 4)(x + 4) = 0$$
so $x = -4$. Since $-3 > -4$, Column B > Column A.

2. **D** If $x(x + p) = 0$, then $x = 0$ *or* $x = p$. Since $x$ may or may not be equal to $p$, two different answers are possible and the correct choice is (D).

3. **A** If $x^2 + 2x - 3 = 0$, then
$$(x + 3)(x - 1) = 0$$
so $x = -3$ or $x = 1$. If $a$ and $b$ are the solutions, let $a = -3$ and $b = 1$. Then
$$a + b = -3 + 1 = -2$$
and
$$ab = (-3)(1) = -3$$
Since $-2 > -3$, Column A > Column B.

4. **D** Solve $\frac{x-1}{5} = \frac{7}{x+1}$ for $x$ by setting the cross-products equal:
$$\frac{x-1}{5} = \frac{7}{x+1}$$

*(Lesson 4-5   Quantitative Comparison continued)*

$$(x-1)(x+1) = (5)(7)$$
$$x^2 - 1 = 35$$
$$x^2 = 35 + 1 = 36$$

Since $6^2 = 36$ and $(-6)^2 = (-6) \times (-6) = 36$, then, $x = 6$ or $x = -6$. If $x = 6$, Column A = Column B. If, however, $x = -6$, Column B > Column A.
Since two different answers are possible, the correct choice is (D).

5. **D** If $y^2 + ky - 9 = 0$ and $y$ is a nonzero integer, you need to find possible factorizations of $y^2 + ky - 9$ that have the form $(y + a)(y + b)$, where $ab = -9$, $k = a + b$, and $k$ is a nonzero integer. Since there are three pairs of integers whose product is $-9$, there are three possibilities to consider:

- $(y + 3)(y - 3)$ and $k = -3 + 3 = 0$. However, you are told that $k$ is a nonzero integer, so this cannot be the case.
- $(y - 1)(y + 9)$ and $k = -1 + 9 = 8$.
- $(y - 9)(y + 1)$ and $k = -9 + 1 = -8$.

If $k = 8$, Column A = Column B. If, however, $k = -8$, Column B > Column A.
Since two different answers are possible, the correct choice is (D).

*(Grid In)*

1. **2** If $(4p + 1)^2 = 81$ and $p > 0$, the expression inside the parentheses is either 9 or $-9$. Since $p > 0$, let $4p + 1 = 9$; then $4p = 8$ and $p = 2$. A possible value of $p$ is 2.

2. **2** If $(x - 1)(x - 3) = -1$, then
$$x^2 - 4x + 3 = -1$$
so $x^2 - 4x + 4 = 0$. Factoring this equation gives $(x - 2)(x - 2) = 0$. Hence, a possible value of $x$ is 2.

3. **13** The roots of the equation $(x - 5)(x + 2) = 0$ are the values of $x$ that make the equation a true statement: $x = 5$ or $x = -2$. The sum of the roots is $5 + (-2)$ or 3, and the product of the roots is $(5)(-2)$ or $-10$. Hence, the sum, 3, exceeds the product, $-10$, by $3 - (-10)$ or 13.

## LESSON 4-6 *(Multiple Choice)*

1. **A** First solve the equation that contains one variable. Since $3x + 15 = 0$, then $3x = -15$, so

$$x = \frac{-15}{3} = -5$$

Substituting $-5$ for $x$ in the other equation, $2x - 3y = 11$, gives $2(-5) - 3y = 11$ or $-10 - 3y = 11$. Adding 10 to both sides of the equation makes $-3y = 21$, so
$$y = \frac{-21}{3} = -7$$

2. **B** If $2a = 3b$, then $4a = 6b$. Substituting $6b$ for $4a$ in $4a + b = 21$ gives $6b + b = 21$ or $7b = 21$, so
$$b = \frac{21}{7} = 3$$

3. **B** Add corresponding sides of the two given equations:
$$\begin{array}{r} 2p + q = 11 \\ + \quad p + 2q = 13 \\ \hline 3p + 3q = 24 \end{array}$$
Dividing each member of $3p + 3q = 24$ by 3 gives $p + q = 8$.

4. **D** Since $p = 2m$ and $k = 2p$, then
$$k = 2(2m) = 4m$$
Substituting for $p$ and $k$ in $m + p + k = 70$ gives
$$m + 2m + 4m = 70$$
$$7m = 70$$
$$m = \frac{70}{7} = 10$$

5. **D** Eliminate $y$ by adding corresponding sides of the two equations:
$$\begin{array}{r} x - y = 3 \\ + \quad x + y = 5 \\ \hline 2x + 0 = 8, \end{array} \quad \text{so} \quad x = \frac{8}{2} = 4$$
Since $x = 4$ and $x - y = 3$, then $4 - y = 3$, so $y = 1$.

6. **C** Subtract corresponding sides of the two given equations:
$$\begin{array}{r} 5x + y = 19 \\ - \quad x - 3y = 7 \end{array} \rightarrow \begin{array}{r} 5x + y = 19 \\ + \quad -x + 3y = -7 \\ \hline 4x + 4y = 12 \end{array}$$
Dividing each member of the equation $4x + 4y = 12$ by 4 gives $x + y = 3$.

7. **E** Subtract corresponding sides of the two given equations:
$$\begin{array}{r} x + 3 = 5y \\ - \quad x - 9 = 2y \end{array} \rightarrow \begin{array}{r} x + 3 = 5y \\ + \quad -x + 9 = -2y \\ \hline 0 + 12 = 3y \end{array}$$
$$\frac{12}{3} = y \text{ or } y = 4$$
Since $y = 4$ and $x + 3 = 5y$, then $x + 3 = 5(4) = 20$, so

*(Lesson 4-6   Multiple Choice continued)*

$$x = 20 - 3 = 17$$

8. **D**   Eliminate $y$ by adding corresponding sides of the given equations:

$$\frac{1}{x} + \frac{1}{y} = \frac{1}{4}$$
$$+ \ \frac{1}{x} - \frac{1}{y} = \frac{3}{4}$$
$$\overline{\ \frac{2}{x} + 0 = \frac{4}{4} = 1\ }$$

Since $\frac{2}{x} = 1$, then $x = 2$.

9. **A**   In the equation $\frac{a}{b} = \frac{2}{5}$, cross-multiplying gives $5a = 2b$. Since $5a + 3b = 35$ and $5a = 2b$, then

$$2b + 3b = 35$$
$$5b = 35$$
$$b = 7$$

Since $5a = 2b = 2(7) = 14$,

$$a = \frac{14}{5}$$

10. **C**   If $\frac{x}{y} = 6$ and $x = 36$, then $\frac{36}{y} = 6$, so $y = 6$. Since $\frac{y}{w} = 4$ and $y = 6$,

$$\frac{6}{w} = 4$$
$$4w = 6$$
$$w = \frac{6}{4} = \frac{3}{2}$$

11. **B**   Add corresponding sides of the given equations:

$$4r + 7s = 23$$
$$+ \ \ \ r - 2s = 17$$
$$\overline{5r + 5s = 40}$$

Dividing each member of $5r + 5s = 40$ by 5 gives $r + s = 8$. Since $r + s = 8$, then

$$3r + 3s = 3(8) = 24$$

12. **C**   If $\frac{p-q}{2} = 3$ and $rp - rq = 12$, then

$$p - q = 2(3) = 6$$

and

$$r(p - q) = 12$$

so $r(6) = 12$ or $6r = 12$. Hence,

$$r = \frac{12}{6} = 2$$

13. **C**   If $(a + b)^2 = 9$, then

$$a^2 + 2ab + b^2 = 9$$

If $(a - b)^2 = 49$, then

$$a^2 - 2ab + b^2 = 49$$

Add corresponding sides of the two equations:

$$a^2 + 2ab + \ b^2 = 9$$
$$+ \ \ a^2 - 2ab + \ b^2 = 49$$
$$\overline{2a^2 + 0 \ \ \ + 2b^2 = 58}$$

Dividing each member of $2a^2 + 2b^2 = 58$ by 2 gives

$$a^2 + b^2 = 29$$

14. **D**   Proceed as follows:

• Find the value of $\frac{s}{r}$. If $\frac{r+s}{r} = 3$, then

$$\frac{r}{r} + \frac{s}{r} = 3 \ \ \text{or} \ \ 1 + \frac{s}{r} = 3$$

so $\frac{s}{r} = 2$.

• Find the value of $\frac{r}{t}$. If $\frac{t+r}{t} = 5$, then

$$\frac{t}{t} + \frac{r}{t} = 5 \ \ \text{or} \ \ 1 + \frac{r}{t} = 5$$

so $\frac{r}{t} = 4$.

• Multiply corresponding sides of the equations $\frac{s}{r} = 2$ and $\frac{r}{t} = 4$:

$$\frac{s}{r} \times \frac{r}{t} = 2 \times 4$$
$$\frac{s}{t} = 8$$

15. **B**   Eliminate $y$ by subtracting corresponding sides of the given equations:

$$3x + y = c$$
$$- \ \ \ x + y = b$$
$$\overline{2x + 0 = c - b,} \ \ \text{so} \ \ x = \frac{c-b}{2}$$

16. **E**   If $a + b = 11$ and $a - b = 7$, then adding corresponding sides of the two equations gives $2a = 18$, so

$$a = \frac{18}{2} = 9$$

If $a + b = 11$ and $a = 9$, then $9 + b = 11$, so

$$b = 11 - 9 = 2$$

Hence,

$$ab = (9)(2) = 18$$

17. **D**   For the given system of three equations,

$$x - z = 7$$
$$x + y = 3$$
$$z - y = 6$$

add the equations two at a time to eliminate the variables $y$ and $z$.

• Eliminate $y$ by adding corresponding sides of the second and third equations. The result is $x + z = 9$.

• Eliminate $z$ by adding $x + z = 9$ to the first equation. The result is $2x = 16$.

Hence, $x = \frac{16}{2} = 8$.

18. **A**   If $x$ students each contributed $y$ dollars to buy a \$60 gift for their advisor, then $xy = 60$ (equation 1). If four more, or $x + 4$ students in total, had contributed, each pupil could have contributed 3 dollars less, or $y - 3$ dollars each, to buy the

*(Lesson 4-6 Multiple Choice continued)*

same gift. Hence, $(x + 4)(y - 3) = 60$ (equation 2).

Choice (A) gives these two equations.

19. **A**  Divide corresponding sides of the given equations:

$$\frac{r^8}{r^7} = 5 \div \frac{3}{t}$$

$$r = 5 \times \frac{t}{3}$$

$$= \frac{5}{3}t$$

20. **D**  • If $\frac{a}{b} = \frac{6}{7}$, then $6b = 7a$, so

$$3b = \frac{7}{2}a$$

• If $\frac{a}{c} = \frac{2}{5}$, then $2c = 5a$, so

$$c = \frac{5}{2}a$$

• Hence,

$$3b + c = \frac{7}{2}a + \frac{5}{2}a = \frac{12}{2}a = 6a$$

*(Quantitative Comparison)*

1. **B**  Solve first the equation that contains only one variable. Since $2q - 1 = 7$, then $2q = 8$ and $q = 4$. Substituting 8 for $2q$ in $3p + 2q = 17$ gives $3p + 8 = 17$, so $3p = 9$ and $p = 3$. Since $q > p$, Column B > Column A.

2. **B**  Solve first the equation that contains only one variable. Since $y + 5 = 2$, then $y = 2 - 5 = -3$. Substituting $-3$ for $y$ in $xy - 5 = 7$ gives $-3x - 5 = 7$ or $-3x = 12$, so $x = -4$. Since $-3 > -4$, Column B > Column A.

3. **A**  Since $4a + 4b = 9$ and $3a - 3b = 7$, then

$$a + b = \frac{9}{4} \quad \text{and} \quad a - b = \frac{7}{3}$$

Since $\frac{9}{4} = 2.25$ and $\frac{7}{3} = 2.333...$, Column A > Column B.

4. **B**  Adding corresponding sides of the two given equations gives $3x = 12$, so $x = 4$. Substituting 4 for $x$ in $x + y = 9$ gives $4 + y = 9$, so $y = 5$. Since $y > x$, Column B > Column A.

5. **A**  Eliminate variable $b$ from the given system of equations by adding corresponding sides of the two equations:

$$
\begin{array}{r}
a - b + \phantom{2}c = 3 \\
+ \quad a + b + \phantom{2}c = 7 \\
\hline
2a + 0 + 2c = 10
\end{array}
$$

Since $2a + 2c = 10$, then

$$a + c = \frac{10}{2} = 5$$

Since $a + c > 4$, Column A > Column B.

6. **C**  Adding the corresponding sides of the three given equations gives $2x + 2y + 2z = 22$, so in Column A

$$x + y + z = \frac{22}{2} = 11$$

Hence, Column A = Column B.

7. **A**  Cross-multiplying in the equation $\frac{2}{x} = \frac{1}{y}$ gives $x = 2y$. Substituting $2y$ for $x$ in $2x - y = 3$ gives $2(2y) - y = 3$, or $4y - y = 3$. This simplifies to $3y = 3$, so $y = 1$. Since $x = 2y$, then $x > y$, so Column A > Column B.

*(Grid In)*

1. **3**  Since

$$5 \text{ sips} + 4 \text{ gulps} = 1 \text{ glass}$$

and

$$13 \text{ sips} + 7 \text{ gulps} = 2 \text{ glasses}$$

then

$$13 \text{ sips} + 7 \text{ gulps} = 2\overbrace{(5 \text{ sips} + 4 \text{ gulps})}^{1 \text{ glass}}$$
$$= 10 \text{ sips} + 8 \text{ gulps}$$
$$13 - 10 \text{ sips} = 8 - 7 \text{ gulps}$$
$$3 \text{ sips} = 1 \text{ gulp}$$

2. **1.5**  Write one equation underneath the other, aligning like terms in the same vertical column. Eliminate variable $b$ by multiplying the first equation by 3 and then adding the result to the second equation:

$$
\begin{array}{ll}
\begin{array}{r}
2a = 9 - b \\
\underline{4a = 3b - 12}
\end{array}
&
\overset{3\times}{\to}
\quad
\begin{array}{r}
6a = \phantom{-}27 - 3b \\
+ \quad 4a = -12 + 3b \\
\hline
10a = \phantom{-}15 + 0 \\
a = \frac{15}{10} = 1.5
\end{array}
\end{array}
$$

3. **1.45**  If $p =$ the cost of a pen and $n =$ the cost of a notebook, then

$$
\begin{array}{ll}
\begin{array}{r}
1p + 2n = 3.50 \\
\underline{2p + 3n = 5.55}
\end{array}
&
\overset{2\times}{\to}
\quad
\begin{array}{r}
2p + 4n = 7.00 \\
- \quad 2p + 3n = 5.55 \\
\hline
n = 1.45
\end{array}
\end{array}
$$

The charge for one notebook is \$1.45.

## LESSON 4-7  *(Multiple Choice)*

1. **A**  *Solution 1*: Since $4 + 3p < p + 1$, then

$$3p - p < 1 - 4 \quad \text{or} \quad 2p < -2$$

so $p < -1$. Hence, the largest integer value for $p$ is $-2$.

*(Lesson 4-7   Multiple Choice continued)*

*Solution 2*: Plug each of the answer choices for $p$ into $4 + 3p < p + 1$ until you find one that makes the inequality a true statement. Since choice (A) gives
$$4 + 3(-2) < (-2) + 1$$
there is no need to continue.

2. **E**   *Solution 1*: Solve $-3 < 2x + 5 < 9$ by first subtracting 5 from each member. The result is $-8 < 2x < 4$. Now divide each member of this inequality by 2, obtaining $-4 < x < 2$. Examine each of the answer choices until you find one (E) that is not between $-4$ and 2. Since $x$ is less than 2, 2 is not a possible value of $x$.
*Solution 2*: Plug each of the answer choices for $x$ into $-3 < 2x + 5 < 9$ until you find one (E) that does not make the inequality a true statement.

3. **E**   If the sum of a number, $x$, and the original number increased by 5, $x + 5$, is greater than 11, then $x + (x + 5) > 11$, so $2x + 5 > 11$. Then $2x > 6$, so $x > 3$. The only answer choice that is greater than 3 is (E).

4. **C**   Determine whether each Roman numeral statement is always true when $0 < a^2 < b$.
- I. From the given inequality, you know that $a^2 < b$. Although $a^2$ is positive, $a$ may or may not be positive. If $a > 0$, then $a < \frac{b}{a}$. If $a < 0$, then dividing each side of $a^2 < b$ by $a$ reverses the inequality sign, so $a > \frac{b}{a}$. Hence, statement I is not always true.
- II. Multiplying both sides of $a^2 < b$ by $a^2$ gives $a^4 < a^2 b$, so statement II is always true.
- III. Since $b > 0$, dividing both sides of $a^2 < b$ by $b$ gives $\frac{a^2}{b} < 1$, so statement III is always true.

Only Roman numeral statements II and III are always true.

5. **A**   *Solution 1*: If $4 - 3x < 11$, then $-3x < 7$, so $x > -\frac{7}{3}$. Since $-\frac{7}{3}$ is between $-2$ and $-3$, the smallest integer value of $x$ that satisfies this inequality is $-3$.
*Solution 2*: Plug each of the answer choices for $x$, starting with (A), into $4 - 3x < 11$ until you find one that makes the inequality a true statement. Choice (A) gives

$$4 - 3(-3) < 11$$

so there is no need to continue.

6. **B**   Determine whether each Roman numeral statement is always true when $a > b > c > 0$.
- I. Since $a > b$, then $a - c > b - c$, and $b - a$ represents a negative number. Hence, when both sides of the inequality $a - c > b - c$ are divided by $b - a$, a true inequality results only if the direction of the inequality is reversed. Thus,
$$\frac{a-c}{b-a} < \frac{b-c}{b-a}$$
so statement I is not always true.
- II. Since $b > c$ and $a > 0$, multiplying both sides of the inequality $b > c$ by $a$ results in the true inequality $ab > ac$, so statement II is always true.
- III. Since $a > c$, then $\frac{1}{a} < \frac{1}{c}$. Since $b > 0$, multiplying both sides of the inequality $\frac{1}{a} < \frac{1}{c}$ by $b$ produces the true inequality $\frac{b}{a} > \frac{b}{c}$, so statement II is not always true.

Hence, only Roman numeral statement II is always true.

7. **B**   If $-4 \le 3n \le 87$, then
$$-\frac{4}{3} \le n \le \frac{87}{3}$$
or, equivalently,
$$-1\frac{1}{3} \le n \le 29$$

Since $n$ is an integer, $n$ can be any integer from $-1$ to 29, inclusive. Including 0, there are 31 integers in this interval.

8. **E**   If $a^2 < b^2$ and $a$ and $b$ are not 0, then $a$ and $b$ may be either positive or negative numbers. Determine whether each Roman numeral statement must be true.
- I. If $a > 0$, then $\frac{a^2}{a} < \frac{b^2}{a}$. Since dividing both sides of an inequality by a negative number reverses the inequality sign, if $a < 0$, then $\frac{a^2}{a} < \frac{b^2}{a}$. Hence, statement I is not always true.
- II. Since $a^2 < b^2$, their reciprocals have the opposite size relationship, so $\frac{1}{a^2} > \frac{1}{b^2}$. Statement II is always true.
- III. Since $a^2 < b^2$, then $a^2 - b^2 < 0$. Factoring the left side of this inequality gives $(a + b)(a - b) < 0$, so statement III is always true.

*(Lesson 4-7   Multiple Choice continued)*

Only Roman numeral statements II and III must be true.

9. **D**  If $b + 3 > 0$, then $b > -3$ . Since $1 > 2b - 9$, then $10 > 2b$, so $5 > b$ or $b < 5$. Since $b$ is an integer, $b$ may be equal to any of these seven integers: $-2, -1, 0, 1, 2, 3,$ or $4$.

10. **C**  Determine whether each Roman numeral statement is always true when $xy > 1$ and $z < 0$.

   • I. If $x > 0$, then $x > z$. However, the fact that $x$ may be a negative number could mean that $x < z$, so statement I is not always true.

   • II. Multiplying an inequality reverses the direction of the inequality, so $(xy)z < (1)z$, or $xyz < z$. Since $z$ may or may not be greater than or equal to $-1$, the inequality $xyz < -1$ may or may not be true. Hence, statement II is not always true.

   • III. Dividing $xy > 1$ by a negative quantity reverses the direction of the inequality, so $\frac{xy}{z} < \frac{1}{z}$. Statement III is always true.

Only Roman numeral statement III is always true.

*(Quantitative Comparison)*

1. **B**  Since $\frac{1}{k} > 1$, then $k$ must be a number between 0 and 1, so $k^3 < 1$. Since $1^3 = 1$, Column B > Column A.

2. **B**  Since $2x - 1 < 7$, then $2x < 8$, so $x < 4$. Adding 2 to both sides of $x < 4$ gives $x + 2 < 6$. Hence, Column B > Column A.

3. **A**  Since $ab > 0$, multiplying both sides of $\frac{1}{a} < \frac{1}{b}$ by $ab$ produces a true inequality:
$$\frac{1}{a}(ab) < \frac{1}{b}(ab)$$
$$b < a \quad \text{or} \quad a > b$$
Hence, Column A > Column B.

4. **D**  Since $-5 < 2x - 3 < 7$, then
$$-5 + 3 < 2x < 7 + 3$$
$$-2 < 2x < 10$$
Dividing each member of $-2 < 2x < 10$ by 2 gives $-1 < x < 5$. Adding 2 to each member of $-1 < x < 5$ gives
$$-1 + 2 < x + 2 < 5 + 2$$
$$1 < x + 2 < 7$$
Since $x + 2$ represents any number from 1

to 7, it may or may not be greater than 6. Since more than one answer is possible, choice (D) is correct.

5. **D**  In general, if $a < b$ and $x < y$, the size relationship between $a - x$ and $b - y$ is unpredictable. In Column A,
$$\frac{p-1}{4} = \frac{p}{4} - \frac{1}{4}$$
In Column B,
$$\frac{q-9}{12} = \frac{q}{12} - \frac{9}{12} = \frac{q}{12} - \frac{3}{4}$$
Since $\frac{p}{4} < \frac{q}{12}$ and $\frac{1}{4} < \frac{3}{4}$, the size relationship between $\frac{p}{4} - \frac{1}{4}$ and $\frac{q}{12} - \frac{3}{4}$ will depend on the particular values of the variables and, as a result, cannot be determined.

Since more than one answer is possible, choice (D) is correct.

6. **D**  Compare the two columns for particular integer values of $a$ and $b$. Since $7 < ab < 21$, let $a = 8$ and $b = 1$. Then $a + b = 9$ and $ab = 8$, so Column A > Column B. If, however, $a = 7$ and $b = 2$, then $a + b = 9$ and $ab = 14$, so Column B > Column A.

Since the size relationship between the columns depends on the values of the variables, the correct choice is (D).

7. **A**  Since $-1 \le x - 3 \le 6$, then
$$-1 + 3 \le x \le 6 + 3$$
$$2 \le x \le 9$$
Also, since $5 \le y + 1 < 9$, then
$$5 - 1 < y \le 9 - 1$$
$$4 \le y \le 8$$
In Column A, the largest possible value of $\frac{x}{y}$ occurs when $x$ has its greatest value, which is 9, and $y$ has its smallest value, which is 4.

Thus, the largest possible value of $\frac{x}{y}$ is $\frac{9}{4}$ or 2.25, so Column A > Column B.

8. **B**  Since $3 \le x \le 10$, then,
$$\frac{1}{10} \le \frac{1}{x} \le \frac{1}{3}$$
Also, since $2 \le y \le 8$, then
$$\frac{1}{8} \le \frac{1}{y} \le \frac{1}{2}$$
Hence, in Column A the largest possible value of $\frac{1}{x} + \frac{1}{y}$ is
$$\frac{1}{3} + \frac{1}{2} = \frac{4}{12} + \frac{6}{12} = \frac{10}{12}$$
so Column B > Column A.

*(Lesson 4-7   Quantitative Comparison continued)*

9. **A**   Since $a + b = 5 = \frac{15}{3}$ and $b < \frac{7}{3}$, then in Column A

$$a > \frac{15}{3} - \frac{7}{3}$$

$$a > \frac{8}{3}$$

so Column A > Column B.

10. **A**   Since $b < j$ and $j < p$, then $b < p$. You are also given that $h < d$. Adding corresponding sides of $b < p$ and $h < d$ gives $b + h < p + d$, so Column A > Column B.

11. **D**   Since $(a + b)^2 = a^2 + 2ab + b^2$ and $(a + b)^2 > 5$, then

$$a^2 + 2ab + b^2 > 5$$

You are also given that $a^2 + b^2 = 2$, so

$$2ab + 2 > 5$$

$$2ab > 3$$

$$ab > \frac{3}{2}$$

Although $ab$ is always greater than $\frac{3}{2}$, it is impossible to tell whether $ab$ is always greater than 3.

Since Column A may or may not be greater than Column B, the correct choice is (D).

*(Grid In)*

1. **4**   If $2y - 3 < 7$, then $2y \leq 10$, so $y < 5$. Since

$$y + 5 > 8 \quad \text{and} \quad 2y - 3 < 7$$

then $y > 3$ and at the same time $y < 5$. The integer for which the quesiton asks must be 4.

2. **6**   When 2 times an integer $x$ is increased by 5, the result is always greater than 16 and less than 29, so $16 < 2x + 5 < 29$. Subtracting 5 from each member of this inequality gives $11 < 2x < 24$. Then

$$\frac{11}{2} < \frac{2x}{2} < \frac{24}{2}$$

so $5\frac{1}{2} < x < 12$. According to this inequality, $x$ is greater than $5\frac{1}{2}$, so the least integer value of $x$ is 6.

# Word Problems

This chapter uses algebraic methods and reasoning to help prepare you to solve the different types of word problems that may appear on the SAT I.

---

## LESSONS IN THIS CHAPTER

# Translating from English to Algebra

## OVERVIEW

*Some* SAT I *algebra questions require that you translate an English sentence into an algebraic sentence.*

## ASSOCIATING ENGLISH PHRASES WITH MATHEMATICAL OPERATIONS

When translating an English phrase into a mathematical expression, look for key words that tell you which arithmetic operation is needed. Table 5.1 lists some common phrases and the symbols for the arithmetic operations that they indicate. For example,

- "A number $n$ *increased by* 3" is translated as $n + 3$.
- "Three *less than* a number $y$" is translated as $y - 3$.
- "Two *times* the *sum* of a number $x$ and 3" is translated as $2(x + 3)$.
- "Twelve *divided by* a number $p$" is translated as $\frac{12}{p}$.
- "The *sum* of a number $x$ and 3 is *at least* 7" is translated as $x + 3 \geq 7$.
- "When 2 is *subtracted from* a number $n$ the difference is *at most* 5" is translated as $n - 2 \leq 5$.

TABLE 5.1   TRANSLATING ENGLISH PHRASES INTO
ARITHMETIC OPERATIONS

| English Phrase | Symbol |
| --- | --- |
| Is; was; will be; has | = |
| Plus; sum; more than; greather than; older than; farther than; years from now; increased by; exceeds by | + |
| Decreased by; minus; less than; diminished by; younger than; years ago; subtracted from | − |
| Times; multiplied by; product; of | × |
| Divided by; quotient; per | ÷ |
| At least | ≥ |
| At most | ≤ |

## TRANSLATING AN ENGLISH SENTENCE INTO AN EQUATION

To translate an English sentence into an equation, identify key phrases that can be translated directly into mathematical terms. For example:

- Two times a number $n$ increased by 5 is 17.

  $2\times$       $n$       $+5$       $= 17$

  The equation is $2n + 15 = 17$.

- One-half of a number $n$ diminished by 5 is 17.

  $\dfrac{1}{2}$   $\times$    $n$       $-5$       $= 17$

  The equation is $\dfrac{1}{2}n - 5 = 17$.

- Five times a number $n$ exceeds 2 times that number by 21.

  $5n$       $=$       $2n$       $+ 21$

  The equation is $5n = 2n + 21$.

## REPRESENTING ONE QUANTITY IN TERMS OF ANOTHER

Sometimes it is necessary to compare two quantities by representing one quantity in terms of another. For example, if Tim weighs 13 pounds more than Sue and $x$ represents Sue's weight, then Tim's weight is $x + 13$. Here are a few more examples.

- The number of dimes exceeds 3 times the number of pennies by 2. If $x$ represents the number of pennies, then $3x + 2$ represents the number of dimes.
- Bill has 7 fewer dollars than twice the number Kim has. If $x$ represents the number of dollars Kim has, then the number of dollars Bill has is $2x - 7$.
- The sum of two integers is 25. If $x$ represents one of these integers, then $25 - x$ represents the other integer.

# LESSON 5-1  TUNE-UP EXERCISES

## Multiple Choice

**1** Carl has 7 fewer than twice the number of course credits that Steve has. Steve has 5 more course credits than Gary. If Gary has 8 course credits, how many does Carl have?

(A) 6
(B) 19
(C) 20
(D) 27
(E) 33

**2** Which of the following expressions represents the phrase "3 less than 2 times $x$"?

(A) $3 - 2x$
(B) $2 - 3x$
(C) $3x - 2$
(D) $2x - 3$
(E) $2(3 - x)$

**3** When 4 times a number $n$ is increased by 9, the result is 21.

Which of the following equations represents the statement above?

(A) $4(n + 9) = 21$
(B) $4n = 9 + 21$
(C) $n + 4 \times 9 = 21$
(D) $4n + 9 = 21$
(E) $4 + 9n = 21$

**4** If $x - 4$ is 2 greater than $y + 1$, then by how much is $x + 6$ greater than $y$?

(A) 7
(B) 8
(C) 13
(D) 14
(E) 15

**5** When 3 is subtracted from 5 times a number $n$, the result is 27.

Which of the following equations represents the statement above?

(A) $5n - 3 = 27$
(B) $3n - 5 = 27$
(C) $5(n - 3) = 27$
(D) $3(n - 5) = 27$
(E) $3 - 5n = 27$

**6** If $\frac{1}{3}$ of a number is 4 less than $\frac{1}{2}$ of the number, the number is

(A) 12
(B) 18
(C) 24
(D) 30
(E) 36

**7** Susan weighs $p$ pounds. If Susan gains 17 pounds, she will weigh as much as Carol, who weighs 8 pounds less than Judy. If Judy weighs $x$ pounds, then Susan's weight, $p$ in terms of $x$ is

(A) $x - 25$
(B) $x - 9$
(C) $x + 9$
(D) $17x - 8$
(E) $x + 25$

**8** When 2 times a number $n$ is decreased by 5, the result is at least 11.

Which of the following expressions represents the sentence above?

(A) $2(n - 5) \geq 11$
(B) $2n - 5 \leq 11$
(C) $2n - 5 \geq 11$
(D) $2n - 5 = 11$
(E) $2(n - 5) \leq 11$

**9** When 3 times a number $n$ is increased by 7, the result is at most 4 times the number decreased by 1.

Which of the following expressions represents the sentence above?

(A) $3n + 7 \le 4n - 1$
(B) $3n + 7 > 4n - 1$
(C) $3n + 7 \le 4(n - 1)$
(D) $3n + 7 \ge 4(n - 1)$
(E) $3(n + 7) \le 4(n - 1)$

**10** If $t$ multiplied by the sum of $x$ and $2y$ is divided by $4y$, the result is

(A) $\dfrac{xt}{2} + \dfrac{yt}{4}$

(B) $\dfrac{xt}{4} + \dfrac{yt}{2}$

(C) $\dfrac{xt}{4y} + \dfrac{t}{2}$

(D) $\dfrac{xt}{2y} + \dfrac{t}{4y}$

(E) $\dfrac{xt + yt}{2y}$

**11** If 4 subtracted from $x$ is 1 more than $y$, what is $x$ in terms of $y$?

(A) $y - 3$
(B) $y + 1$
(C) $y + 3$
(D) $y + 4$
(E) $y + 5$

**12** If the number $a - 5$ is 3 less than $b$, which expression has the same value as 1 more than $b$?

(A) $a - 2$
(B) $a - 1$
(C) $a$
(D) $a + 1$
(E) $a + 2$

**13** If $x + y$ is 4 more than $x - y$, which of the following statements must be true?

  I. $x = 2$
  II. $y = 2$
  III. $xy$ has more than one value.

(A) I only
(B) II only
(C) III only
(D) I and III only
(E) II and III only

**14** Jill's present weight is 14 pounds less than her weight a year ago. If her weight at that time was $\dfrac{9}{8}$ of her present weight, what is her present weight in pounds?

(A) 98
(B) 104
(C) 112
(D) 118
(E) 120

**15** A number $x$ is 3 less than 4 times the number $y$. Two times the sum of $x$ and $y$ is 9. Which of the following pairs of equations could be used to find the values of $x$ and $y$?

(A) $x = 4y - 3$
   $2(x + y) = 9$
(B) $y = 4x - 3$
   $2(x + y) = 9$
(C) $x = 4(y - 3)$
   $2(x + y) = 9$
(D) $y = 4(x - 3)$
   $2x + y = 9$
(E) $x = 4y - 3$
   $2(xy) = 9$

**16** Arthur has 3 times as many marbles as Vladimir. If Arthur gives Valdimir 6 marbles, Arthur will be left with 4 more marbles than Vladimir. What is the total number of marbles that Arthur and Vladimir have?

(A) 36
(B) 32
(C) 30
(D) 24
(E) 21

**17** A video store rents each video tape at the rate of $x$ dollars for the first day and $y$ dollars for each additional day the tape is out. When Sara returns a video tape to this store, she is charged $c$ dollars. In terms of $x$, $y$, and $c$, what is the number of rental days for which Sara is charged?

(A) $x + cy$

(B) $\dfrac{c-x}{y}$

(C) $\dfrac{x+y}{c}$

(D) $x + y(c - 1)$

(E) $1 + \dfrac{c-x}{y}$

**18** A computer program is designed so that, when a number is entered, the computer output is obtained by multiplying the number by 3 and then subtracting 4 from the product. If the output that results from entering a number $x$ is then entered, which expression represents, in terms of $x$, the final output?

(A) $3x - 8$
(B) $3x - 12$
(C) $9x - 8$
(D) $9x - 16$
(E) $6x + 9$

**19** After a certain number of people enter an empty room, $\dfrac{2}{3}$ of the people who entered the room leave. After 2 more people leave, $\dfrac{1}{4}$ of the original number of people who entered the room remain. What was the original number of people who entered the empty room?

(A) 9
(B) 12
(C) 18
(D) 24
(E) 30

**20** In a high school that has a total of 950 students, the number of seniors is $\dfrac{3}{4}$ of the number of juniors, and the number of juniors is $\dfrac{2}{3}$ of the number of sophomores. If this school has the same number of freshmen as sophomores, how many students are seniors?

(A) 120
(B) 150
(C) 180
(D) 200
(E) 300

---

### Quantitative Comparison

Each question consists of two quantities in boxes, one in Column A and one in Column B. You are to compare the two quantities and on the answer sheet fill in

A  if the quantity in Column A is greater;
B  if the quantity in Column B is greater;
C  if the two quantities are equal;
D  if the relationship cannot be determined from the information given

AN E RESPONSE WILL NOT BE SCORED

---

| Column A | Column B |
| --- | --- |

A two-digit positive integer
$N$ has $t$ as its tens digit
and $u$ as its units digit

**1**

| The number formed by inter-changing the tens and units digit of $N$ | $t + 10u$ |
| --- | --- |

When 2 times a
number $n$ is increased
by 7, the result is 3

**2**

| Two times $n$ increased by 3 | One-half of $n$ |
| --- | --- |

| Column A | Column B |
| --- | --- |

The sum of the squares of
two positive odd integers
is 34

**3**

| The product of the two integers | 14 |
| --- | --- |

---

### Grid In

**1**  If $\frac{5}{4}$ of $x$ is 20, what number is $x$ decreased by 1?

**2**  Half the difference of two positive numbers is 10. If the smaller of the two numbers is 3, what is the sum of the two numbers ?

**3**  In a certain college class, each student received a grade of A, B, C, or D or an "Incomplete." In this class, $\frac{1}{6}$ of the students received A's, $\frac{1}{4}$ received B's, $\frac{1}{3}$ received C's, and $\frac{1}{6}$ received D's. If three students received grades of "Incomplete," how many students were in the class?

# Percent Problems

## OVERVIEW

*Simple algebraic equations can be used to help solve different types of percent problems.*

## THE THREE TYPES OF PERCENT PROBLEMS

You can solve each of the three basic types of percent problems by writing and solving an equation.

- **Type 1:** Finding a percent of a given number

**EXAMPLE:**   What is 15% of 80?

$$n = 0.15 \times 80$$

$$= 12$$

15% of 80 is 12.

- **Type 2:** Finding a number when a percent of it is given

**EXAMPLE:**   30% of what number is 12?

$$0.30 \times n = 12$$

$$0.30n = 12$$

$$10(0.3n) = 10(12)$$

$$3n = 120$$

$$n = \frac{120}{3}$$

$$= 40$$

30% of 40 is 12.

- **Type 3:** Finding what percent one number is of another

**EXAMPLE:**

$$\underbrace{\text{What percent}}_{\downarrow} \underbrace{\text{of}}_{\downarrow} \underbrace{30}_{\downarrow} \underbrace{\text{is}}_{} \underbrace{9?}_{\downarrow}$$

$$\frac{p}{100} \times 30 = 9$$

$$\frac{p}{100} \times 30 = 9$$

$$\frac{\overset{3}{\cancel{30}p}}{\underset{10}{\cancel{100}}} = 9$$

$$p = \frac{10 \cdot 9}{3} = 30$$

9 is 30% of 30.

# LESSON 5-2  TUNE-UP EXERCISES

## Multiple Choice

**1** The sum of 25% of 32 and 40% of 15 is what percent of 35?

(A) 20%
(B) 25%
(C) 30%
(D) 40%
(E) 50%

**2** 30% of 150 equals 4.5% of

(A) 10
(B) 100
(C) 250
(D) 1000
(E) 10,000

**3** If $A$ is 125% of $B$, then $B$ is what percent of $A$?

(A) 60%
(B) 75%
(C) 80%
(D) 88%
(E) 90%

**4** By the end of the school year, Terry had passed 80% of his science tests. If Terry failed 4 science tests, how many science tests did Terry pass?

(A) 12
(B) 15
(C) 16
(D) 18
(E) 20

**5** If 30% of $x$ is 21, what is 80% of $x$?

(A) 28
(B) 35
(C) 42
(D)  48
(E) 56

**6** A soccer team has played 25 games and has won 60% of the games it has played. What is the minimum number of additional games the team must win in order to finish the season winning 80% of the games it has played?

(A) 28
(B) 25
(C) 21
(D) 18
(E) 15

**7** What percent of 800 is 5?

(A) $\frac{1}{160}$%
(B) $\frac{5}{8}$%
(C) 1.6%
(D) 62.5%
(E) $\frac{800}{5}$%

**8** $\frac{1}{2}$ is what percent of $\frac{1}{5}$ ?

(A) 250%
(B) 210%
(C) 140%
(D) 40%
(E) 20%

**9** In an opinion poll of 50 men and 40 women, 70% of the men and 25% of the women said that they preferred fiction to nonfiction books. What percent of the number of people polled preferred to read fiction?

(A) 40%
(B) 45%
(C) 50%
(D) 60%
(E) 75%

**10**  If 25% of $x$ is 12.5, what is 12.5% of $2x$?

(A) 6.25
(B) 12.5
(C) 25
(D) 37.5
(E) 50

**11**  If $\frac{1}{8}$ of a number is 9, what is 75% of the same number?

(A) 36
(B) 48
(C) 54
(D) 72
(E) 81

**12**  300% of 6 is what percent of 24?

(A) 40%
(B) 50%
(C) 60%
(D) 75%
(E) 80%

**13**  $\sqrt{2}$ is what percent of $\sqrt{8}$?

(A) 20%
(B) 25%
(C) $33\frac{1}{3}$%
(D) 50%
(E) 75%

**14**  The price of a stock falls 25%. By what percent of the new price must the stock price rise in order to reach its original value?

(A) 25%
(B) 30%
(C) $33\frac{1}{3}$%
(D) 40%
(E) 75%

**15**  If $a$ is 30% greater than $A$ and $b$ is 20% greater than $B$, then $ab$ is what percent greater than $AB$?

(A) 25%
(B) 50%
(C) 56%
(D) 60%
(E) 75%

**16**  A number $a$ increased by 20% of $a$ results in a number $b$. When $b$ is decreased by $33\frac{1}{3}$% of $b$, the result is $c$. The number $c$ is what percent of $a$?

(A) 40%
(B) 60%
(C) 80%
(D) 120%
(E) 150%

---

## Grid In

**1**  A high school tennis team is scheduled to play 28 matches. If the team wins 60% of the first 15 matches, how many additional matches must the team win in order to finish the season winning 75% of its scheduled matches?

**2**  In a club of 35 boys and 28 girls, 80% of the boys and 25% of the girls have been members for more than 2 years. If $n$ percent of the club have been members for more than 2 years, what is the value of $n$?

# Some Special Types of Word Problems

## OVERVIEW

*This lesson applies algebra translation skills to solving consecutive-integer, and age problems*

## CONSECUTIVE INTEGER PROBLEMS

Consecutive integers differ by 1, and consecutive even or odd integers differ by 2.

- A list of consecutive integers that begins with $x$ is

$$x, x + 1, x + 2, x + 3, \ldots$$

- A list of consecutive even or odd integers that begins with $x$ is

$$x, x + 2, x + 4, x + 6, \ldots$$

**EXAMPLE:** In a set of 3 consecutive odd integers, twice the sum of the second and the third integers is 43 more than 3 times the first integer. What is the smallest of the 3 integers?

**SOLUTION:**

- Represent the unknown quantities in mathematical terms:

Let $x$ = the first of the three consecutive odd integers.
Then $x + 2$ = the second of the three consecutive odd integers.
and $x + 4$ = the third of the three consecutive odd integers.

- Write an equation by directly translating the key words of the English sentence into mathematical terms:

| Twice the sum of the second and the third | is | 43 more than | 3 times the first |
|:---:|:---:|:---:|:---:|
| $2\big[(x + 2) + (x + 4)\big]$ | $=$ | $43 +$ | $3x$ |

- Solve the equation:

$$2\big[(x + 2) + (x + 4)\big] = 43 + 3x$$

$$2\big[2x + 6\big] = 43 + 3x$$

$$4x + 12 = 43 + 3x$$

$$4x - 3x = 43 - 12$$

$$x = 31$$

Hence, the smallest of the three consecutive odd integers is 31.

## AGE PROBLEMS

If $x$ represents a person's present age, then subtracting a number of years from $x$ gives the person's age in the past. Similarly, adding a number of years to $x$ gives the person's age in the future. For example:

- $x - 5$ represents a person's age 5 years ago.
- $x + 5$ represents a person's age 5 years from now.

**EXAMPLE:**   If 7 years from now Susan will be 2 times as old as she was 3 years ago, what is Susan's present age?

**SOLUTION:**

Let $x$ = Susan's present age.
Then $x + 7$ = Susan's age 7 years from now,
and $x - 3$ = Susan's age 3 years ago.

$$\underbrace{\text{7 years from now}}_{x + 7} \text{ Susan } \underbrace{\text{will be}}_{=} \underbrace{\text{2 times as old as she was 3 years ago}}_{2(x - 3)}$$

$$x + 7 = 2(x - 3)$$

$$x + 7 = 2x - 6$$

$$13 = x$$

Susan is now 13 years old.

# LESSON 5-3 TUNE-UP EXERCISES

| **Multiple Choice** |
|---|

**1** In an ordered set of 4 consecutive odd integers, the sum of 3 times the second integer and the greatest integer is 104. Which is the least integer in the set?

(A) 11
(B) 17
(C) 19
(D) 23
(E) 27

**2** Four years ago Jim was one-half of the age he will be 7 years from now.

If $x$ represents Jim's present age, which of the following equations represents the above statement ?

(A) $\frac{1}{2}(x - 4) = x + 7$

(B) $x - 4 = \frac{1}{2}(x + 7)$

(C) $\frac{1}{2}(x + 4) = x + 7$

(D) $\frac{1}{2}(x + 4) = x - 7$

(E) $(x - 4) = 2(x + 7)$

**3** The sum of $n$ consecutive integers is 11. If the least of these integers is −10, what is the value of $n$?

(A) 11
(B) 12
(C) 20
(D) 21
(E) 22

**4** How old is David if his age 6 years from now will be twice his age 7 years ago?

(A) 13
(B) 15
(C) 17
(D) 20
(E) 23

**5** If $x$ is the least of 4 consecutive even integers whose sum is $S$, what is $x$ in terms of $S$?

(A) $\frac{S}{4} - 3$

(B) $\frac{3(S + 1)}{4}$

(C) $\frac{S + 3}{4}$

(D) $\frac{4S}{3}$

(E) $3S - 4$

**6** The greatest of 4 consecutive odd integers is 11 more than twice the sum of the first and the second odd integers.

If $x$ is the least of the 3 consecutive odd integers, which of the following equations represents the above statement?

(A) $x + 4 = 2[(x + 1) + (x + 2)] + 11$
(B) $x + 8 = 2[(x + 2) + (x + 4)] + 11$
(C) $(x + 6) + 11 = 2[x + (x + 2)]$
(D) $x + 5 = 2[x + (x + 1)] + 11$
(E) $x + 6 = 2[x + (x + 2)] + 11$

**7** If the value of $n$ nickels plus $d$ dimes is $c$ cents, what is $n$ in terms of $d$ and $c$?

(A) $\frac{c}{5} - 2d$

(B) $5c - 2d$

(C) $\frac{c - d}{10}$

(D) $\frac{cd}{10}$

(E) $\frac{c + 10d}{5}$

**8** Ellen paid $10.82 for 40 postage stamps. If some of the stamps were 29-cent stamps and the rest were 23-cent stamps, how many 29-cent stamps did she buy?

(A) 13
(B) 17
(C) 22
(D) 25
(E) 27

**9** If John's age is increased by Mary's age, the result is 2 times John's age 3 years ago. If Mary is now $M$ years old, what is John's present age in terms of $M$?

(A) $M + 3$
(B) $M + 6$
(C) $2M - 3$
(D) $2M - 6$
(E) $2M + 3$

**10** If $3x$ years from today Reyna will be $(3y + 4)$ times her present age, what is Reyna's present age in terms of $x$ and $y$?

(A) $\dfrac{3x}{4(y-1)}$

(B) $\dfrac{3x}{3y+4}$

(C) $\dfrac{x}{y+1}$

(D) $\dfrac{3x}{4y}$

(E) $\dfrac{x}{4y-1}$

# 5-4 Ratio and Variation

## OVERVIEW

*If Mary is 16 years old and her brother Gary is 8 years old, then Mary is 2 times as old as Gary. The ratio of Mary's age to Gary's age is 2 : 1 (read as "2 to 1") since*

$$\frac{\text{Mary's age}}{\text{Gary's age}} = \frac{16 \text{ years}}{8 \text{ years}} = \frac{2}{1} \text{ or } 2 : 1$$

*A **ratio** is a comparison by division of two quantities that are measured in the same units.*

*One quantity may be related to another quantity so that either the ratio or product of these quantities always remains the same.*

## RATIO OF *a* TO *b*

The ratio of $a$ to $b$ ($b \neq 0$) is the fraction $\frac{a}{b}$, which can be written as $a : b$ (read as "$a$ is to $b$").

**EXAMPLE:** The ratio of the number of girls to the number of boys in a certain class is 3 : 5. If there is a total of 32 students in the class, how many girls are in the class?

**SOLUTION:** Since the number of girls is a multiple of 3 and the number of boys is the same multiple of 5, let

$$3x = \text{the number of girls in the class}$$
$$\text{and } 5x = \text{the number of boys in the class}$$

Then

$$3x + 5x = 32$$

$$8x = 32$$

$$x = \frac{32}{8} = 4$$

The number of girls = $3x = 3(4) = 12$.

**200**

## RATIO OF *a* TO *b* TO *c*

If $a : b$ represents the ratio of $A$ to $B$ and $b : c$ represents the ratio of $B$ to $C$, then the ratio of $A$ to $C$ is $a : c$, provided that $b$ stands for the same number in both ratios. For example, if the ratio of $A$ to $B$ is $3 : 5$ and the ratio of $B$ to $C$ is $5 : 7$, then the ratio of $A$ to $C$ is $3 : 7$. In this case $B$ represents the number 5 in both ratios.

**EXAMPLE:**   If the ratio of $A$ to $B$ is $3 : 5$ and the ratio of $B$ to $C$ is $2 : 7$, what is the ratio of $A$ to $C$?

**SOLUTION:**   Change each ratio into an equivalent ratio in which the term that corresponds to $B$ is the same number.

- The ratio of $A$ to $B$ is $3 : 5$, so the term corresponding to $B$ in this ratio is 5. The ratio of $B$ to $C$ is $2 : 7$, so the term corresponding to $B$ in this ratio is 2.
- The least common multiple of 5 and 2 is 10. You need to change each ratio into an equivalent ratio in which the term corresponding to $B$ is 10.
- Multiplying each term of the ratio $3 : 5$ by 2 gives the equivalent ratio $6 : 10$. Multiplying each term of the ratio $2 : 7$ by 5 gives $10 : 35$.
- Since the ratio of $A$ to $B$ is equivalent to $6 : 10$ and the ratio of $B$ to $C$ is equivalent to $10 : 35$, the ratio of $A$ to $C$ is $6 : 35$.

## DIRECT VARIATION

If two variables change in value so that their ratio always remains the same, then one variable is said to vary **directly** with the other variable. When one variable varies directly with another variable, a change in one variable causes a change in the other variable in the same direction—both increase or both decrease.

**EXAMPLE:**   If 28 pennies weigh 42 grams, what is the weight in grams of 50 pennies ?

**SOLUTION:**   The number of pennies and their weight vary directly since multiplying one of the two quantities of pennies by a constant causes the other to be multiplied by the same constant. If $x$ represents the weight in grams of 50 pennies, then

$$\frac{\text{pennies}}{\text{grams}} = \frac{28}{42} = \frac{50}{x}$$

Cross-multiply:   $28x = 42(50)$

$$x = \frac{2100}{28} = 75$$

The weight of 50 pennies is 75 grams.

## INVERSE VARIATION

If two variables change in opposite directions, so that their product always remains the same, then one variable is said to vary **inversely** with the other variable.

**EXAMPLE:** Four men working together can build a house in 9 days. How many days would it take 3 men working together to build the same house?

**SOLUTION:** As the number of men working on the house *decreases*, the number of days needed to build the house *increases*. Since this is an inverse variation, the number of men working times the number of days needed to build the house stays constant.

If $d$ represents the number of days that 3 men take to build the house, then

$$3 \times d = 4 \times 9$$

$$3d = 36$$

$$d = \frac{36}{3} = 12$$

Three men working together would take 12 days to build the house.

## LESSON 5-4  TUNE-UP EXERCISES

| Multiple Choice |
|---|

**1** A recipe for 4 servings requires salt and pepper to be added in the ratio of 2 : 3. If the recipe is adjusted from 4 to 8 servings, what is the ratio of the salt and pepper that must now be added?

(A) 4 : 3
(B) 2 : 6
(C) 2 : 3
(D) 3 : 2
(E) 8 : 4

**2** On a certain map, $\frac{3}{8}$ of an inch represents 120 miles. How many miles does $1\frac{3}{4}$ inches represent?

(A) 300
(B) 360
(C) 400
(D) 480
(E) 560

**3** The population of a bacteria culture doubles in number every 12 minutes. The ratio of the number of bacteria at the end of 1 hour to the number of bacteria at the beginning of that hour is

(A) 64 : 1
(B) 60 : 1
(C) 32 : 1
(D) 16 : 1
(E) 8 : 1

**4** At the end of the season, the ratio of the number of games a team has won to the number of games it lost is 4 : 3. If the team won 12 games and each game played ended in either a win or a loss, how many games did the team play during the season?

(A) 9
(B) 15
(C) 18
(D) 21
(E) 24

**5** The ratio $3\frac{1}{2}$ : 8 is equivalent to which of the following ratios?

(A) $\frac{1}{2} : \frac{8}{3}$
(B) 7 : 16
(C) 2 : 5
(D) $8 : 3\frac{1}{2}$
(E) 7 : 8

**6** A school club includes only sophomores, juniors, and seniors, in the ratio of 1 : 3 : 2. If the club has 42 members, how many seniors are in the club?

(A) 6
(B) 7
(C) 12
(D) 14
(E) 21

**7** If $\frac{c-3d}{4} = \frac{d}{2}$, what is the ratio of $c$ to $d$?

(A) 5 : 1
(B) 3 : 2
(C) 4 : 3
(D) 3 : 4
(E) 2 : 3

**8** If 4 pairs of socks costs $10.00, how many pairs of socks can be purchased for $22.50?

(A) 5
(B) 7
(C) 8
(D) 9
(E) 10

**9** Two boys can paint a fence in 5 hours. How many hours would it take 3 boys to paint the same fence?

(A) $\frac{3}{2}$
(B) 3
(C) $3\frac{1}{3}$
(D) $7\frac{1}{2}$
(E) $4\frac{2}{3}$

*(handwritten) 2 boys → 5 hrs.*
*3 boys → x hrs.*
*$\frac{3}{5} = \frac{2}{x}$*
*2(5) = 3x*

**10** A car moving at a constant rate travels 96 miles in 2 hours. If the car maintains this rate, how many miles will the car travel in 5 hours?

(A) 480
(B) 240
(C) 210
(D) 192
(E) 144

**11** The number of kilograms of corn needed to feed 5000 chickens is 30 less than twice the number of kilograms needed to feed 2800 chickens. How many kilograms of corn are needed to feed 2800 chickens?

(A) 70
(B) 110
(C) 140
(D) 190
(E) 250

**12** In an ordered list of 5 consecutive positive even integers, the ratio of the greatest integer to the least integer is 2 to 1. Which of the following is the middle integer in the list?

(A) 10
(B) 12
(C) 14
(D) 16
(E) 18

**13** If the ratio of $p$ to $q$ is 3 to 2, what is the ratio of $2p$ to $q$?

(A) 1 : 3
(B) 2 : 3
(C) 3 : 3
(D) 3 : 1
(E) 3 : 4

**14**
$$\frac{x}{z} = \frac{1}{3}$$

If in the equation above $x$ and $z$ are integers, which are possible values of $\frac{x^2}{z}$?

I. $\frac{1}{9}$

II. $\frac{1}{3}$

III. 3

(A) II only
(B) III only
(C) I and III only
(D) II and III only
(E) None

**15** If $a - 3b = 9b - 7a$, then ratio of $a$ to $b$ is

(A) 3 : 2
(B) 2 : 3
(C) 3 : 4
(D) 4 : 3
(E) 1 : 2

**16** The ratio of $A$ to $B$ is $a : 8$, and the ratio of $B$ to $C$ is $12 : c$. If the ratio of $A$ to $C$ is $2 : 1$, what is the ratio of $a$ to $c$?

(A) $2 : 3$
(B) $3 : 2$
(C) $4 : 3$
(D) $3 : 4$
(E) $1 : 3$

**17** If $8^r = 4^t$, the ratio of $r$ to $t$ is

(A) $2 : 3$
(B) $3 : 2$
(C) $4 : 3$
(D) $3 : 4$
(E) $1 : 2$

**18** If $\dfrac{a+b}{b} = 4$ and $\dfrac{a+c}{c} = 3$, what is the ratio of $c$ to $b$?

(A) $2 : 3$
(B) $3 : 2$
(C) $2 : 1$
(D) $3 : 1$
(E) $6 : 1$

**19** Jars $A$, $B$, and $C$ each contain eight marbles. What is the minimum number of marbles that must be transferred among the jars so that the ratio of the number of marbles in jar $A$ to the number in jar $B$ to the number in jar $C$ is 1 to 2 to 3?

(A) 2
(B) 3
(C) 4
(D) 6
(E) 8

**20** At a college basketball game, the ratio of the number of freshmen who attended to the number of juniors who attended is 3 to 4. The ratio of the number of juniors who attended to the number of seniors who attended is 7 to 6. What is the ratio of the number of freshmen to the number of seniors who attended the basketball game?

(A) $7 : 8$
(B) $3 : 4$
(C) $2 : 3$
(D) $1 : 2$
(E) $1 : 3$

---

## Quantitative Comparison

Each question consists of two quantities in boxes, one in Column A and one in Column B. You are to compare the two quantities and on the answer sheet fill in

A  if the quantity in Column A is greater;
B  if the quantity in Column B is greater;
C  if the two quantities are equal;
D  if the relationship cannot be determined from the information given

AN E RESPONSE WILL NOT BE SCORED

---

| <u>Column A</u> | <u>Column B</u> | | <u>Column A</u> | <u>Column B</u> |
|---|---|---|---|---|
| $\dfrac{4}{x} = \dfrac{x}{9}$ | | | $a : b = 3 : 5$ $a = 6$ | |

**1** | $x$ | 6 |     **2** | $b$ | 9 |

| Column A | Column B |
| --- | --- |

**3** | The ratio of $\frac{3}{8}$ to $\frac{1}{4}$ | The ratio of 1 to $\frac{2}{3}$ |

| Column A | Column B |
| --- | --- |

$$3h - k = k - 5h$$

**4** | The ratio of $h$ to $k$ | $\frac{1}{2}$ |

## Grid In

**1** A string is cut into 2 pieces that have lengths in the ratio of 2 : 9. If the difference between the lengths of the 2 pieces of string is 42 inches, what is the length in inches of the shorter piece?

**2** The ratio of $a$ to $b$ is 5 : 9, and the ratio of $x$ to $y$ is 10 : 3. The ratio of $ay$ to $bx$ is equivalent to the ratio of 1 to what number?

**3** The ratio of dimes to pennies in a purse is 3 to 4. If 3 pennies are taken out of the purse, the ratio of dimes to pennies becomes 1 to 1. How many dimes are in the purse?

**4** For integer values of $a$ and $b$, $b^a = 8$. The ratio of $a$ to $b$ is equivalent to the ratio of $c$ to $d$, where $c$ and $d$ are integers. What is the value of $c$ when $d = 10$?

**5** If $6a - 8b = 0$ and $c = 12b$, the ratio of $a$ to $c$ is equivalent to the ratio of 1 to what number?

# 5-5 Rate Problems

## OVERVIEW

*A ratio of 2 quantities that have different units of measurement is called a* **rate**. *For example, if a car travels a total distance of 150 miles in 3 hours, the average rate of speed is the distance traveled divided by the amount of time required to travel that distance:*

$$\text{Rate} = \frac{\text{Distance}}{\text{Time}} = \frac{150 \text{ miles}}{3 \text{ hours}} = 50 \text{ miles per hour}$$

*Rate problems are usually solved using the general relationship*

$$\text{Rate (of } A \text{ per unit } B) \times B = A$$

*You are using the correct rate relationship if the units check, as in*

$$\text{Rate} \quad \times \quad \text{Time} \quad = \quad \text{Distance}$$

$$\frac{\text{miles}}{\text{hour}} \quad \times \quad \text{hours} \quad = \quad \text{miles}$$

## UNIT COST PROBLEMS

Unit cost problems require you to figure out a rate by calculating the cost per item. Multiplying this rate by a given number of items gives the total cost of those items.

**EXAMPLE:**   If 5 cans of soup cost $1.95, how much do 3 cans of soup cost?

**SOLUTION 1:**   Since five cans of soup cost $1.95, the cost of one can is

$$\frac{\$1.95}{5 \text{ cans}} = 0.39 \text{ dollars per can}$$

To find the cost of three cans, multiply the rate of 0.39 dollars per can by 3:

$$\text{cost of 3 cans} = 3 \text{ cans} \times 0.39 \frac{\$}{\text{can}} = \$1.17$$

**SOLUTION 2:** The number of cans of soup varies directly with the cost of the cans. Form a proportion in which $x$ represents the cost of three cans of soup:

$$\frac{\text{Cost}}{\text{Number of cans}} = \frac{x}{3} = \frac{1.95}{5}$$

$$5x = 3(1.95)$$

$$x = 5.85$$

$$x = \frac{5.85}{5} = 1.17$$

The cost of three cans of soup is $1.17.

## MOTION PROBLEMS

The solution of motion problems depends on the relationship

$$\text{Rate} \times \text{Time} = \text{Distance}$$

**EXAMPLE:**  John rode his bicycle to town at the rate of 15 miles per hour. He left the bicycle in town for minor repairs and walked home along the same route at the rate of 3 miles per hour. Excluding the time John spent in taking the bike into the repair shop, the trip took 3 hours. How many hours did John take to walk back?

**SOLUTION:**  Since the trip took a total of 3 hours,

let $x$ = the number of hours John took to ride his bicycle to town,
and $3 - x$ = the number of hours John took to walk back from town.

| | Rate | × | Time | = | Distance |
|---|---|---|---|---|---|
| To Town | 15 mph | | $x$ hours | | $15x$ |
| Return Trip | 3 mph | | $3 - x$ hours | | $3(3 - x)$ |

Since John traveled over the same route, the two distances must be equal. Hence,

$$15x = 3(3 - x)$$

$$x = 9 - 3x$$

$$18x = 9$$

$$x = \frac{9}{18} = \frac{1}{2} \text{ hour}$$

John took $3 - \frac{1}{2} = 2\frac{1}{2}$ hours to walk back.

## WORK PROBLEMS

Work problems are solved using the relationship

$$\begin{pmatrix} \text{Rate of} \\ \text{work} \end{pmatrix} \times \begin{pmatrix} \text{Time} \\ \text{worked} \end{pmatrix} = \begin{pmatrix} \text{Part of job} \\ \text{completed} \end{pmatrix}$$

The rate at which work is done is the reciprocal of the amount of time needed to complete the whole job. For example, if a car mechanic takes 3 hours to repair a car, his average rate of work is $\frac{1}{3} \frac{\text{job}}{\text{hour}}$. Thus, the mechanic completes $\frac{1}{3}$ of the job in 1 hour, $\frac{2}{3}$ of the job in 2 hours, and $\frac{3}{3}$ of the job in 3 hours. In $x$ hours the mechanic completes $\frac{1}{3}(x)$, or $\frac{x}{3}$, of the job.

**EXAMPLE:**    Jim can wax a car in 3 hours, and Sue can wax the same car in 2 hours. If Jim and Sue work together, how long will it take them to wax the car ?
(A) 60 minutes
(B) 72 minutes
(C) 75 minutes
(D) 105 minutes
(E) 120 minutes

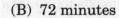
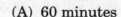
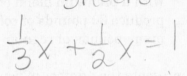

**SOLUTION:**    Since Jim and Sue start and finish at the same time, they work the same number of hours, say $x$. Make a table as shown below.

|  | **Rate** | $\times$ | **Time** | $=$ | **Part of job** |
|---|---|---|---|---|---|
| Jim | $\frac{1}{3}$ |  | $x$ |  | $\frac{x}{3}$ |
| Sue | $\frac{1}{2}$ |  | $x$ |  | $\frac{x}{2}$ |

- For the whole job to be completed, the fractional parts of the job done by Jim and Sue must add up to 1. Thus:

$$\frac{x}{3} + \frac{x}{2} = 1$$

- The fractions can be eliminated by multiplying each member of the equation by 6, the LCD of 3 and 2:

$$6\left(\frac{x}{3}\right) + 6\left(\frac{x}{2}\right) = 6(1)$$

$$2x + 3x = 6$$

$$5x = 6$$

$$x = \frac{6}{5}$$

- Working together, Jim and Sue take $\frac{6}{5}$ hours to wax the car. Since 1 hour is equivalent to 60 minutes, $\frac{6}{5}$ hours is equivalent to

$$\frac{6}{5} \times 60 = 6 \times 12 = 72 \text{ minutes}$$

The correct choice is (B).

## DRY MIXTURE PROBLEMS

Sometimes two or more dry ingredients that cost different amounts of money per unit of weight are mixed together. For the mixture,

$$\left(\begin{array}{c}\text{Cost of ingredient} \\ \text{per unit of amount}\end{array}\right) \times \left(\begin{array}{c}\text{Amount of} \\ \text{ingredient}\end{array}\right) = \left(\begin{array}{c}\text{Value of} \\ \text{ingredient}\end{array}\right)$$

**EXAMPLE:**   Regular blend coffee that sells for $3.00 a pound is to be mixed with a premium blend of coffee that sells for $4.50 a pound to produce 30 pounds of coffee that sells for $4.00 a pound. How many pounds of regular coffee should be in the mixture?

**SOLUTION:** If $x$ represents the number of pounds of regular blend coffee in the mixture, then $30 - x$ is the number of pounds of premium blend in the mixture. Make a table as shown below.

| | Price per pound $\times$ | Number of pounds | = Value |
|---|---|---|---|
| Regular blend | $3.00 | $x$ | $3.00x$ |
| Premium blend | $4.50 | $30 - x$ | $4.50(30 - x)$ |
| Mixture | $4.00 | 30 | $4.00(30)$ |

- Find the total value of the mixture, which is equal to the sum of the values of both of the ingredients of the mixture. Thus,

$$3.00x + 4.50(30 - x) = 4.00(30)$$

- Multiply each side of the equation by 100 by moving the decimal point in each term two places to the right:

$$300x + 450(30 - x) = 400(30)$$

$$300x + 13{,}500 - 450x = 12{,}000$$

$$-150x + 13{,}500 = 12{,}000$$

$$-150x = -1500$$

$$x = \frac{-1500}{-150} = 10$$

Hence, 10 pounds of regular blend coffee should be in the mixture.

## LESSON 5-5  TUNE-UP EXERCISES

| **Multiple Choice** |
|---|

**1** If four pens cost $1.96, what is the greatest number of pens that can be purchased for $7.68?

(A) 12
(B) 14
(C) 15
(D) 16
(E) 17

*handwritten: $x = 4$ pens*
*handwritten: $1.96x = 7.68$, $\frac{1.96}{1.96}$ = $\frac{7.68}{1.96}$*

**2** If a car is traveling at a constant rate of 45 miles per hour, how many miles does it travel from 10:40 A.M. to 1:00 P.M. of the same day?

(A) 165
(B) 150
(C) 120
(D) 105
(E) 90

**3** If $k$ pencils cost $c$ cents, what is the cost in cents of $p$ pencils?

(A) $\dfrac{pc}{k}$

(B) $\dfrac{kc}{p}$

(C) $\dfrac{c}{kp}$

(D) $c - kp$

(E) $\dfrac{c}{p-k}$

**4** A freight train left a station at 12 noon, going north at a rate of 50 miles per hour. At 1 P.M. a passenger train left the same station, going south at a rate of 60 miles per hour. At what time were the trains 380 miles apart?

(A) 3:00 P.M.
(B) 4:00 P.M.
(C) 4:30 P.M.
(D) 5:00 P.M.
(E) 5:30 P.M.

*handwritten: $2 = 160$, $3 = 270$, $4 = 380$*

**5** Julie can type a manuscript in 4 hours. Pat takes 6 hours to type the same manuscript. If Julie and Pat begin working together at 12 noon, at what time will they complete the typing of the manuscript?

(A) 2:24 P.M.
(B) 2:30 P.M.
(C) 2:40 P.M.
(D) 3:00 P.M.
(E) 3:30 P.M.

*handwritten: $12\left(\dfrac{x}{4} + \dfrac{x}{6}\right)$  2.4 hrs.*
*handwritten: $3x + 2x = 12$, $\dfrac{5x}{5} = \dfrac{12}{5}$*

**6** If $x$ men working together at the same rate can complete a job in $h$ hours, what part of the same job can one man working alone complete in $k$ hours?

(A) $\dfrac{k}{xh}$

(B) $\dfrac{h}{xk}$

(C) $\dfrac{k}{x+h}$

(D) $\dfrac{kh}{x}$

(E) $\dfrac{kx}{h}$

**7** If 1 cup of milk is added to a 3-cup mixture that is $\frac{2}{5}$ flour and $\frac{3}{5}$ milk, what percent of the 4-cup mixture is milk?

(A) 80%
(B) 75%
(C) 70%
(D) 65%
(E) 60%

*handwritten: $50(x+1) + 60(x) = 380$*
*handwritten: $50x + 50 + 60x = 380$*
*handwritten: $\dfrac{110x}{110} = \dfrac{330}{110}$  $x = 3$ hrs.*

**8** A freight train and a passenger train start toward each other at the same time from two towns that are 500 miles apart. After 3 hours the trains are still 80 miles apart. If the average rate of speed of the passenger train is 20 miles per hour faster than the average rate of speed of the freight train, what is the average rate of speed, in miles per hour, of the freight train?

(A) 40
(B) 45
(C) 50
(D) 55
(E) 60

**9** One machine can seal 360 packages per hour, and an older machine can seal 140 packages per hour. How many MINUTES will the two machines working together take to seal a total of 700 packages?

(A) 48
(B) 72
(C) 84
(D) 90
(E) 108

**10** Carrie can inspect a case of watches in 5 hours. James can inspect the same case of watches in 3 hours. After working alone for 1 hour, Carrie stops for lunch. After taking a 40-minute lunch break, Carrie and James work together to inspect the remaining watches. How long do Carrie and James work together to complete the job?

(A) 1 hour and 30 minutes
(B) 1 hour and 45 minutes
(C) 2 hours
(D) 2 hours and 15 minutes
(E) 2 hours and 30 minutes

**11** A man who can complete a job in $h$ hours stops working before the job is finished. If the man did not stop, he could have finished the job by working $p$ additional hours. What part of the job had the man completed when he stopped working?

(A) $\dfrac{p}{h}$

(B) $\dfrac{h-p}{h}$

(C) $\dfrac{h+p}{h}$

(D) $\dfrac{h}{h+p}$

(E) $\dfrac{p}{h+p}$

**12** A motor boat traveling at 18 miles per hour traveled the length of a lake in one-quarter of an hour less time than it took when traveling at 12 miles per hour. What was the length in miles of the lake?

(A) 6
(B) 9
(C) 12
(D) 15
(E) 21

**13** A man went on a trip of 120 miles, traveling at an average of $x$ miles per hour. Several days later he returned over the same route at a rate that was 5 miles per hour faster than his previous rate. If the time for the return trip was one-third of an hour less than the time for the outgoing trip, which equation can be used to find the value of $x$?

(A) $\dfrac{120}{x+5} = \dfrac{1}{3}$

(B) $\dfrac{x}{120} = \dfrac{x+5}{120} - \dfrac{1}{3}$

(C) $\dfrac{120}{x+(x+5)} = \dfrac{1}{3}$

(D) $\dfrac{120}{x} = \dfrac{120}{x+5} + \dfrac{1}{3}$

(E) $120(x+5) - 120x = \dfrac{1}{3}$

## Quantitative Comparison

Each question consists of two quantities in boxes, one in Column A and one in Column B. You are to compare the two quantities and on the answer sheet fill in

    A  if the quantity in Column A is greater;
    B  if the quantity in Column B is greater;
    C  if the two quantities are equal;
    D  if the relationship cannot be determined from the information given

AN E RESPONSE WILL NOT BE SCORED

| Column A | Column B |
|---|---|

Bob works twice as fast as Vincent. Working alone, Bob can complete a job in $x$ hours.

**1**

| The part of the job completed in 2 hours when Bob and Vincent work together | $\dfrac{4}{x}$ |
|---|---|

**2**

| The number of hours a hiker takes to walk 14 kilometers at an average rate of 3.5 kilometers per hour | The number of hours 2 people working together take to complete a job that each of them takes 8 hours to complete when working alone |
|---|---|

| Column A | Column B |
|---|---|

$x$ pounds of \$3.00-per-pound coffee are mixed with $y$ pounds of \$7.00-per-pound coffee to make a blend of coffee that sells for \$6.00 per pound.

**3**

| $\dfrac{x}{y}$ | 3 |
|---|---|

*(handwritten)*
$$48 - 2x + 6x = 120$$
$$48 + 4x = 120$$
$$4x = 72$$
$$x = \frac{72}{4} = 18$$

## Grid In

**1** Fruit needed for a dessert costs \$1.20 a pound. If 5 pounds of fruit are needed to make a dessert that serves 18 people, what is the cost of the fruit needed to make enough of the same dessert to serve 24 people?

**2** Candy selling for \$2.00 per pound will be mixed with candy selling for \$6.00 per pound. How many pounds of the more expensive candy will be needed to produce a 24-pound mixture that will sell for \$5.00 per pound?

*(handwritten)*
candy1   \$2.00   24-x   2(24-x)
candy2   \$6.00   x   6x
2.00 + 6(24) = 5.00(24)
mixture   5.00 → 24 lbs   24(5)

# ANSWERS TO CHAPTER 5 TUNE-UP EXERCISES

**LESSON 5-1**  *(Multiple Choice)*

1. **B**  Since Steve has 5 more course credits than Gary and Gary has 8 course credits, Steve has 5 + 8 or 13 course credits. Carl has 7 fewer than twice the number of course credits that Steve has, so Carl has $(2 \times 13) - 7$ or 19 course credits.

2. **D**  The phrase "3 less than 2 times $x$" means $2x$ minus 3 or $2x - 3$.

3. **D**  When 4 times a number $n$ (= $4n$) is increased by 9 (= $4n + 9$), the result is 21. Hence, the equation is $4n + 9 = 21$.

4. **C**  If $x - 4$ is 2 greater than $y + 1$, then
$$x - 4 = (y + 1) + 2$$
$$= y + 3$$
$$x = y + 7$$
Since $x + 6 = (y + 7) + 6$ or $x + 6 = y + 13$, $x + 6$ is greater than $y$ by 13.

5. **A**  When 3 is subtracted from 5 times a number $n$ (= $5n - 3$), the result is 27. Hence, the equation is $5n - 3 = 27$.

6. **C**  *Solution 1:* Let $x$ represent the unknown number. Since $\frac{1}{3}$ of $x$ is 4 less than $\frac{1}{2}$ of $x$, $\frac{x}{3} = \frac{x}{2} - 4$. Multiplying each member of this equation by 6, the LCD of its denominators, gives $2x = 3x - 24$, so
$$-x = -24 \quad \text{or} \quad x = 24$$
*Solution 2:* Plug each of the answer choices into the statement of the problem until you find one (C) that works: $\frac{1}{3}$ of 24 (= 8) is 4 less than $\frac{1}{2}$ of 24 (= 12).

7. **A**  Since Carol weighs 8 pounds less than Judy and Judy weighs $x$ pounds, Carol weighs $x - 8$ pounds. If Susan, who weighs $p$ pounds, gains 17 pounds, she will weigh $p + 17$ pounds, which is as much as Carol weighs. Hence, $p + 17 = x - 8$, so $p = x - 25$.

8. **C**  The expression "at least" in the given sentence means greater than or equal to. If two times a number $n$ is decreased by 5 (= $2n - 5$), the result is at least ($\geq$) 11. Hence, $2n - 5 \geq 11$.

9. **A**  The expression "at most" in the given sentence means less than or equal to. When 3 times a number $n$ is increased by 7 (= $3n + 7$), the result is at most ($\leq$)

4 times the number decreased by 1 (= $4n - 1$). Hence, $3n + 7 \leq 4n - 1$.

10. **C**  The sum of $x$ and $2y$ is $x + 2y$, and $t$ multiplied by this sum is $t(x + 2y)$. When this expression is divided by $4y$, the result is
$$\frac{t(x + 2y)}{4y} = \frac{xt + 2yt}{4y}$$
$$= \frac{xt}{4y} + \frac{\overset{1}{\cancel{2yt}}}{\underset{2}{\cancel{4y}}}$$
$$= \frac{xt}{4y} + \frac{t}{2}$$

11. **E**  If 4 subtracted from $x$ is 1 more than $y$, then $x - 4 = y + 1$, so $x = y + 1 + 4$ or $x = y + 5$.

12. **B**  If the number $a - 5$ is 3 less than $b$, then
$$a - 5 = b - 3$$
$$a - 5 + 3 = b$$
$$a - 2 = b$$
Since 1 more than $b$ is $b + 1$, add 1 to each side of the equation $a - 2 = b$, obtaining $a - 2 + 1 = b + 1$ or $a - 1 = b + 1$. Hence, $a - 1$ has the same value as 1 more than $b$.

13. **E**  If $x + y$ is 4 more than $x - y$, then $x + y = x - y + 4$ or $y + y = x - x + 4$, so $2y = 4$. Determine whether each Roman numeral statement is true or false.
- I. The value of $x$ cannot be determined, so statement I is false.
- II. Statement II is true since $2y = 4$, so $y = 2$.
- III. Since the value of $x$ is not fixed, $xy$ can have more than one value for different values of $x$. Hence, statement III is true.

Only Roman numeral statements II and III are true.

14. **C**  Jill's weight 1 year ago was $\frac{9}{8}$ of her present weight, so Jill lost $\frac{1}{8}$ of her weight. Since her present weight is 14 pounds less than her weight a year ago, 14 pounds represents $\frac{1}{8}$ of her present weight. Hence, Jill's present weight is 8 × 14 or 112 pounds.

15. **A**  If the number $x$ is 3 less than 4 times the number $y$, then $x = 4y - 3$. If 2 times the sum of $x$ and $y$ is 9, then $2(x + y) = 9$.

*(Lesson 5-1 Multiple Choice continued)*

The two equations in choice (A) can be used to find the values of $x$ and $y$.

16. **B** Since Arthur has 3 times as many marbles as Vladimir, let $x$ represent the number of marbles Vladimir has, and let $3x$ represent the number of marbles Arthur has. If Arthur gives Valdimir 6 marbles, Arthur now has $3x - 6$ marbles and Vladimir has $x + 6$ marbles. Since Arthur is left with 4 more marbles than Vladimir,

$$3x - 6 = (x + 6) + 4$$
$$= x + 10$$
$$2x = 16$$
$$x = 8$$

Since $x = 8$, $3x = 24$. The total number of marbles that Arthur and Vladimir have is $x + 3x = 8 + 24$ or 32.

17. **E** Let $n$ represent the number of rental days. Since the video store charges $x$ dollars for the first day and $y$ dollars for each additional day that the tape is out, for $n$ days the charge is $1x + (n - 1)y$. Since Sara is charged $c$ dollars, $c = x + (n - 1)y$. Now solve this equation for $n$ by multiplying each term inside the parentheses by $y$:

$$c = x + ny - y$$
$$y + c - x = ny$$
$$\frac{y}{y} + \frac{c}{y} - \frac{x}{y} = \frac{ny}{y}$$
$$1 + \frac{c - x}{y} = n$$

18. **D** If $x$ is the number entered, then multiplying $x$ by 3 gives $3x$ and subtracting 4 from that product gives $3x - 4$. Since $3x - 4$ is then entered, the final output is obtained by multiplying $3x - 4$ by 3 and then subtracting 4. Thus, the final output is $3(3x - 4) - 4$, which simplifies to $9x - 12 - 4$ or $9x - 16$.

19. **D** If the original number of people in the room is represented by $x$, then, after $\frac{2}{3}$ of the $x$ people in the room leave, $\frac{1}{3}x$ people remain. After two more people leave, $\frac{1}{3}x - 2$ people remain. Since this number represents $\frac{1}{4}$ of the $x$ people who originally entered the room, $\frac{1}{3}x - 2 = \frac{1}{4}x$. Multiplying each member of this equation by 12 gives $4x - 24 = 3x$, so $4x - 3x = 24$ and $x = 24$.

20. **B** Let $x$ represent the number of sophomores. Then $\frac{2}{3}x$ represents the number of juniors, and $\frac{3}{4}\left(\frac{2}{3}x\right)$ or $\frac{1}{2}x$ represents the number of seniors. Since this school has the same number of freshmen as sophomores, the number of freshmen is $x$. The total number of students is 950 so

$$x + x + \frac{2}{3}x + \frac{1}{2}x = 950$$
$$2x + \frac{2}{3}x + \frac{1}{2}x = 950$$
$$6(2x) + 6\left(\frac{2}{3}x\right) + 6\left(\frac{1}{2}x\right) = 6(950)$$
$$12x + 4x + 3x = 5700$$
$$19x = 5700$$
$$x = \frac{5700}{19} = 300$$

Since $\frac{1}{2}x = \frac{1}{2}(300) = 150$, 150 students are seniors.

*(Quantitative Comparison)*

1. **C** Any two-digit positive integer can be represented as $10t + u$, where $t$ is the tens digit and $u$ is the units digit. For example, the tens digit of 53 is 5, and the units digit is 3, so $53 = 5(10) + 3$. Interchanging the tens and units digits of 53 gives 35 which can be written as $5 + 3(10)$. Interchanging the tens and units digit of $10t + u$ gives $10u + t$ or, equivalently, $t + 10u$. Since Column B represents the integer formed by interchanging the tens digit $t$ and the units digit $u$ of $N$ ($= 10t + u$), Column A = Column B.

2. **C** You are given that, when 2 times a number $n$ is increased by 7, the result is 3. Hence, $2n + 7 = 3$ or $2n = 3 - 7 = -4$, so $n = \frac{-4}{2} = -2$. In Column A, two times $n$ increased by 3 is

$$2(-2) + 3 = -4 + 3 = -1$$

In Column B, one-half of $n$ is

$$\frac{1}{2}(-2) = -1$$

Hence, Column A = Column B.

3. **A** If the sum of the squares of two positive odd integers is 34, then these two integers must be 3 and 5 since

$$3^2 + 5^2 = 9 + 25 = 34$$

In Column A, the product of the two integers is $3 \times 5 = 15$, so Column A > Column B.

*(Lesson 5-1   Grid In)*

1. **15**  If $\frac{5}{4}$ of $x$ is 20, then $\frac{5}{4}x = 20$, so
$$x = \frac{4}{5}(20) = 16$$
Hence, $x$ decreased by 1 is $x - 1$ or 15.

2. **26**  If half the difference of two positive numbers is 10, then the difference of the two positive numbers is 20. If the smaller of the two numbers is 3, then the other positive number must be 23 since $23 - 3 = 20$. Hence, the sum of the two numbers is $3 + 23$ or 26.

3. **36**  You are given that in a certain college class each student received a grade of A, B, C, D or an "Incomplete." Since $\frac{1}{6} + \frac{1}{4} + \frac{1}{3} + \frac{1}{6}$ or $\frac{11}{12}$ of the number of students in the class received letter grades, $\frac{1}{12}$ of the students in the class received grades of "Incomplete." You are told that 3 students received this grade. Since these 3 students represent $\frac{1}{12}$ of the class, there were $3 \times 12$ or 36 students in the class.

## LESSON 5-2  *(Multiple Choice)*

1. **D**  First find the sum of 25% of 32 and 40% of 15:
$$25\% \text{ of } 32 = \frac{1}{4} \times 32 = 8$$
$$+ \quad 40\% \text{ of } 15 = \frac{2}{5} \times 15 = 6$$
$$\overline{\phantom{+ \quad 40\% \text{ of } 15 = } 14}$$
Then answer the question "14 is what percent of 35?" by solving the equation $14 = \frac{p}{100} \times 35$, where $p$ is the unknown percent. Since $14 = 0.35\,p$,
$$p = \frac{14}{0.35} = 40$$
so 14 is 40% of 35.

2. **D**  30% of 150 = $0.30 \times 150 = 45$. Answer the question "4.5% of what number is 45?" by solving the equation $0.045 \times n = 45$. Since $0.045n = 45$,
$$n = \frac{45}{0.045} = 1000$$

3. **C**  If $A$ is 125% of $B$, then $A = 1.25 \times B$. Since $A = 1.25B$,
$$B = \frac{1}{1.25} \times A = 0.80 \times A$$
so $B$ is 80% of $A$.

4. **C**  Since Terry passed 80% of his science tests, he failed 20% or $\frac{1}{5}$ of the tests taken.

Terry failed 4 science tests, so 4 represents $\frac{1}{5}$ of the total number of science tests. Hence, Terry took $5 \times 4$ or 20 science tests. Since he failed 4 tests, he passed $20 - 4$ or 16 science tests.

5. **E**  If 30% of $x$ is 21, then $\frac{3}{10}x = 21$, so
$$\frac{1}{10}x = \frac{1}{3}(21) = 7$$
Since $\frac{1}{10}x = 7$, $\frac{8}{10}x$, or 80% of $x$, is $8 \times 7 = 56$.

6. **B**  Since the soccer team has won 60% of the 25 games it has played, it has won $0.6 \times 25$ or 15 games.

*Solution 1:* If $x$ represents the minimum number of additional games the team must win in order to finish the season winning 80% of the games it has played, then $15 + x = 80\%$ of $(25 + x)$, so $15 + x = 0.80(25 + x)$. Remove the parentheses by multiplying each term inside the parentheses by 0.80:
$$15 + x = 20 + 0.8x$$
$$x - 0.8x = 20 - 15$$
$$0.2x = 5$$
$$x = \frac{5}{0.2} = 25$$

*Solution 2:* For each answer choice, form and then evaluate the fraction
$$\frac{\text{wins}}{\text{total games}} = \frac{\text{answer choice} + 15}{\text{answer choice} + 25}$$
The correct answer choice is (B), which gives a fraction equal to 0.80:
$$\frac{15}{25} = \frac{25 + 15}{25 + 25} = \frac{40}{50} = 0.80$$

7. **B**  Answer the question "What percent of 800 is 5?" by solving the equation $\frac{p}{100} \times 800 = 5$, where $p$ is the unknown percent. Since
$$\frac{p}{100} \times 800 = p \times 8 = 5$$
then $p = \frac{5}{8}\%$.

8. **A**  Answer the question "$\frac{1}{2}$ is what percent of $\frac{1}{5}$?" by solving the equation $\frac{1}{2} = \frac{p}{100} \times \frac{1}{5}$, where $p$ is the unknown percent. Since $\frac{1}{2} = \frac{p}{500}$, then
$$p = \frac{500}{2} = 250\%$$

9. **C**  In the opinion poll, 70% of the 50 men or $0.70 \times 50 = 35$ men preferred fiction to nonfiction books. In the same poll, 25% of the 40 women or $0.25 \times 40 = 10$ women

*(Lesson 5-2  Multiple Choice continued)*

preferred fiction to nonfiction books. Thus, 45 (= 35 + 10) of the 90 (= 50 + 40) people polled preferred fiction to nonfiction books. Since $\frac{45}{90} = \frac{1}{2} = 50\%$, 50% of the people polled preferred to read fiction.

10. **B**  12.5% of $2x = 12.5\% \times 2x$
$$= (12.5\% \times 2) \times x$$
$$= 25\% \times x$$
Hence, 12.5% of $2x$ = 25% of $x$ = 12.5.

11. **C**  If $\frac{1}{8}$ of a number is 9, then $\frac{8}{8}$ or the whole number is $8 \times 9$ or 72. Hence, 75% of the same number is $0.75 \times 72 = 54$.

12. **D**  300% of 6 is $3.00 \times 6$ or 18. Now answer the question "18 is what percent of 24?" by solving the equation $18 = \frac{p}{100} \times 24$, where $p$ is the unknown percent. Hence, $18 = 0.24p$, so
$$p = \frac{18}{0.24} = 75\%$$

13. **D**  Answer the question "$\sqrt{2}$ is what percent of $\sqrt{8}$?" by solving the equation $\sqrt{2} = \frac{p}{100} \times \sqrt{8}$, where $p$ is the unknown percent:
$$\frac{\sqrt{2}}{\sqrt{8}} = \frac{p}{100}$$
$$\frac{1}{\sqrt{4}} = \frac{p}{100}$$
$$\frac{100}{2} = p$$
$$50 = p$$
Hence, $\sqrt{2}$ is 50% of $\sqrt{8}$.

14. **C**  Suppose the original price of the stock was $100. If the price falls 25%, the new price of the stock is $0.75 \times \$100 = \$75$. To reach its original value, the price of the stock must rise $25, which is $\frac{1}{3}$ or $33\frac{1}{3}\%$ of its $75 price.

15. **C**  If $a$ is 30% greater than $A$, then
$$a = A + 30\% \text{ of } A$$
$$= A + 0.30A$$
$$= 1.3A$$
Similarly, if $b$ is 20% greater than $B$, then $b = 1.2B$. Thus,
$$ab = (1.3A)(1.2B)$$
$$= 1.56(AB)$$
$$= AB + 0.56AB$$
Since $0.56 = 56\%$, $ab$ is 56% greater than $AB$.

16. **C**  If $a$ increased by 20% of $a$ results in $b$, then

$$a + (20\% \text{ of } a) = b$$
$$a + 0.2a = b$$
$$1.2a = b$$
When $b$ is decreased by $33\frac{1}{3}\%$ of $b$, the result is $c$, so
$$b - \frac{1}{3}b = \frac{2}{3}b = c$$
Since
$$c = \frac{2}{3}b = \frac{2}{3}(1.2a) = 0.80a$$
then $c$ is 80% of $a$.

*(Grid In)*

1. **12**  Since the team wins 60% of the first 15 matches, it wins $0.60 \times 15$ or 9 of its first 15 matches. Let $x$ represent the number of additional matches it must win to finish the 28-match season winning 75% of its scheduled matches. Since $75\% = \frac{3}{4}$,
$$\frac{\text{total wins}}{\text{total matches}} = \frac{9 + x}{28} = \frac{3}{4}$$
Solve the equation by cross-multiplying:
$$4(9 + x) = 3(28)$$
$$36 + 4x = 84$$
$$4x = 48$$
$$x = \frac{48}{4}$$
$$= 12$$

2. **55.5**  Since
$$80\% \text{ of } 35 = 0.80 \times 35 = 28$$
and
$$25\% \text{ of } 28 = 0.25 \times 28 = 7$$
then 35 (= 28 + 7) of the 63 (35 + 28) boys and girls have been club members for more than 2 years. Since $\frac{35}{63} = 0.5555...$, 55.5% of the club have been members for more than 2 years.

**LESSON 5-3**  *(Multiple Choice)*

1. **D**  If $x$ is an odd integer, then $x + 2$, $x + 4$, and $x + 6$ represent the next three consecutive odd integers. Since the sum of 3 times the second integer, $x + 2$, and the largest integer, $x + 6$, is 104,
$$3(x + 2) + (x + 6) = 104$$
$$3x + 6 + x + 6 = 104$$
$$4x + 12 = 104$$
$$4x = 92$$
$$x = \frac{92}{4} = 23$$

*(Lesson 5-3  Multiple Choice continued)*

2. **B** If $x$ is Jim's present age, then $x - 4$ represents Jim's age 4 years ago and $x + 7$ represents his age 7 years from now. Since 4 years ago Jim was one-half of the age he will be 7 years from now,

$$x - 4 = \frac{1}{2}(x + 7)$$

3. **E** If the sum of $n$ consecutive integers starting with $-10$ is 11, then the consecutive integers must range from $-10$ to $+11$, inclusive, so that number-pair opposites from $-10$ to $+10$ cancel each other. Including 0, there are 22 integers in this interval.

4. **D** If $x$ is David's present age, then $x + 6$ is his age 6 years from now and $x - 7$ was his age 7 years ago. Since 6 years from now David will be twice his age 7 years ago,

$$x + 6 = 2(x - 7)$$
$$= 2x - 14$$
$$x = 20$$

5. **A** If $x$ is the least of 4 consecutive even integers whose sum is $S$, then

$$x + (x + 2) + (x + 4) + (x + 6) = S$$
$$4x + 12 = S$$
$$4x = S - 12$$
$$x = \frac{S - 12}{4} = \frac{S}{4} - \frac{\overset{3}{\cancel{12}}}{\underset{1}{\cancel{4}}} = \frac{S}{4} - 3$$

6. **E** If $x$ is the least of the 4 consecutive odd integers, then the next 3 consecutive odd integers are $x + 2$, $x + 4$, and $x + 6$. The greatest of these 4 integers is $x + 6$, and 2 times the sum of the first and the second of these integers is $2[x + (x + 2)]$. Hence, the required equation is

$$x + 6 = 2[x + (x + 2)] + 11$$

7. **A** The value in cents of $n$ nickels plus $d$ dimes is $5n + 10d$, which you are told is equal to $c$ cents. Hence, $5n + 10d = c$ or $5n = c - 10d$, so

$$n = \frac{c}{5} - \frac{10d}{5} = \frac{c}{5} - 2d$$

8. **E** Suppose the number of 29-cent stamps purchased is $x$. A total of 40 stamps was purchased, so $40 - x$ is the number of 23-cent stamps. Since the total value of the stamps is \$10.82,

$$0.29x + 0.23(40 - x) = 10.82$$
$$29x + 23(40 - x) = 1082$$

$$29x + 920 - 23x = 1082$$
$$6x = 1082 - 920$$
$$x = \frac{162}{6} = 27$$

9. **B** Let $J$ represent John's age. If John's age is increased by Mary's age, the result is $J + M$, which equals 2 times John's age 3 years ago or $2(J - 3)$. Thus, $J + M = 2(J - 3)$ or $J + M = 2J - 6$, so

$$M + 6 = 2J - J \quad \text{and} \quad J = M + 6$$

10. **C** Suppose $R$ represents Reyna's present age. In $3x$ years, Reyna will be $R + 3x$, which equals $(3y + 4)$ times her present age $R$. Hence, $R + 3x = (3y + 4)R$. Solve this equation for $R$:

$$R + 3x = 3yR + 4R$$
$$3x = 3yR + 4R - R$$
$$= 3yR + 3R$$
$$= R(3)(y + 1)$$
$$\frac{3x}{3(y + 1)} = R$$
$$\frac{x}{y + 1} = R$$

## LESSON 5-4 *(Multiple Choice)*

1. **C** When the recipe is adjusted from 4 to 8 servings, the amounts of salt and pepper are *each* doubled, so the ratio of 2 : 3 remains the same.

2. **E** *Solution 1*: If $x$ represents the unknown number of miles, then

$$\frac{\text{inches}}{\text{miles}} = \frac{\frac{3}{8}}{120} = \frac{1\frac{3}{4}}{x}$$

Cross-multiplying gives

$$\frac{3}{8}x = 1\frac{3}{4}(120) = \frac{7}{4}(120) = 210$$

Then

$$x = \frac{8}{3}(210) = 560 \text{ miles}$$

*Solution 2*: Since $\frac{3}{8}$ of an inch represents 120 miles, $\frac{1}{8}$ of an inch represents $\frac{1}{3} \times 120$ = 40 miles. Since $1\frac{3}{4} = \frac{14}{8}$, $1\frac{3}{4}$ inches represent $14 \times 40$ or 560 miles.

3. **C** Let $p$ represent the initial population of bacteria. After 12 minutes the population is $2p$, after 24 minutes it is $4p$, after 36 minutes it is $8p$, after 48 minutes it is $16p$, and after 60 minutes or 1 hour the population is $32p$. Since

*(Lesson 5-4   Multiple Choice continued)*

$$\frac{32p}{p} = \frac{32}{1}$$

the ratio of the number of bacteria at the end of 1 hour to the number of bacteria at the beginning of that hour is 32 : 1.

4. **D**  If $x$ represents the number of losses, then

$$\frac{\text{wins}}{\text{losses}} = \frac{4}{3} = \frac{12}{x}$$
$$4x = 36$$
$$x = 9$$

The total number of games played was 12 + 9 or 21.

5. **B**  Since

$$3\frac{1}{2} : 8 = \frac{7}{2} \div 8 = \frac{7}{2} \times \frac{1}{8} = \frac{7}{16}$$

then $3\frac{1}{2} : 8$ is equivalent to the ratio 7 : 16.

6. **D**  If $x$ represents the number of sophomores in the school club, then $3x$ represents the number of juniors, and $2x$ represents the number of seniors. Since the club has 42 members,

$$x + 3x + 2x = 42$$
$$6x = 42$$
$$x = \frac{42}{6} = 7$$

The number of seniors in the club is $2x = 2(7) = 14$.

7. **A**  If $\frac{c-3d}{4} = \frac{d}{2}$, then

$$2(c - 3d) = 4d$$
$$2c - 6d = 4d$$
$$2c = 10d$$

Since $\frac{c}{d} = 5$, the ratio of $c$ to $d$ is 5:1.

8. **D**  If $x$ represents the number of pairs of sock that can be purchased for \$22.50, then

$$\frac{\text{pairs of socks}}{\text{cost}} = \frac{4}{10} = \frac{x}{22.50}$$

So

$$10x = 4(22.50) = 90$$
$$x = \frac{90}{10} = 9$$

9. **C**  The number of boys working and the time needed to complete the job are inversely related since, as one of these quantities increases, the other decreases. Let $x$ represent the time three boys take to paint a fence; then

$$3x = 2(5) = 10$$
$$x = \frac{10}{3} = 3\frac{1}{3}$$

10. **B**  If $x$ represents the number of miles the car travels in 5 hours, then

$$\frac{\text{distance}}{\text{time}} = \frac{96}{2} = \frac{x}{5}$$

so

$$2x = 5(96) = 480$$
$$x = \frac{480}{2} = 240$$

11. **C**  If $x$ represents the number of kilograms of corn needed to feed 2800 chickens, then $2x - 30$ is the number of kilograms needed to feed 5000 chickens. If the amount of feed needed for each chicken is assumed to be constant,

$$\frac{x}{2800} = \frac{2x-30}{5000}$$

so

$$5000x = 2800(2x - 30)$$
$$50x = 28(2x - 30)$$
$$= 56x - 840$$
$$6x = 840$$
$$x = \frac{840}{6} = 140$$

12. **B**  If $x$ is the least integer in an ordered list of five consecutive positive even integers, then the other integers are $x + 2$, $x + 4$, $x + 6$, and $x + 8$. Since the ratio of the greatest integer to the least integer is 2 to 1,

$$\frac{x+8}{x} = \frac{2}{1}$$
$$2x = x + 8$$
$$x = 8$$

The middle integer in the list is $x + 4 = 8 + 4 = 12$.

13. **D**  If the ratio of $p$ to $q$ is 3 to 2, then $\frac{p}{q} = \frac{3}{2}$, so

$$2\left(\frac{p}{q}\right) = 2\left(\frac{3}{2}\right) \quad \text{and} \quad \frac{2p}{q} = \frac{3}{1}$$

Hence, the ratio of $2p$ to $q$ is 3 : 1.

14. **D**  Since $\frac{x}{z} = \frac{1}{3}$, then

$$\frac{x^2}{z} = x\left(\frac{x}{z}\right) = x\left(\frac{1}{3}\right)$$

Determine whether each Roman numeral value can be a possible value of $\frac{x^2}{z}$ when $x$ and $z$ are integers.

- I. If $\frac{x^2}{z} = x\left(\frac{1}{3}\right) = \frac{1}{9}$, then $x = \frac{1}{3}$. Since $x$ must be an integer, $\frac{1}{9}$ is not a possible value of $\frac{x^2}{z}$.

- II. $\frac{x^2}{z} = x\left(\frac{1}{3}\right) = \frac{1}{3}$, then $x = 1$. Hence, $\frac{1}{3}$ is

*(Lesson 5-4 Multiple Choice continued)*

a possible value of $\frac{x^2}{z}$.

- III. If $\frac{x^2}{z} = x\left(\frac{1}{3}\right) = 1$, then $x = 3$. Hence, 3 is a possible value of $\frac{x^2}{z}$.

Only Roman numeral values II and III are possible values of $\frac{x^2}{z}$.

15. **A** If $a - 3b = 9b - 7a$, then
$$a + 7a = 9b + 3b$$
$$8a = 12b$$
$$\frac{a}{b} = \frac{12}{8} = \frac{3}{2}$$

Hence, the ratio of $a$ to $b$ is 3 : 2.

16. **C** Since the lowest common multiple of 8 and 12 is 24, multiply each term of the ratio $a$ : 8 by 3 and multiply each term of the ratio 12 : $c$ by 2. The ratio of $A$ to $B$ is then $3a$ : 24, and the ratio of $B$ to $C$ is 24 : $2c$, so the ratio of $A$ to $C$ is $3a$ : $2c$. Since you are given that the ratio of $A$ to $C$ is 2 : 1, then
$$\frac{2}{1} = \frac{3a}{2c}, \text{ so}$$
$$\frac{a}{c} = \frac{2(2)}{3(1)} = \frac{4}{3}$$

Hence, the ratio of $a$ to $c$ is 4 : 3.

17. **A** If $8^r = 4^t$, then $2^{3r} = 2^{2t}$ so $3r = 2t$. Dividing both sides of this equation by $3t$ gives
$$\frac{3r}{3t} = \frac{2t}{3t} \quad \text{or} \quad \frac{r}{t} = \frac{2}{3}$$

The ratio of $r$ to $t$ is 2 : 3.

18. **B** Rewrite each fraction by dividing each term of the numerator by the denominator:
$$\frac{a+b}{b} = \frac{a}{b} + \frac{b}{b} = \frac{a}{b} + 1 = 4, \quad \frac{a}{b} = 3$$
$$\frac{a+c}{c} = \frac{a}{c} + \frac{c}{c} = \frac{a}{c} + 1 = 3, \quad \frac{a}{c} = 2 \text{ or } \frac{c}{a} = \frac{1}{2}$$

Multiply corresponding sides of the two proportions:
$$\frac{a}{b} \times \frac{c}{a} = 3 \times \frac{1}{2}$$
$$\frac{c}{b} = \frac{3}{2}$$

The ratio of $c$ to $b$ is 3 : 2.

19. **C** Since jars $A$, $B$, and $C$ each contain 8 marbles, there are 24 marbles in the three jars. Let $x$, $2x$, and $3x$ represent the new numbers of marbles in jars $A$, $B$, and $C$, respectively. Hence, $x + 2x + 3x = 24$ or $6x = 24$, so

$$x = \frac{24}{6} = 4$$

To achieve a ratio of 1 to 2 to 3, jars $A$, $B$, and $C$ must contain 4, 8, and 12 marbles, respectively. Since jar $B$ already contains 8 marbles, 4 of the 8 marbles originally in jar $A$ must be transferred to jar $C$.

20. **A** The ratio of freshmen to juniors who attended the game is 3 : 4, and the ratio of juniors to seniors is 7 : 6. Since the least common multiple of 4 and 7 is 28, multiply the terms of the first ratio by 7 and multiply the terms of the second ratio by 4. Since 21 : 28 is then the ratio of freshmen to juniors and 28 : 24 is the ratio of juniors to seniors, 21 : 24 is the ratio of freshmen to seniors.
Since
$$21 : 24 = \frac{21}{24} = \frac{7}{8}$$
the correct choice is (A).

*(Quantitative Comparison)*

1. **D** Since $\frac{4}{x} = \frac{x}{9}$, then $x^2 = 36$, so $x = 6$ or $x = -6$. If $x = 6$, Column A = Column B. If, however, $x = -6$, Column B > Column A. Since two answer choices are possible, the correct choice is (D).

2. **A** Since $a : b = 3 : 5$ and $a = 6$, then $\frac{6}{b} = \frac{3}{5}$. Since 3 is one-half of 6, 5 must be one half of $b$, so $b = 10$. Hence, Column A > Column B.

3. **C** In column A, the ratio of $\frac{3}{8}$ to $\frac{1}{4}$ is
$$\frac{3}{8} \div \frac{1}{4} = \frac{3}{8} \times \frac{4}{1} = \frac{3}{2}$$
In Column B, the ratio of 1 to $\frac{2}{3}$ is the reciprocal of $\frac{2}{3}$, which is $\frac{3}{2}$. Hence, Column A = Column B.

4. **B** Since $3h - k = k - 5h$, then $3h + 5h = k + k$ or $8h = 2k$, so
$$\frac{h}{k} = \frac{2}{8} = \frac{1}{4}$$
Since $\frac{1}{2} > \frac{1}{4}$, Column B > Column A.

*(Grid In)*

1. **12** Since the lengths of the two pieces of string are in the ratio of 2 : 9, let $2x$ and $9x$ represent their lengths. Hence:
$$9x - 2x = 42$$
$$7x = 42$$
$$x = \frac{42}{7} = 6$$

*(Lesson 5-4 Grid In continued)*

Since $2x = 2(6) = 12$, the length of the shorter piece of string is 12 inches.

2. **1/6** Since the ratio of $a$ to $b$ is $5 : 9$ and the ratio of $x$ to $y$ is $10 : 3$, $\frac{a}{b} = \frac{5}{9}$ and $\frac{x}{y} = \frac{10}{3}$, so $\frac{y}{x} = \frac{3}{10}$. Hence,

$$\frac{ay}{bx} = \frac{a}{b} \cdot \frac{y}{x} = \frac{5}{9} \cdot \frac{3}{10} = \frac{15}{90} = \frac{1}{6}$$

Grid in as 1/6.

3. **9** Since the ratio of dimes to pennies in the purse is 3 to 4, let $3x$ and $4x$ represent the numbers of dimes and pennies, respectively. If three pennies are taken out of the purse, then $4x - 3$ pennies remain.

Since the ratio of dimes to pennies is now 1 to 1,

$$\frac{3x}{4x-3} = \frac{1}{1} \quad \text{or} \quad 4x - 3 = 3x$$

so $x = 3$. Hence, the number of dimes in the purse is $3x = 3(3) = 9$.

4. **15** Since $2^3 = 2 \times 2 \times 2 = 8$, then $a = 3$ and $b = 2$, so $\frac{a}{b} = \frac{3}{2}$. You are told that $\frac{a}{b} = \frac{c}{d}$. Substituting $\frac{3}{2}$ for $\frac{a}{b}$ and 10 for $d$ gives $\frac{3}{2} = \frac{c}{10}$. Since 10 is 5 times 2, $c$ must be 5 times 3 or 15.

5. **9** To find the value of the ratio of $\frac{a}{c}$, multiply $\frac{a}{b}$ by $\frac{b}{c}$. If $6a - 8b = 0$, then $6a = 8b$, so

$$\frac{a}{b} = \frac{8}{6} = \frac{4}{3}$$

Since $c = 12b$, then $\frac{b}{c} = \frac{1}{12}$, so

$$\frac{a}{c} = \frac{a}{b} \cdot \frac{b}{c} = \frac{\overset{1}{\cancel{4}}}{3} \cdot \frac{1}{\underset{3}{\cancel{12}}} = \frac{1}{9}$$

Hence, the ratio of $a$ to $c$ is equivalent to the ratio of 1 to 9.

## LESSON 5-5 *(Multiple Choice)*

1. **C** If 4 pens cost $1.96, the cost of 1 pen is $\frac{\$1.96}{4} = \$0.49$. The greatest number of pens that can be purchased for $7.68 is the greatest integer that is equal to or less than $\frac{\$7.68}{\$0.49}$. Since $\frac{\$7.68}{\$0.49} = 15.67$, the greatest number of pens that can be purchased is 15.

2. **D** From 10:40 A.M. to 1:00 P.M., 2 hours and 20 minutes elapse. Since the car travels at a constant rate of 45 miles per hour, the car travels $2 \times 45$ or 90 miles in 2

hours. Also, since 20 minutes is $\frac{1}{3}$ of an hour, the car travels $\frac{1}{3} \times 45$ or 15 miles in 20 minutes. The total distance the car travels is $90 + 15$ or 105 miles.

3. **A** If $k$ pencils cost $c$ cents, then 1 pencil costs $\frac{c}{k}$, so $p$ pencils cost $p \times \frac{c}{k}$ or $\frac{pc}{k}$.

4. **B** Since the freight train leaves 1 hour before the passenger train, it travels 1 hour longer. Therefore, when the passenger train has traveled $x$ hours, the freight train has traveled $x + 1$ hours. Make a table.

| Rate | × | Time | = | Distance |
|---|---|---|---|---|
| 50 mph | | $x + 1$ hours | | $50(x + 1)$ |
| 60 mph | | $x$ hours | | $60x$ |

Since the two trains are traveling in opposite directions, the sum of their distances must equal 380 miles:

$$50(x + 1) + 60x = 380$$
$$50x + 50 + 60x = 380$$
$$110x = 330$$
$$x = \frac{330}{110} = 3$$

Since the passenger train leaves at 1:00 P.M., the two trains are 380 miles apart 3 hours later or at 4:00 P.M.

5. **A** Julie can type a manuscript in 4 hours so her rate of work is $\frac{1}{4}$ of the job per hour. Pat needs 6 hours to type the same manuscript, so her rate of work is $\frac{1}{6}$ of the job per hour. If Julie and Pat work $x$ hours to complete the whole job, then

$$x\left(\frac{1}{4}\right) + x\left(\frac{1}{6}\right) = 1 \quad \text{or} \quad \frac{x}{4} + \frac{x}{6} = 1$$

Multiplying each member of the equation by 12, the LCD of 4 and 6, gives

$$3x + 2x = 12$$
$$5x = 12$$
$$x = \frac{12}{5} = 2.4 \text{ hours}$$

Since Julie and Pat start working at 12:00 P.M. and 0.4 hour equals $0.4 \times 60 = 24$ minutes, the job is completed in 2 hours and 24 minutes past noon or at 2:24 P.M.

6. **E** If $x$ men working together at the same rate can complete a job in $h$ hours, then each man working alone can complete the same job in $\frac{h}{x}$ hours, so his rate of

*(Lesson 5-5    Multiple Choice continued)*

work is the reciprocal of $\frac{h}{x}$ or $\frac{x}{h}$. Hence, the part of the job one man working alone can complete in $k$ hours is $k\left(\frac{x}{h}\right)$ or $\frac{kx}{h}$.

7. **C**  Since $\frac{3}{5} \times 3 = \frac{9}{5}$, the original 3-cup mixture contains $\frac{9}{5}$ cups of milk. After 1 cup of milk is added to the 3-cup mixture, the 4-cup mixture contains

$$\frac{9}{5} + 1 = \frac{9}{5} + \frac{5}{5} = \frac{14}{5} \text{ cups of milk}$$

Since

$$\frac{\frac{14}{5}}{4} = \frac{14}{5} \times \frac{1}{4} = \frac{14}{20} = 0.70$$

70% of the 4-cup mixture is milk.

8. **E**  Make a table.

| Rate | × | Time | = | Distance |
|---|---|---|---|---|
| $x$ mph | | 3 hours | | $3x$ |
| $x + 20$ mph | | 3 hours | | $3(x + 20)$ |

The total distance the two trains travel in 3 hours is $500 - 80$ or 420 miles. Hence,

$$3x + 3(x + 20) = 420$$
$$3x + 3x + 60 = 420$$
$$6x = 360$$
$$x = \frac{360}{6} = 60$$

9. **C**  If a machine can seal 360 packages per hour, then it seals $\frac{360}{60}$ or 6 packages per minute. If another machine can seal 140 packages per hour, then it seals $\frac{140}{60}$ or $\frac{7}{3}$ packages per minute. If the two machines working together can seal a total of 700 packages in $x$ minutes, then

$$6x + \frac{7}{3}x = 700$$
$$18x + 7x = 2100$$
$$25x = 2100$$
$$x = \frac{2100}{25} = 84$$

10. **A**  If $x$ represents the number of hours Carrie and James work together, then $x + 1$ is the total number of hours Carrie works to complete the job. Hence,

Part of job Carrie completes + Part of job James completes

$$\left(\frac{1}{5}\right)(x + 1) \quad + \quad \left(\frac{1}{3}\right)x \qquad = 1$$

Multiplying each member of the equation by 15, the LCD of 3 and 5, gives $3(x + 1) + 5x = 15$. Then $8x + 3 = 15$, so

$$x = \frac{12}{8} = \frac{3}{2} = 1\frac{1}{2} \text{ hours}$$

or 1 hour and 30 minutes.

11. **D**  A man's rate of work is the reciprocal of the total number of hours he takes to complete the job working alone, which, in this case, is $\frac{1}{h+p}$. Hence, after working $h$ hours, the part of the job the man completes is

$$h\left(\frac{1}{h + p}\right) \text{ or } \frac{h}{h + p}$$

12. **B**  Make a table.

| Rate | × | Time | = | Distance |
|---|---|---|---|---|
| 18 mph | | $x - \frac{1}{4}$ hours | | $18\left(x - \frac{1}{4}\right)$ |
| 12 mph | | $x$ hours | | $12x$ |

Since the length of the lake remains constant,

$$18\left(x - \frac{1}{4}\right) = 12(x)$$
$$18x - \frac{18}{4} = 12x$$
$$6x = \frac{9}{2}$$
$$x = \frac{9}{6(2)} = \frac{9}{12} = \frac{3}{4}$$

Hence,

$$\text{Rate} \times \text{Time} = \text{Distance}$$
$$12 \quad \times \quad \frac{3}{4} \quad = 9 \text{ miles}$$

13. **D**  Make a table.

| Rate | × | Time | = | Distance |
|---|---|---|---|---|
| $x$ mph | | $\frac{120}{x}$ hours | | 120 |
| $x + 5$ mph | | $\frac{120}{x+5}$ hours | | 120 |

Since the time for the return trip was one-third of an hour less than the time for the outgoing trip,

$$\frac{120}{x} = \frac{120}{x + 5} + \frac{1}{3}$$

*(Quantitative Comparison)*

1. **B**  Bob works twice as fast as Vincent. If Bob can complete a job working alone in $x$ hours, then Vincent can complete the same job working alone in $2x$ hours.

*(Lesson 5-5   Quantitative Comparison continued)*

Working together for 2 hours, Bob and Vincent complete

$$2\left(\frac{1}{x}\right) + 2\left(\frac{1}{2x}\right) \quad \text{or} \quad \frac{2}{x} + \frac{1}{x} = \frac{3}{x}$$

of this job. Since $x$ must be a positive number, $\frac{4}{x} > \frac{3}{x}$, so Column B > Column A.

2. **C**  In Column A, the number of hours a hiker takes to walk 14 kilometers at an average rate of 3.5 kilometers per hour is 4 since

$$\text{Time} = \frac{\text{Distance}}{\text{Rate}} = \frac{14}{3.5} = 4$$

For Column B, let $x$ be the number of hours the two people working together take to complete the same job. Then

$$\frac{x}{8} + \frac{x}{8} = 1$$

so $x = 4$. Since the number of hours in each column is 4, Column A = Column B.

3. **B**  You are told that $x$ pounds of $3.00-per-pound coffee are mixed with $y$ pounds of $7.00-per-pound coffee to make a blend of coffee that sells for $6.00 per pound. Hence,

$$3x + 7y = 6(x + y)$$
$$= 6x + 6y$$
$$7y - y = 6x - 3x$$
$$6y = 3x$$

Dividing both sides of this equation by $3y$ gives

$$\frac{6y}{3y} = \frac{3x}{3y} \quad \text{or} \quad 2 = \frac{x}{y}$$

Hence, Column B > Column A.

*(Grid In)*

1. **8.0 (or 8)**   If 5 pounds of fruit serve 18 people, then $\frac{5}{18}$ pound serves one person, so

$$24 \times \frac{5}{18} = 4 \times \frac{5}{3} = \frac{20}{3} \text{ pounds}$$

serve 24 people. Since the fruit costs $1.20 a pound, the cost of the fruit needed to serve 24 people is

$$\frac{20}{3} \times \$1.20 = 20 \times \$0.40 \text{ or } \$8.00.$$

2. **6**   Let $x$ represent the number of pounds of the more expensive candy. Since the total mixture is 24 pounds, $24 - x$ represents the number of pounds of the less expensive candy. Make a table.

| $\begin{pmatrix}\text{Price per}\\\text{pound}\end{pmatrix}$ | $\times$ | $\begin{pmatrix}\text{Number of}\\\text{pounds}\end{pmatrix}$ | $=$ Total value |
|---|---|---|---|
| $2.00 | | $x$ | $2.00x$ |
| $6.00 | | $24 - x$ | $6.00(24 - x)$ |
| $5.00 | | $24$ | $5.00(24)$ |

$$2.00x + 6.00(24 - x) = 5.00(24)$$
$$2x + 144 - 6x = 120$$
$$-4x = -24$$
$$x = \frac{-24}{-4} = 6$$

Six pounds of the more expensive candy will be needed.

# CHAPTER 6

# Geometric Concepts and Reasoning

This chapter reviews the geometric relationships and properties of figures that are tested on the SAT I. Keep in mind that you will not be required to complete formal geometric proofs or to memorize key formulas.

Unless otherwise stated, figures on the SAT I are drawn to scale. If you find a note that tells you the accompanying figure is *not* drawn to scale, try redrawing the figure so that it looks in scale. You may then be able to arrive at the answer by estimating or by eliminating answer choices that clearly do not fit the diagram.

---

## LESSONS IN THIS CHAPTER

---

# LESSON 6-1

# Angle Relationships

## OVERVIEW

*This lesson reviews the relationships between*

- *angles formed when two lines intersect;*
- *angles formed when two parallel lines are cut by a third line;*
- *the angles of a triangle.*

## NAMING ANGLES

The **vertex** of an angle is the point at which the two sides of an angle intersect. The symbol $\angle$ is used as an abbreviation for the word *angle*. An angle may be named by using a number or a letter, as shown below.

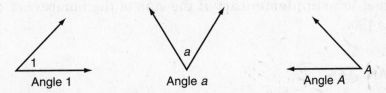

When more than one angle has the same vertex, three letters can be used to name a particular angle, provided that the middle letter is the vertex of the angle. In the accompanying figure, $\angle a$, $\angle BAC$, and $\angle CAB$ all name the same angle.

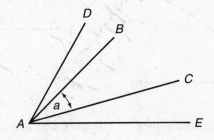

**225**

## CLASSIFYING ANGLES

Angles can be classified according to their degree measures (see Figure 6.1).

- An **acute angle** is an angle that measures less than 90°.
- A **right angle** is an angle that measures 90°.
- An **obtuse angle** is an angle that measures more than 90° and less than 180°.
- A **straight angle** is an angle that measures 180°, so its sides lie on the same straight line.

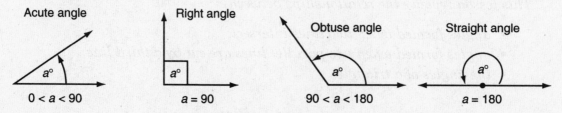

**Figure 6.1** *Classifying Angles*

Pairs of angles whose measures add up to 90° or 180° are given special names.

- Two angles are **complementary** if the sum of the numbers of degrees in the angles is 90°.
- Two angles are **supplementary** if the sum of the numbers of degrees in the angles is 180°.

## ADJACENT ANGLES

In Figure 6.2, angles 1 and 2 are **adjacent angles** because they have the same vertex, share a common side, and do not overlap.

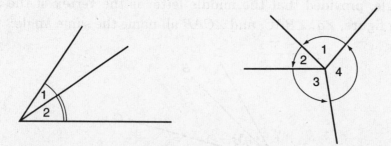

**Figure 6.2** *Adjacent Angles*      **Figure 6.3** *Adding Angles About a Point*

The sum of the measures of the adjacent angles about a point is 360°. In Figure 6.3,

$$\angle 1 + \angle 2 + \angle 3 + \angle 4 = 360°$$

In the figure at the right, the measure
of angle $a$ is $360° - 70°$ or $290°$.

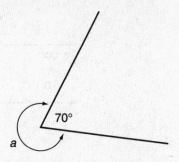

## ANGLES FORMED BY INTERSECTING LINES

When two lines intersect, **vertical** or opposite angles have equal measures and non-adjacent angles are supplementary. In the figure below:

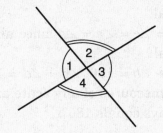

- $\angle 1 = \angle 3$ and $\angle 2 = \angle 4$
- $\angle 1 + \angle 2 = 180°$ and $\angle 1 + \angle 4 = 180°$
  $\angle 3 + \angle 4 = 180°$ and $\angle 2 + \angle 3 = 180°$

## BISECTING ANGLES AND LINE SEGMENTS

A line segment or an angle is **bisected** when it is divided into two parts that have equal measures.

- In the figure below left, segment $AB$ is bisected at point $M$ since line segments $AM$ and $MB$ have the same length.

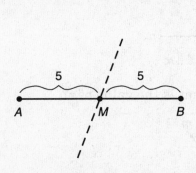

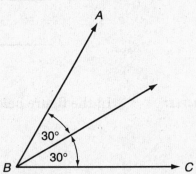

- In the figure above right, $\angle ABC$ is bisected since it is divided into two angles that have the same number of degrees.

## ANGLES FORMED BY PARALLEL LINES

**Parallel lines** are always the same distance apart and, as a result, do not intersect. In Figure 6.4, lines $\ell$ and $m$ are parallel. The notation $\ell \parallel m$ is read as "$\ell$ is parallel to $m$." When two parallel lines are cut by a third line:

- four equal acute angles and four equal obtuse angles are formed.
- any acute angles and any obtuse angles are supplementary.

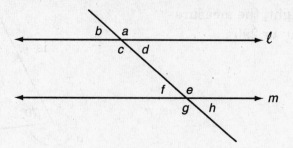

**Figure 6.4** *Parallel-Line Angle Relationships*

In Figure 6.4:

- $\angle b = \angle d = \angle f = \angle h$ since all of the acute angles formed by parallel lines are equal.
- $\angle a = \angle c = \angle e = \angle g$ since all of the obtuse angles formed by parallel lines are equal.
- $\angle a + \angle b = 180$, $\angle a + \angle d = 180$, $\angle a + \angle h = 180$, and so forth since the sum of the measures of any acute angle and any obtuse angle formed by parallel lines always equals 180°.

If a line is perpendicular to one of two parallel lines, it must also be perpendicular to the other parallel line. In the accompanying figure, line $l$ is parallel to line $m$. If line $p$ is perpendicular to line $l$, then line $p$ must also be perpendicular to line $m$.

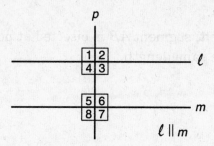

**EXAMPLE:**   In the figure below, if $l \parallel m$, what is the value of $x$?

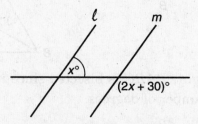

**SOLUTION:**   The angle marked $x$ is acute, and the angle marked $2x + 30$ is obtuse. Since these angles are formed by parallel lines, their degree measures add up to 180. Hence:

$$x + (2x + 30) = 180$$

$$3x = 150$$

$$x = \frac{150}{3} = 50$$

## ANGLES OF A TRIANGLE

The sum of the measure of the three angles of a triangle is 180° (see Figure 6.5).

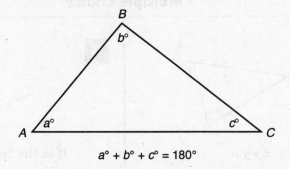

$$a° + b° + c° = 180°$$

**Figure 6.5**   *Sum of the Angles of a Triangle*

An **exterior angle** of a triangle, as shown in Figure 6.6, can be formed at any vertex of the triangle by extending one of its sides.

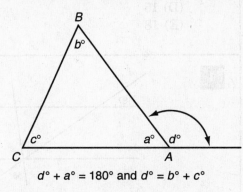

$$d° + a° = 180° \text{ and } d° = b° + c°$$

**Figure 6.6**   *Exterior Angle of a Triangle*

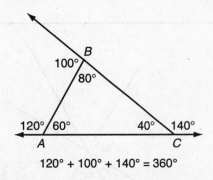

$$120° + 100° + 140° = 360°$$

**Figure 6.7**   *Adding the Exterior Angles of a Triangle*

In any triangle,

- An exterior angle and interior angle that have the same vertex are supplementary.
- An exterior angle is equal to the sum of the measures of the two remote (nonadjacent) interior angles.
- The sum of the measures of the exterior angles, one drawn at each vertex, is 360° (see Figure 6.7).

If a triangle contains a right angle, then the measures of the two remaining angles of the triangle must add up to 90°. In the figure at the right, since $\angle C$ is a right angle it measures 90, so $a + b = 90°$.

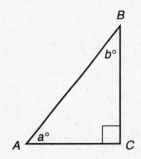

# LESSON 6-1 TUNE-UP EXERCISES

| **Multiple Choice** |
|---|

**1**

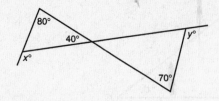

In the figure above, $x + y =$

(A) 270
(B) 230
(C) 210
(D) 190
(E) 180

**2**

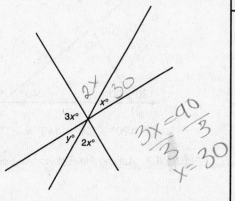

In the figure above, what is the value of $y$?

(A) 20
(B) 30
(C) 45
(D) 60
(E) It cannot be determined from the information given.

**3**

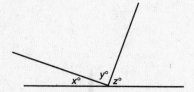

If in the figure above $\frac{y}{x} = 5$ and $\frac{z}{x} = 4$, what is the value of $x$?

(A) 8
(B) 10
(C) 12
(D) 15
(E) 18

**4**

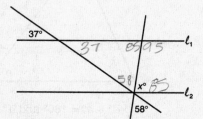

In the figure above, if $\ell_1 \parallel \ell_2$, what is the value of $x$?

(A) 90
(B) 85
(C) 75
(D) 70
(E) 63

**5** If the measures of the angles of a triangle are in the ratio of 3 : 4 : 5, what is the measure of the smallest angle of the triangle?

(A) 25
(B) 30
(C) 45
(D) 60
(E) 75

**6**

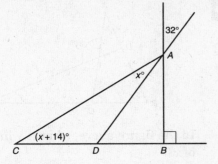

In the figure above, if $AB \perp BC$, what is the value of $x$?

(A) 18
(B) 22
(C) 25
(D) 27
(E) 32

**7**

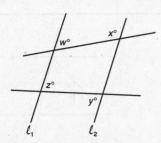

In the figure above, if $\ell_1 \parallel \ell_2$, what is $x + y$ in terms of $w$ and $z$?

(A) $180 - w + z$
(B) $180 + w - z$
(C) $180 - w - z$
(D) $180 + w + z$
(E) $w + z$

**8**

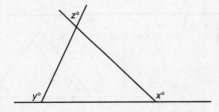

In the figure above, what is $z$ in terms of $x$ and $y$?

(A) $x + y + 180$
(B) $x + y - 180$
(C) $180 - (x + y)$
(D) $x + y + 360$
(E) $360 - (x + y)$

**9**

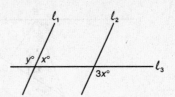

In the figure above, if $\ell_1 \parallel \ell_2$, what is the value of $y$?

(A) 90
(B) 100
(C) 120
(D) 135
(E) 150

**10**

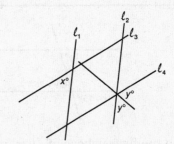

In the figure above, if $\ell_1 \parallel \ell_2$ and $\ell_3 \parallel \ell_4$, what is $y$ in terms of $x$?

(A) $90 + x$
(B) $90 + 2x$
(C) $90 - \frac{x}{2}$
(D) $90 - 2x$
(E) $180 - 2x$

**11**

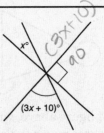

In the figure above, what is the value of $x$?

(A) 12
(B) 15
(C) 20
(D) 24
(E) 30

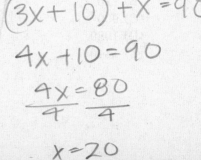

$(3x+10)+x=90$

$4x+10=90$

$4x=80$

$x=20$

**12**

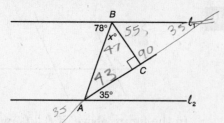

In the figure above, for which value of $x$ is $\ell_1 \parallel \ell_2$?

(A) 37
(B) 43
(C) 45
(D) 47
(E) 55

**13**

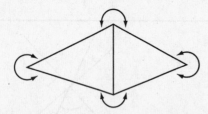

In the figure above, if $\ell \parallel m$, what is the value of $x$?

(A) 60
(B) 50
(C) 45
(D) 30
(E) 25

**14**

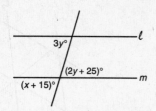

In the figure above, what is the sum of the degree measures of all of the angles marked?

(A) 540
(B) 720
(C) 900
(D) 1080
(E) 1440

**15**

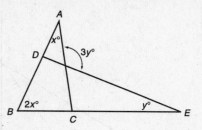

In the figure above, what is $y$ in terms of $x$?

(A) $\frac{4}{3}x$

(B) $\frac{3}{2}x$

(C) $\frac{3}{4}x$

(D) $\frac{2}{3}x$

(E) $x$

**16**

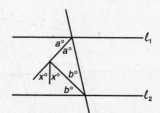

In the figure above, if $\ell_1 \parallel \ell_2$, the value of $x$ is

(A) $22\frac{1}{2}$
(B) 30
(C) 45
(D) 60
(E) It cannot be determined from the information given.

**17**

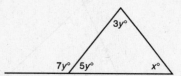

In the figure above, what is the value of $x$?

(A) 35
(B) 45
(C) 50
(D) 60
(E) 75

**18**

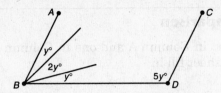

In the figure above, if line segment $AB$ is parallel to line segment $CD$, what is the value of $y$?

(A) 12
(B) 15
(C) 18
(D) 20
(E) 24

**19**

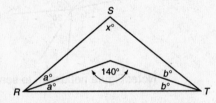

Note: Figure is not drawn to scale.

In $\triangle RST$ above, what is the value of $x$?

(A) 40
(B) 60
(C) 80
(D) 90
(E) 100

**20**

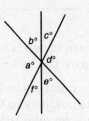

In the figure above, which of the following statements must be true?

  I. $a + b = d + c$
 II. $a + c + e = 180$
III. $b + f = c + e$

(A) I only
(B) II only
(C) III only
(D) I and II only
(E) II and III only

## Quantitative Comparison

Each question consists of two quantities in boxes, one in Column A and one in Column B. You are to compare the two quantities and on the answer sheet fill in

A if the quantity in Column A is greater;
B if the quantity in Column B is greater;
C if the two quantities are equal;
D if the relationship cannot be determined from the information given

AN E RESPONSE WILL NOT BE SCORED

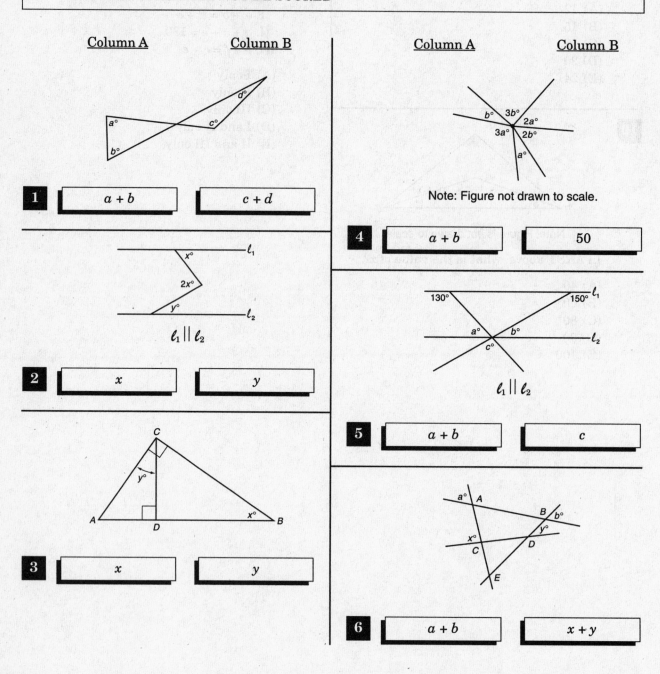

Column A          Column B

**1** $\boxed{a + b}$        $\boxed{c + d}$

$\ell_1 \| \ell_2$

**2** $\boxed{x}$        $\boxed{y}$

**3** $\boxed{x}$        $\boxed{y}$

Column A          Column B

Note: Figure not drawn to scale.

**4** $\boxed{a + b}$        $\boxed{50}$

$\ell_1 \| \ell_2$

**5** $\boxed{a + b}$        $\boxed{c}$

**6** $\boxed{a + b}$        $\boxed{x + y}$

| Grid In |
|---|

**1**

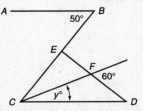

In the figure above, if line segment $AB$ is parallel to line segment $CD$ and $BE \perp ED$, what is the value of $y$?

**2**

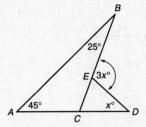

In the figure above, what is the value of $x$?

**Questions 3 and 4.**

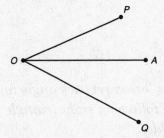

Line segments $OP$, $OA$, and $OQ$, in the figure above, coincide at the time $t = 0$ second. At the same instant of time, $OP$ and $OQ$ rotate in the plane about point $O$ in opposite directions while $OA$ remains fixed. With respect to $OA$, segment $OP$ rotates at a constant rate of $4°$ per second and segment $OQ$ rotates at a constant rate of $5°$ per second.

**3**  What is the smallest positive value of $t$, in seconds, for which segments $OA$ and $OQ$ will coincide?

**4**  When $t = 1$ hour, how many more revolutions does segment $OP$ complete than segment $OQ$?

# Special Triangles

## OVERVIEW

*This lesson reviews angle and side relationships in special types of triangles. The following right triangle reference formulas are supplied in the SAT I test booklet:*

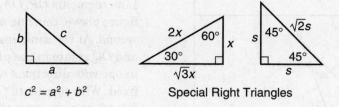

$$c^2 = a^2 + b^2$$

Special Right Triangles

## CLASSIFYING TRIANGLES

A triangle can be classified according to the number of equal sides that it contains, as shown in Figure 6.8.

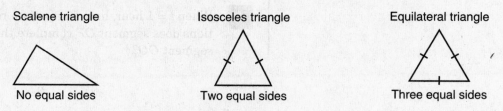

Scalene triangle
No equal sides

Isosceles triangle
Two equal sides

Equilateral triangle
Three equal sides

**Figure 6.8** *Classifying Triangles by Side Lengths*

A triangle can also be classified according to the measure of its greatest angle, as shown in Figure 6.9.

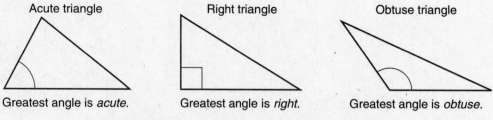

Acute triangle
Greatest angle is *acute.*

Right triangle
Greatest angle is *right.*

Obtuse triangle
Greatest angle is *obtuse.*

**Figure 6.9** *Classifying Triangles by Angle Measures*

## ANGLE AND SIDE RELATIONSHIPS IN AN ISOSCELES TRIANGLE

In an **isosceles triangle,** equal angles are opposite equal sides, as shown in Figure 6.10.

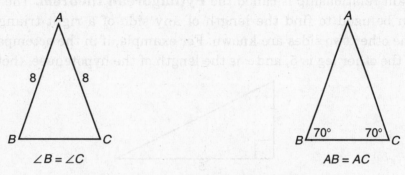

**Figure 6.10**  *Isosceles Triangle Relationships*

## ANGLE AND SIDE RELATIONSHIPS IN AN EQUILATERAL TRIANGLE

In an **equilateral triangle,** all three sides are equal *and* all three angles each measure 60°.

If the three sides of a triangle have the same length, you may conclude that the triangle contains three 60° angles (see Figure 6.11).

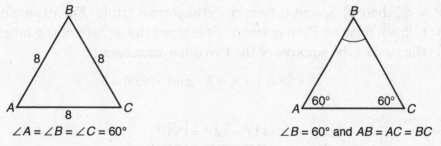

**Figure 6.11**  *Equilateral Triangles Are Equiangular*    **Figure 6.12**  *Equiangular Triangles are Equilateral*

If two angles of a triangle measure 60°, the third angle also measures 60° and, as a result, the three sides have the same length (see Figure 6.12).

## SIDE RELATIONSHIPS IN A RIGHT TRIANGLE

In a **right triangle,** the side opposite the right angle is called the **hypotenuse** and the other two sides are called **legs** (see Figure 6.13).

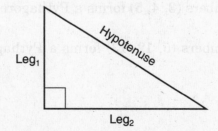

**Figure 6.13**  *Right Triangle*

In any right triangle:

$$(\text{leg}_1)^2 + (\text{leg}_2)^2 = (\text{hypotenuse})^2$$

This important relationship is called the **Pythagorean theorem.** The Pythagorean theorem can be used to find the length of any side of a right triangle when the lengths of the other two sides are known. For example, if in the accompanying figure one leg is 3, the other leg is 5, and $x$ is the length of the hypotenuse, then

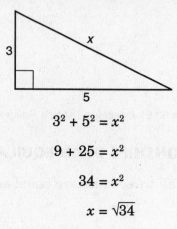

$$3^2 + 5^2 = x^2$$

$$9 + 25 = x^2$$

$$34 = x^2$$

$$x = \sqrt{34}$$

## PYTHAGOREAN TRIPLES

If $a^2 + b^2 = c^2$, then $a$, $b$, and $c$ form a Pythagorean triple. For example, the set of numbers $(1, 2, \sqrt{5})$ forms a Pythagorean triple since the square of the largest number, $\sqrt{5}$, equals the sum of the squares of the two other numbers:

$$1^2 + 2^2 = 1 + 4 = 5 \quad \text{and} \quad (\sqrt{5})^2 = 5$$

so

$$\underbrace{(1)^2}_{a^2} + \underbrace{(2)^2}_{b^2} = \underbrace{(\sqrt{5})^2}_{c^2}$$

If the lengths of the sides of a triangle form a Pythagorean triple, then the triangle is a right triangle in which the longest side is the hypotenuse of the triangle. For example, the triangle whose sides measure 1, 2, and $\sqrt{5}$ is a right triangle in which the length of the hypotenuse is $\sqrt{5}$.

Some Pythagorean triples occur so frequently that you should memorize them. For example:

- The ordered set of numbers $(3, 4, 5)$ forms a Pythagorean triple since $3^2 + 4^2 = 5^2$.
- The ordered set of numbers $(5, 12, 13)$ forms a Pythagorean triple since $5^2 + 12^2 = 13^2$.

Any multiple of a Pythagorean triple is also a Pythagorean triple. Thus, ($3k$, $4k$, $5k$) and ($5k$, $12k$, $13k$) are Pythagorean triples for all positive integer values of $k$. For example, if the lengths of the two legs of a right triangle are 9 and 12, then the length of the hypotenuse is 15 since (9, 12, 15) is a multiple of a (3, 4, 5) Pythagorean triple.

**Example:**    In $\triangle ABC$ below, what is the length of line segment $BD$?

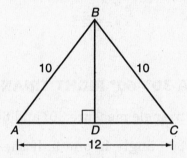

**Solution:**    If two sides of a triangle are equal, a perpendicular drawn to the third side bisects that side. Hence, $AD = DC = 6$. In right triangle $ADB$, the lengths of the sides are members of a (6, $x$, 10) Pythagorean triple where $AD = 6$, hypotenuse $AB = 10$, and $x = BD = 8$.

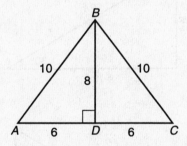

## SIDE RELATIONSHIPS IN A 45°–45° RIGHT TRIANGLE

In an isosceles right triangle:

- Each acute angle measures 45°.
- The length of the hypotenuse is √2 times the length of a leg.
- The length of each leg is $\frac{1}{\sqrt{2}}(= \frac{\sqrt{2}}{2})$ times the length of the hypotenuse.

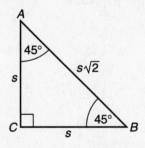

For example,

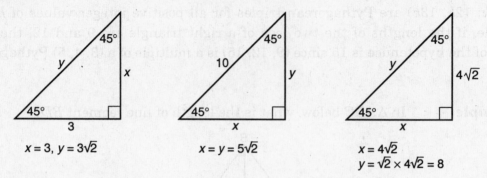

$$x = 3, y = 3\sqrt{2} \qquad x = y = 5\sqrt{2} \qquad \begin{aligned} x &= 4\sqrt{2} \\ y &= \sqrt{2} \times 4\sqrt{2} = 8 \end{aligned}$$

## SIDE RELATIONSHIPS IN A 30°–60° RIGHT TRIANGLE

If the acute angles of a right triangle measure 30° and 60°, then

- The leg opposite the 30° angle is one-half the length of the hypotenuse. Equivalently, the length of the hypotenuse is 2 times the length of the leg opposite the 30° angle.
- The leg opposite the 60° angle is √3 times the length of the other leg.

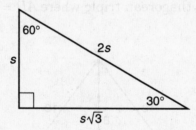

For example:

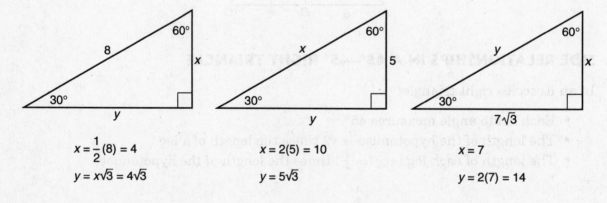

$$\begin{aligned} x &= \frac{1}{2}(8) = 4 \\ y &= x\sqrt{3} = 4\sqrt{3} \end{aligned} \qquad \begin{aligned} x &= 2(5) = 10 \\ y &= 5\sqrt{3} \end{aligned} \qquad \begin{aligned} x &= 7 \\ y &= 2(7) = 14 \end{aligned}$$

# LESSON 6-2   TUNE-UP EXERCISES

| **Multiple Choice** |
|---|

**1**

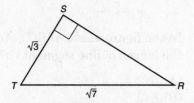

In the figure above, what is the length of RS?

(A) 2
(B) $\sqrt{7} - \sqrt{3}$
(C) 4
(D) $7 + \sqrt{3}$
(E) 10

**2**

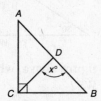

In $\triangle ABC$, $\angle C = 90°$, $CD = BD$, and the ratio of the measure of $\angle A$ to the measure of $\angle B$ is 2 to 3. What is the value of $x$?

(A) 18
(B) 36
(C) 40
(D) 54
(E) 72

**3**

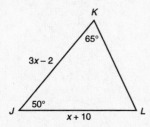

In $\triangle JKL$ above, what is the value of $x$?

(A) 2
(B) 3
(C) 4
(D) 6
(E) It cannot be determined from the information given.

**4**

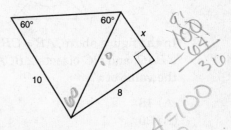

In the figure above, $x =$

(A) 4
(B) 6
(C) $4\sqrt{2}$
(D) $4\sqrt{3}$
(E) $8\sqrt{3}$

**5**

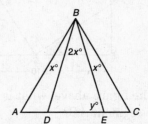

In equilateral triangle $ABC$ above, $y =$

(A) 40
(B) 60
(C) 70
(D) 75
(E) 80

**6**

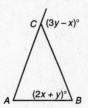

In the figure above, if $AC = BC$ what is $y$ in terms of $x$?

(A) $x$
(B) $\frac{3}{2}x$
(C) $2x$
(D) $3x$
(E) $5x$

**7**

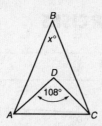

In the figure above, $AB = CB$, $DA$ bisects $\angle BAC$, and $DC$ bisects $\angle BCA$. What is the value of $x$?

(A) 18
(B) 30
(C) 36
(D) 72
(E) 108

**8**

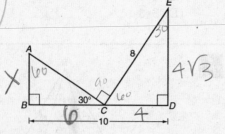

In the figure above, what is the length of line segment $AB$?

(A) $2\sqrt{3}$
(B) 3
(C) 4
(D) $3\sqrt{3}$
(E) 6

$$\frac{a\sqrt{3}=6}{\sqrt{3}} \quad \frac{6}{\sqrt{3}}$$

$$\frac{6}{\sqrt{3}} \cdot \frac{\sqrt{3}}{\sqrt{3}}$$

$$\frac{6\sqrt{3}}{3}$$

$$2\sqrt{3}$$

**9**

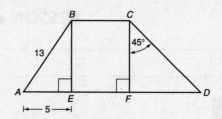

In the figure above, if $BC \parallel AD$, what is the length of line segment $CD$?

(A) $6\sqrt{2}$
(B) $8\sqrt{2}$
(C) $12\sqrt{2}$
(D) 15
(E) 17

**10**   Note: Figure is not drawn to scale.

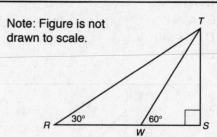

In the figure above, what is the ratio of $RW$ to $WS$?

(A) $\sqrt{2}$ to 1
(B) $\sqrt{3}$ to 1
(C) 2 to 1
(D) 3 to 1
(E) 3 to 2

## Quantitative Comparison

Each question consists of two quantities in boxes, one in Column A and one in Column B. You are to compare the two quantities and on the answer sheet fill in

A  if the quantity in Column A is greater;
B  if the quantity in Column B is greater;
C  if the two quantities are equal;
D  if the relationship cannot be determined from the information given

AN E RESPONSE WILL NOT BE SCORED

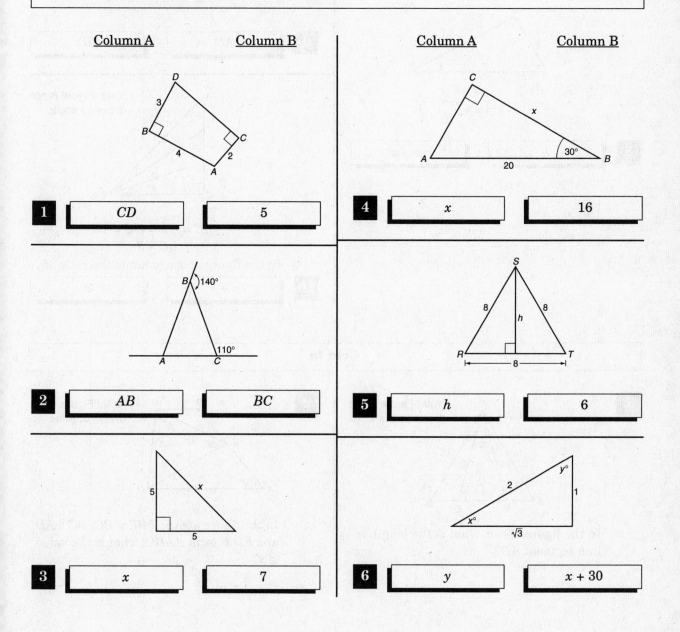

<table>
<tr><td colspan="2">Column A</td><td colspan="2">Column B</td></tr>
</table>

**1**  | $CD$ | 5

**2**  | $AB$ | $BC$

**3**  | $x$ | 7

**4**  | $x$ | 16

**5**  | $h$ | 6

**6**  | $y$ | $x + 30$

| Column A | Column B |
|---|---|

In $\triangle PQR$, $\dfrac{PR}{PQ} = 1$ and $\dfrac{RQ}{PR} = 1$

**7**  | 45 | The degree measure of $\angle P$ |

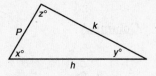

$z = x + y$

**8**  | $h$ | $\sqrt{k^2 + p^2}$ |

| Column A | Column B |
|---|---|

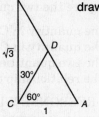

Note: Figure is not drawn to scale.

**9**  | $AD$ | $BD$ |

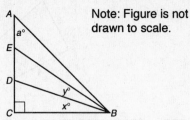

Note: Figure is not drawn to scale.

$AE = ED = DC = 1$
$BC = 3$

**10**  | $x + y$ | $a$ |

---

## Grid In

**1**

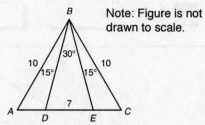

Note: Figure is not drawn to scale.

In the figure above, what is the length of line segment $AD$?

**2**

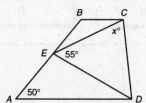

Note: Figure is not drawn to scale.

In the figure above, if $BE = BC$, $BC \parallel AD$, and $ED$ bisects $\angle ADC$, what is the value of $x$?

# LESSON 6-3

# Triangle Inequality Relationships

## OVERVIEW

*In any triangle:*

- *The length of each side is less than the sum of the lengths of the other two sides.*
- *If side lengths or angle measures are unequal, then the longest side is opposite the greatest angle and the shortest side is opposite the smallest angle.*

## TESTING THREE NUMBERS AS POSSIBLE TRIANGLE SIDE LENGTHS

To determine whether a set of three positive numbers can represent the lengths of the sides of a triangle, check that *each* of the three numbers is less than the sum of the other two numbers.

**EXAMPLE:** Which of the following sets of numbers CANNOT represent the lengths of the sides of a triangle?
(A) 9, 40, 41
(B) 7, 7, 3
(C) 4, 5, 1
(D) 1.6, 1.4, 2.9
(E) 6, 6, 6

**SOLUTION:** For each choice, check that each of the three given numbers is less than the sum of the other two numbers. In choice (C), $4 < 5 + 1$ and $1 < 4 + 5$, but 5 is *not* less than $1 + 4$.

The correct choice is (C).

**EXAMPLE:** If the lengths of two sides of an isosceles triangle are 3 and 7, what is the length of the third side?

**SOLUTION:** Since two sides of an isosceles triangle have the same length, the length of the third side is either 3 or 7. The length of the third side cannot be 3 since 7 is *not* less than $3 + 3$. If the third side is 7, then $3 < 7 + 7$ and $7 < 3 + 7$. Hence, the length of the third side must be 7.

## UNEQUAL SIDES ARE OPPOSITE UNEQUAL ANGLES

If two sides (or angles) of a triangle are unequal, then the measures of the angles (or sides) opposite them are also unequal, with the larger angle lying opposite the longer side. These relationships are illustrated in the figures below.

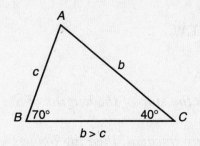

$b > c$

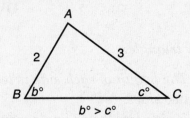

$b° > c°$

# LESSON 6-3 TUNE-UP EXERCISES

**Multiple Choice**

**1** In obtuse triangle $RST$, $RT = TS$. Which of the following statements must be true?

  I. $\angle R = \angle S$
  II. $\angle T$ is obtuse.
  III. $RS > TS$

(A) I only
(B) I and II only
(C) I and III only
(D) II and III only
(E) I, II, and III

**2** In a triangle in which the lengths of two sides are 5 and 9, the length of the third side is represented by $x$. Which statement is always true?

(A) $x > 5$
(B) $x < 9$
(C) $5 \le x \le 9$
(D) $4 < x < 14$
(E) $5 \le x < 14$

**3** In $\triangle ABC$, $BC > AB$ and $AC < AB$. Which statement is always true?

(A) $\angle A > \angle B > \angle C$
(B) $\angle A > \angle C > \angle B$
(C) $\angle B > \angle A > \angle C$
(D) $\angle C > \angle B > \angle A$
(E) $\angle C > \angle A > \angle B$

**4** 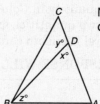 Note: Figure is not drawn to scale.

In $\triangle ABC$, if $AB = BD$ which of the following statements must be true?

  I. $x > z$
  II. $y > x$
  III. $AB > BC$

(A) II only
(B) I and II only
(C) I and III only
(D) II and III only
(E) I, II, and III

**5** How many different triangles are there for which the lengths of the sides are 3, 8, and $n$, where $n$ is an integer and $3 < n < 8$?

(A) Two
(B) Three
(C) Four
(D) Five
(E) Six

**6** In $\triangle RST$, $RS \perp TS$, $\angle T$ measures 40°, and $W$ is a point on side $RT$ such that $\angle RWS$ measures 100°. Which of the following segments has the shortest length?

(A) $RS$
(B) $ST$
(C) $RW$
(D) $TW$
(E) $SW$

## Quantitative Comparison

Each question consists of two quantities in boxes, one in Column A and one in Column B. You are to compare the two quantities and on the answer sheet fill in

A if the quantity in Column A is greater;
B if the quantity in Column B is greater;
C if the two quantities are equal;
D if the relationship cannot be determined from the information given

AN E RESPONSE WILL NOT BE SCORED

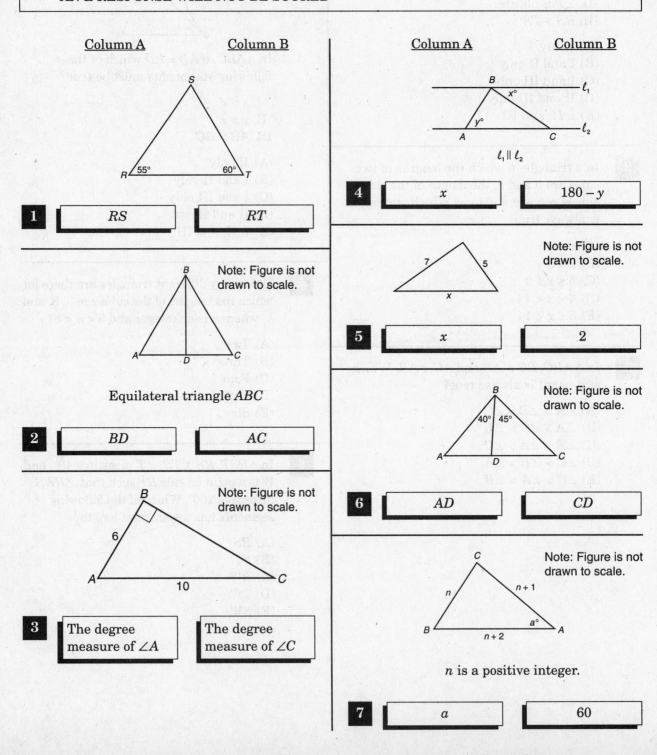

| Column A | Column B |
|---|---|

**1** RS | RT

Equilateral triangle *ABC*

**2** BD | AC

Note: Figure is not drawn to scale.

**3** The degree measure of ∠A | The degree measure of ∠C

| Column A | Column B |
|---|---|

$\ell_1 \parallel \ell_2$

**4** $x$ | $180 - y$

Note: Figure is not drawn to scale.

**5** $x$ | 2

Note: Figure is not drawn to scale.

**6** AD | CD

Note: Figure is not drawn to scale.

*n* is a positive integer.

**7** $a$ | 60

## Grid In

**1** If the lengths of two sides of an isosceles triangle are 7 and 15, what is the perimeter of the triangle?

**2** The perimeter of a triangle in which the lengths of all of the sides are integers is 21. If the length of one side of the triangle is 8, what is the shortest possible length of another side of the triangle?

# Polygons and Parallelograms

## OVERVIEW

*A triangle is a three-sided polygon. This lesson considers some angle and side relationships in polygons with three or more sides.*

## TERMS RELATED TO POLYGONS

The accompanying figure shows a polygon with five sides. The corner points $A$, $B$, $C$, $D$, and $E$ are called **vertices.** Line segments $AC$ and $AD$ are diagonals. A **diagonal** of a polygon is a line segment that connects any two nonconsecutive vertices of the polygon.

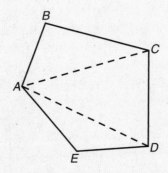

An **equilateral polygon** is a polygon in which all the sides have the same length. An **equiangular polygon** is a polygon in which all the angles are equal.

## ANGLES OF A QUADRILATERAL

A quadrilateral is a polygon with four sides. Since a quadrilateral can be divided into two triangles, as illustrated in Figure 6.14, the sum of the four interior angles of a quadrilateral is $2 \times 180°$ or $360°$.

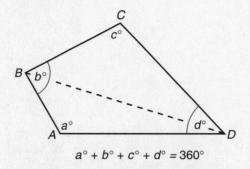

$$a° + b° + c° + d° = 360°$$

**Figure 6.14** *Angles of a Quadrilateral*

## SUM OF THE INTERIOR ANGLES OF A POLYGON

The sum $S$ of the interior angles of a polygon with $n$ sides is given by this formula:

$$S = (n - 2) \times 180°$$

For example, to find the sum of the five interior angles of a five-sided polygon, substitute 5 for $n$ in the above formula:

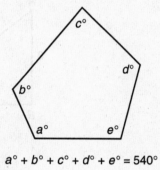

$$S = (n - 2) \times 180°$$
$$= (5 - 2) \times 180°$$
$$= \quad 3 \quad \times 180°$$
$$= 540°$$

$$a° + b° + c° + d° + e° = 540°$$

## SUM OF THE EXTERIOR ANGLES OF A POLYGON

The sum of the exterior angles of any polygon, with one angle drawn at each vertex of the polygon, is 360° (see Figure 6.15).

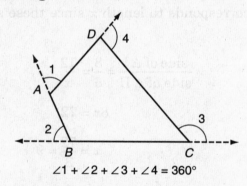

$$\angle 1 + \angle 2 + \angle 3 + \angle 4 = 360°$$

**Figure 6.15**   *Exterior Angles of a Polygon*

## SIMILAR POLYGONS

When a photograph is enlarged, the original and enlarged figures are *similar* since they have the same shape. If two polygons are **similar,** corresponding angles are equal and the ratios of the lengths of corresponding sides are equal. For example, the quadrilaterals below are similar.

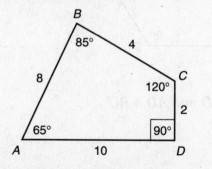
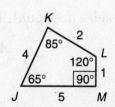

## SIMILAR TRIANGLES

To find out whether two triangles are similar, compare the measures of their angles. If two angles of one triangle are equal to two angles of the second triangle, the two triangles are similar. For example, the triangles in Figure 6.16 are similar since $\angle B = \angle D$ and the vertical angles at $C$ are equal.

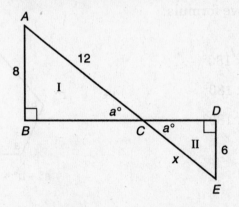

**Figure 6.16** *Similar Triangles*

Since the triangles in Figure 6.16 are similar, the lengths of corresponding sides are in proportion. Length 8 corresponds to length 6 since these sides are opposite equal angles. Also, length 12 corresponds to length $x$ since these sides are opposite equal angles. Hence:

$$\frac{\text{side of } \Delta \text{ I}}{\text{side of } \Delta \text{ II}} = \frac{8}{6} = \frac{12}{x}$$

$$8x = 72$$

$$x = \frac{72}{8} = 9$$

## PARALLELOGRAMS

A **parallelogram** is a special type of quadrilateral in which both pairs of opposite sides are parallel. A parallelogram has these three additional properties:

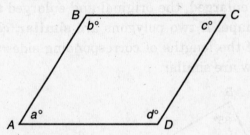

- Opposite sides are equal. Thus:

$$AB = CD \text{ and } AD = BC$$

- Opposite angles are equal. Thus:

$$a = c \text{ and } b = d$$

- Consecutive angles are supplementary. Thus:

$$a + b = b + c = c + d = d + a = 180$$

## SPECIAL TYPES OF PARALLELOGRAMS

A **rectangle** is a parallelogram with four right angles. A **rhombus** is a parallelogram with four equal sides. A **square** is a parallelogram with four right angles and four equal sides.

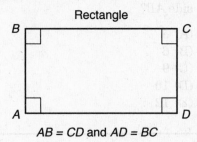

Rectangle

$AB = CD$ and $AD = BC$

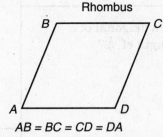

Rhombus

$AB = BC = CD = DA$

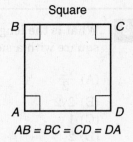

Square

$AB = BC = CD = DA$

## DIAGONAL PROPERTIES OF SPECIAL TYPES OF PARALLELOGRAMS

In any parallelogram, each diagonal cuts the other diagonal in half.

- In a rectangle, the diagonals have the same length.
- In a rhombus, the diagonals intersect at right angles.
- In a square, the diagonals have the same length and intersect at right angles.

## LESSON 6-4  TUNE-UP EXERCISES

| **Multiple Choice** |
|---|

**1** What is the sum of the lengths of the two diagonals in a 9 by 12 rectangle?

(A) 20
(B) 25
(C) 30
(D) 35
(E) 40

**2** What is the length of a diagonal of a square with a side length of √2?

(A) $\frac{\sqrt{2}}{2}$
(B) $2\sqrt{2}$
(C) 1
(D) 2
(E) 4

**3** A diagonal of a rectangle forms a 30° angle with each of the longer sides of the rectangle. If the length of the shorter side is 3, what is the length of the diagonal?

(A) $3\sqrt{2}$
(B) $3\sqrt{3}$
(C) 4
(D) 5
(E) 6

**4** If the degree measures of the angles of a quadrilateral are $4x$, $7x$, $9x$, and $10x$, what is the sum of the measures of the smallest angle and the largest angle?

(A) 140
(B) 150
(C) 168
(D) 180
(E) 192

**5** 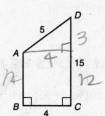  Note: Figure is not drawn to scale.

In the figure above, what is the length of side $AB$?

(A) 5
(B) 6
(C) 9
(D) 10
(E) 12

**6**

If, in the figure above, $AC = 3$, $DB = 4$, and $AB = 14$, then $AE =$

(A) 19
(B) 12
(C) 10.5
(D) 8
(E) 6

**7** Which of the following CANNOT represent the degree measure of an equiangular polygon?

(A) 165
(B) 162
(C) 140
(D) 125
(E) 90

**8**

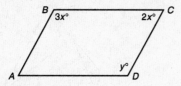

If figure *ABCD* above is a parallelogram, what is the value of *y*?

(A) 108
(B) 72
(C) 54
(D) 45
(E) 36

**9**

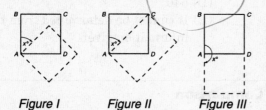

*Figure I*      *Figure II*      *Figure III*

Which of the figures above could represent the rotation of square *ABCD* $x°$ about point *A*?

(A) I only
(B) II only
(C) III only
(D) I and III only
(E) I, II, and III

**Questions 10 and 11.**

If a polygon has *n* sides, then the sum, *S*, of the degree measures of the interior angles is given by the formula
$S = (n - 2)180$.

**10**   If $S = 540t$, what is *n* in terms of *t*?

(A) $3t + 2$
(B) $2t - 3$
(C) $\dfrac{t+2}{3}$
(D) $\dfrac{t-3}{2}$
(E) $\dfrac{t-2}{180}$

**11**   What is the number of sides of a polygon in which the sum of the degree measures of the interior angles is 4 times the sum of the degree measures of the exterior angles?

(A) 4
(B) 6
(C) 8
(D) 10
(E) 12

**12**   In quadrilateral *ABCD*, $\angle A + \angle C$ is 2 times $\angle B + \angle D$. If $\angle A = 40$, then $\angle B =$

(A) 60
(B) 80
(C) 120
(D) 240
(E) It cannot be determined from the information given.

**13**

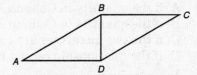

For parallelogram *ABCD* above, if $AB > BD$, which of the following statements must be true?

  I. $CD < BD$
 II. $\angle ADB > \angle C$
III. $\angle CBD > \angle A$

(A) None
(B) I only
(C) II and III only
(D) I and III only
(E) I, II, and III

**14**

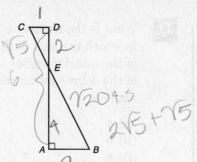

If, in the figure above, $CD = 1$, $AB = 2$, and $AD = 6$, then $BC =$

(A) 5
(B) 9
(C) $2 + \sqrt{5}$
(D) $3\sqrt{5}$
(E) $3\sqrt{2} + 2\sqrt{3}$

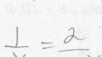

**15**

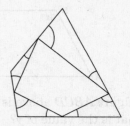

In the figure above, what is the sum of the degree measures of the marked angles?

(A) 120
(B) 180
(C) 360
(D) 540
(E) It cannot be determined from the information given.

---

## Quantitative Comparison

Each question consists of two quantities in boxes, one in Column A and one in Column B. You are to compare the two quantities and on the answer sheet fill in

A  if the quantity in Column A is greater;
B  if the quantity in Column B is greater;
C  if the two quantities are equal;
D  if the relationship cannot be determined from the information given

AN E RESPONSE WILL NOT BE SCORED

---

| <u>Column A</u> | <u>Column B</u> | <u>Column A</u> | <u>Column B</u> |
|---|---|---|---|

**1**

| The length of a diagonal of a square of side length 10 | The length of a diagonal of a 6 by 8 rectangle |
|---|---|

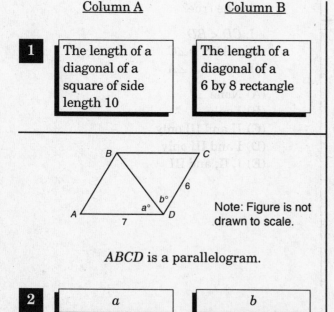

*ABCD* is a parallelogram.

**2**

| $a$ | $b$ |
|---|---|

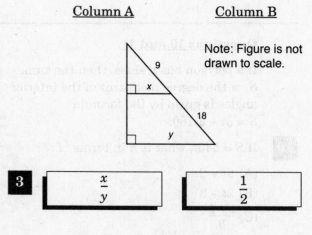

Note: Figure is not drawn to scale.

**3**

| $\dfrac{x}{y}$ | $\dfrac{1}{2}$ |
|---|---|

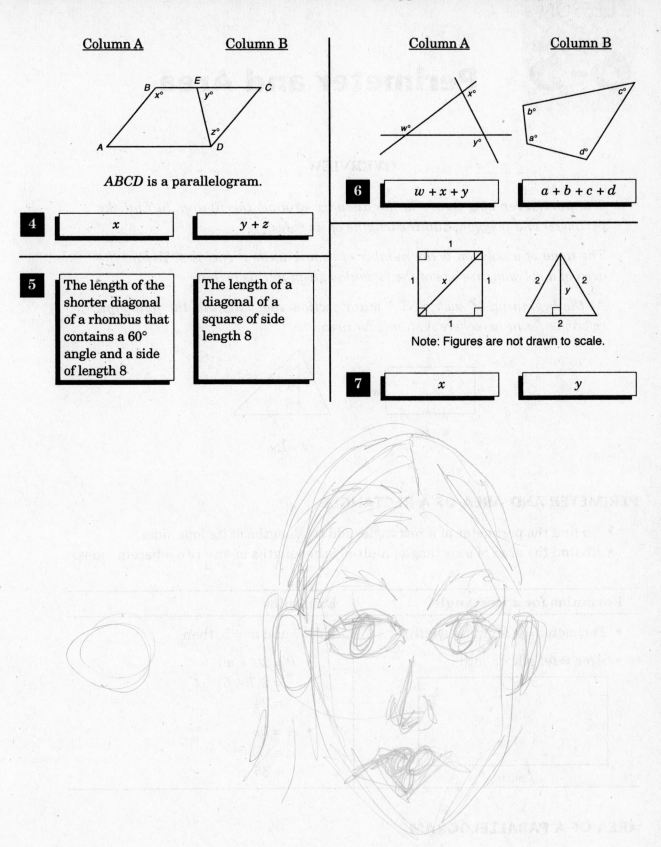

| Column A | Column B |
| --- | --- |

*ABCD* is a parallelogram.

**4** | $x$ | $y + z$

**5** | The length of the shorter diagonal of a rhombus that contains a 60° angle and a side of length 8 | The length of a diagonal of a square of side length 8

| Column A | Column B |
| --- | --- |

**6** | $w + x + y$ | $a + b + c + d$

Note: Figures are not drawn to scale.

**7** | $x$ | $y$

# Perimeter and Area

## OVERVIEW

The **perimeter** of a figure is the distance around the figure. To find the perimeter of a polygon, add the lengths of its sides.

The **area** of a polygon is the number of square units it encloses. To find the area of a polygon, use one of the formulas given in this lesson.

At the beginning of each SAT I math section you will find the following reference formulas where A stands for area:

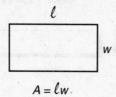

$A = \ell w$

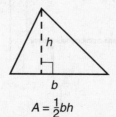

$A = \frac{1}{2}bh$

## PERIMETER AND AREA OF A RECTANGLE

- To find the perimeter of a rectangle, add the lengths of its four sides.
- To find the area of a rectangle, multiply the lengths of any two adjacent sides.

| Formulas for a Rectangle | Examples |
|---|---|
| • Perimeter = 2(*l*ength + *w*idth) <br><br> • Area = *l*ength × *w*idth | If $\ell = 7$ and $w = 5$, then <br><br> • $P = 2(\ell + w)$ <br> $\quad = 2(7 + 5)$ <br> $\quad = 24$ <br><br> • $A = \ell w$ <br> $\quad = (7)(5)$ <br> $\quad = 35$ |

## AREA OF A PARALLELOGRAM

To find the area of a parallelogram, multiply the length of a side (called the **base**) by the length of the perpendicular (called the **height**) drawn to it from the opposite side.

| Area of a Parallelogram Formula | Example |
|---|---|
| Area = $base \times height$ | If $b = 8$ and $h = 5$, then $$A = bh$$ $$= (8)(5)$$ $$= 40$$ |

## AREA OF A RHOMBUS

To find the area of a rhombus, multiply the lengths of the two diagonals and then divide the product by 2.

| Area of a Rhombus Formula | Example |
|---|---|
| Area = $\dfrac{diagonal_1 \times diagonal_2}{2}$ | If, in parallelogram $ABCD$, diagonal $d_1 = AC = 8$ and diagonal $d_2 = BD = 6$, then $$A = \frac{AC \times BD}{2}$$ $$= \frac{8 \times 6}{2}$$ $$= 24$$ |

## AREA OF A SQUARE

To find the area of a square, multiply the length of a side by itself. Since the diagonals of a square have the same length, you can also find the area of a square by multiplying the length of a diagonal by itself and then dividing the product by 2.

| Area of a Square Formula | Example |
|---|---|
| Area = $(side)^2$ or Area = $\dfrac{(diagonal)^2}{2}$ | If $d = AC = 10$, then $$A = \frac{d^2}{2}$$ $$= \frac{10^2}{2}$$ $$= 50$$ |

## AREA OF A TRIANGLE

The height (or **altitude**) of a triangle is the length of the perpendicular segment drawn from *any* vertex to the opposite side, called the **base**. To find the area of a triangle, multiply the product of its base and height by $\frac{1}{2}$.

| Area of a Triangle Formula | Example |
|---|---|
| Area $= \frac{1}{2}$ (base $\times$ height) | If $b = 9$ and $h = 4$, then $$A = \frac{1}{2}bh$$ $$= \frac{1}{2}(9)(4)$$ $$= \frac{1}{2}(36)$$ $$= 18$$ |

## USING AREA FORMULAS

Sometimes it is necessary to figure out the base or height of a figure before an area formula can be used.

**EXAMPLE:** To find the area of rectangle *ABCD*, note that the diagonal forms a (5-12-13) right triangle so the width of the rectangle is 5. Hence:

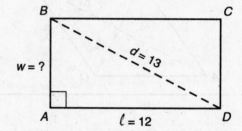

Area of rectangle $= \ell w = 12 \times 5 = 60$

**EXAMPLE:** To find the area of parallelogram *ABCD*, draw perpendicular segment *BH*, as shown. Since *BH* is the side opposite a 45° angle in a right triangle:

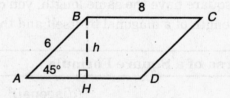

$$h = 6 \times \frac{\sqrt{2}}{2} = 3\sqrt{2}$$

Opposite sides of a parallelogram are equal, so $AD = 8$. Hence,

Area of parallelogram $ABCD = bh$

$$= AD \times h$$

$$= 8 \times 3\sqrt{2}$$

$$= 24\sqrt{2}$$

**EXAMPLE:** To find the area of $\triangle ABC$, note that the lengths of the sides of $\triangle ABC$ form a (*3-4-5*) Pythagorean triple, where $AC = 4$. Hence,

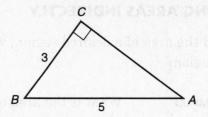

$$\text{Area of } \triangle ABC = \frac{1}{2}bh = \frac{1}{2}(3)(4) = 6$$

Note that in a right triangle either leg is the base and the other leg is the altitude.

**EXAMPLE:** To find the area of $\triangle JKL$, drop a perpendicular segment from vertex $J$ to side $KL$, extending it as necessary. Since $\angle JKH$ measures 30°:

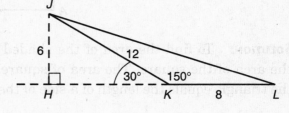

$$h = JH = \frac{1}{2} \times 12 = 6$$

and

$$\text{Area of } \triangle JKL = \frac{1}{2}bh = \frac{1}{2}(8)(6) = 24$$

## AREA OF AN EQUILATERAL TRIANGLE

If the length of a side of an equilateral triangle is known, the area of the triangle can be determined using a special formula.

| Area of an Equilateral Triangle | Example |
|---|---|
| $\text{Area} = \dfrac{(\text{side})^2 \times \sqrt{3}}{4}$ <br><br> (triangle with all sides $s$ and all angles 60°) | If $s = 8$, then <br><br> $A = \dfrac{s^2 \times \sqrt{3}}{4}$ <br> $= \dfrac{(8)^2 \times \sqrt{3}}{4}$ <br> $= \dfrac{64 \times \sqrt{3}}{4}$ <br> $= 16\sqrt{3}$ |

## FINDING AREAS INDIRECTLY

To find the area of a desired region, you may need to subtract the area of an overlapping region.

**EXAMPLE:**      What is the area of the region shaded in square *ABCD* below?

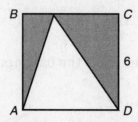

**SOLUTION:**   To find the area of the shaded region, subtract the area of △*AED* from the area of the square. The area of square *ABCD* is 6 × 6 = 36. Since the height of the triangle equals the length of a side of the square, the area of △*AED* is $\frac{1}{2}(6)(6) = 18$).

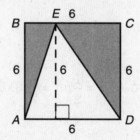

Hence, the area of the shaded region is 36 − 18 = 18.

# LESSON 6-5 TUNE-UP EXERCISES

## Multiple Choice

**1** What is the perimeter of a square that has an area of 25?

(A) 15
(B) 20
(C) 25
(D) $20\sqrt{2}$
(E) $25\sqrt{2}$

**2**

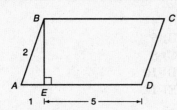

In the figure above, what is the area of parallelogram *ABCD*?

(A) $4\sqrt{2}$
(B) $4\sqrt{3}$
(C) $6\sqrt{2}$
(D) $6\sqrt{3}$
(E) 12

**3**

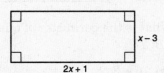

If the perimeter of the rectangle above is 44, then $x =$

(A) 2
(B) 4
(C) 7
(D) 8
(E) 9

**4**

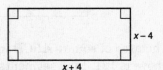

The rectangle above has an area of 65 when $x =$

(A) 9
(B) 8
(C) 6
(D) 5
(E) 3

**5** What is the area of a square with a diagonal of $\sqrt{2}$?

(A) $\frac{1}{2}$
(B) 1
(C) $\sqrt{2}$
(D) 2
(E) 4

**6**

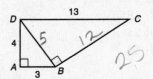

In the figure above, what is the area of quadrilateral *ABCD*?

(A) 28
(B) 32
(C) 36
(D) 42
(E) 60

**7**

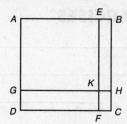

The area of square *ABCD* in the figure above is 121. Line segments *EF* and *GH* are drawn parallel to the sides of the square. If the area of rectangle *FKHC* is 6 and the area of rectangle *EKHB* is 16, what is the area of rectangle *GDFK*?

(A) 21
(B) 24
(C) 27
(D) 32
(E) 36

**8**

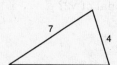

If the perimeter of the triangle above is 18, what is the area of the triangle?

(A) $6\sqrt{5}$
(B) $9\sqrt{5}$
(C) $18\sqrt{5}$
(D) $2\sqrt{33}$
(E) 14

**9**

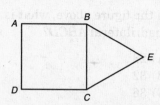

In the figure above, if the area of square *ABCD* is 64, what is the area of equilateral triangle *BEC*?

(A) 8
(B) $8\sqrt{3}$
(C) $12\sqrt{3}$
(D) $16\sqrt{3}$
(E) 32

**10** If the area of an isosceles right triangle is 8, what is the perimeter of the triangle?

(A) $8 + \sqrt{2}$
(B) $8 + 4\sqrt{2}$
(C) $4 + 8\sqrt{2}$
(D) $12\sqrt{2}$
(E) 10

**11** The lengths of the sides of △*ABC* are consecutive integers. If △*ABC* has the same perimeter as an equilateral triangle with a side length of 9, what is the length of the shortest side of △*ABC*?

(A) 4
(B) 6
(C) 8
(D) 10
(E) 12

Questions 12 and 13 refer to the figure below.

Note: Figure is not drawn to scale.

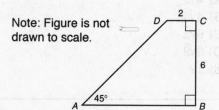

**12** What is the perimeter of quadrilateral *ABCD*?

(A) $16 + 3\sqrt{2}$
(B) $16 + 6\sqrt{2}$
(C) $22 + 6\sqrt{2}$
(D) $22\sqrt{2}$
(E) 28

**13** What is the area of quadrilateral *ABCD*?
(A) 20
(B) 24
(C) 30
(D) 36
(E) 40

**14**

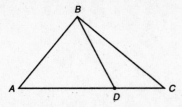

In the figure above, the ratio of *AD* to *DC* is 3 to 2. If the area of △*ABC* is 40, what is the area of △*BDC*?

(A) 16
(B) 24
(C) 30
(D) 36
(E) It cannot be determined from the information given.

**15** The perimeter of a rectangle with adjacent side lengths of *x* and *y*, where *x* > *y*, is 8 times as great as the shorter side of the rectangle. What is the ratio of *y* to *x*?

(A) 1 : 2
(B) 1 : 3
(C) 1 : 4
(D) 2 : 3
(E) 3 : 4

**16** A square has the same area as a rectangle whose longer side is 2 times the length of its shorter side. If the perimeter of the rectangle is 24, what is the perimeter of the square?

(A) $8\sqrt{2}$
(B) $16\sqrt{2}$
(C) $32\sqrt{2}$
(D) 32
(E) 64

**17**

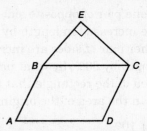

In the figure above, *ABCD* is a parallelogram with *AB* = *BE* = *EC*. If the area of right triangle *BEC* is 8, what is the perimeter of polygon *ABECD*?

(A) 20
(B) 16
(C) $16 + 8\sqrt{2}$
(D) $12 + 8\sqrt{2}$
(E) $16 + 4\sqrt{2}$

**18**

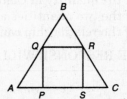

In the figure above, the vertices of square *PQRS* lie on the sides of equilateral triangle *ABC*. If the area of the square is 3, what is the perimeter of △*ABC*?

(A) $6\sqrt{3}$
(B) $3 + 6\sqrt{3}$
(C) $6 + 3\sqrt{3}$
(D) 9
(E) 12

**19**

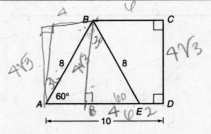

In the figure above, what is the area of quadrilateral *BCDE*?

(A) $8\sqrt{3}$
(B) $16\sqrt{3}$
(C) $8 + 4\sqrt{3}$
(D) $4 + 12\sqrt{3}$
(E) 24

**20** If one pair of opposite sides of a square are increased in length by 20% and the other pair of sides are increased in length by 50%, by what percent is the area of the rectangle that results greater than the area of the original square?

(A) 10%
(B) 50%
(C) 70%
(D) 75%
(E) 80%

---

## Quantitative Comparison

Each question consists of two quantities in boxes, one in Column A and one in Column B. You are to compare the two quantities and on the answer sheet fill in

A  if the quantity in Column A is greater;
B  if the quantity in Column B is greater;
C  if the two quantities are equal;
D  if the relationship cannot be determined from the information given

AN E RESPONSE WILL NOT BE SCORED

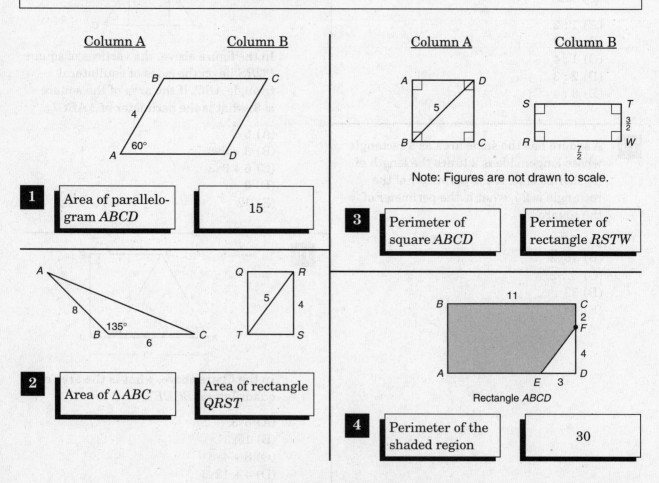

|  | Column A | Column B |
|---|---|---|
| **1** | Area of parallelogram *ABCD* | 15 |
| **2** | Area of △*ABC* | Area of rectangle *QRST* |

Note: Figures are not drawn to scale.

|  | Column A | Column B |
|---|---|---|
| **3** | Perimeter of square *ABCD* | Perimeter of rectangle *RSTW* |
| **4** | Perimeter of the shaded region | 30 |

Rectangle *ABCD*

|  Column A  |  Column B  |
|---|---|

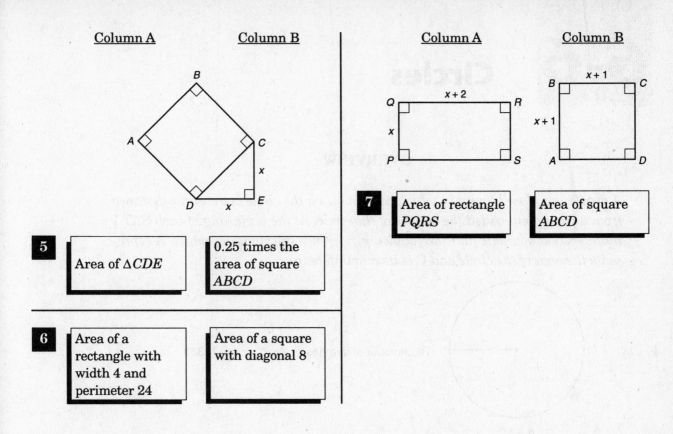

**5**

| Area of △CDE | 0.25 times the area of square ABCD |

**6**

| Area of a rectangle with width 4 and perimeter 24 | Area of a square with diagonal 8 |

|  Column A  |  Column B  |
|---|---|

**7**

| Area of rectangle PQRS | Area of square ABCD |

## Grid In

**1** Brand $X$ paint costs \$14 per gallon, and 1 gallon provides coverage of an area of at most 150 square feet. What is the minimum cost of the amount of brand $X$ paint needed to cover the four walls of a rectangular room that is 12 feet wide, 16 feet long, and 8 feet high?

**2**

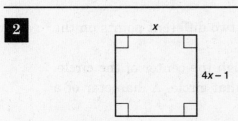

What is the area of the square above?

**3**

In the figure above, $P$ and $Q$ are the midpoints of sides $AB$ and $BC$, respectively, of square $ABCD$. Line segment $PB$ is extended by its own length to point $E$, and line segment $PQ$ is extended to point $F$ so that $FE \perp PE$. If the area of square $ABCD$ is 9, what is the area of quadrilateral $QBEF$?

## OVERVIEW

*A **circle** is a closed figure in which each point on the circle is the same distance from a fixed point called the **center** of the circle. At the beginning of each SAT I math section you will find the following reference information, where A represents the area of the circle and C is its circumference:*

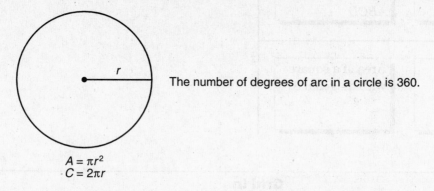

The number of degrees of arc in a circle is 360.

$A = \pi r^2$
$C = 2\pi r$

## SEGMENTS OF A CIRCLE

A circle is usually denoted by a single capital letter that identifies the center of the circle. Figure 6.17 shows key segments related to circle $O$, where point $O$ is the center of this circle.

- A **radius** is any line segment drawn from the center of a circle to any point on the circle. The plural of the word *radius* is *radii*. All radii of the same circle are equal in length.
- A **chord** of a circle is a line segment that connects two different points on the circle.
- A **diameter** of a circle is a chord that passes through the center of the circle. Every diameter of a circle is the longest chord of that circle. A diameter of a circle is 2 times the length of a radius of that circle.
- A **tangent** of a circle is a line that intersects a circle in exactly one point.

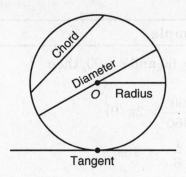

**Figure 6.17**  *Segments Related to a Circle*

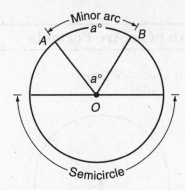

**Figure 6.18**  *Angles and Arcs of a Circle*

## ANGLES AND ARCS OF A CIRCLE

Figure 6.18 names important angle and arcs of circle.

- An **arc** of a circle is a curved section of the circle. A diameter of a circle divides a circle into two equal arcs, each of which is called a semicircle. An arc smaller than a **semicircle** is called a **minor arc**. A **major arc** is an arc that is greater than a semicircle.
- A **central angle** of a circle is an angle whose vertex is the center of the circle and whose two sides are radii. In Figure 6.18, $\angle AOB$ is a central angle. A central angle has the same degree measure as the minor arc between its sides.

## CIRCUMFERENCE AND ARC LENGTH OF A CIRCLE

- To find the circumference of a circle, multiply the radius by 2 and by $\pi$.
- To find the length of an arc of a circle, multiply the circumference of the circle by the fraction $\frac{n°}{360°}$, where $n$ is the number of degrees in the arc or in the central angle that intercepts the arc.

| Circumference Formula | Example |
|---|---|
| Circumference $= 2\pi r$ | If $r = 5$, then |
| | $$C = 2\pi(5) = 10\pi$$ |

| Length of an Arc Formula | Example |
|---|---|
| $\text{Length} = \dfrac{n^\circ}{360^\circ} \times 2\pi r$ | If $n = 60$ and $r = 90$, then $$L = \frac{60^\circ}{360^\circ} \times 2\pi\,(9)$$ $$= \frac{1}{6} \times 18\pi$$ $$= 3\pi$$ |

## AREA OF A CIRCLE AND SECTOR

- To find the area of a circle, square the radius and multiply by $\pi$.
- To find the area of any fractional part or **sector** of a circle, multiply the area of the circle by the fraction $\frac{n^\circ}{360^\circ}$, where $n$ is the number of degrees in the central angle of the sector.

| Area of a Circle Formula | Example |
|---|---|
| $\text{Area} = \pi r^2$ | If $r = 6$, then <br> • $A = \pi(6^2) = 36\pi$ <br> • The area of any semicircle of this circle is $$\frac{1}{2} \times 36\pi = 18\pi$$ |

| Area of a Sector Formula | Example |
|---|---|
| $\text{Area}_{\text{sector}} = \dfrac{n^\circ}{360^\circ} \times \pi r^2$ | If $n = 40$ and $r = 6$, then $$A_{\text{sector}} = \frac{40^\circ}{360^\circ} \times \pi(6)^2$$ $$= \frac{1}{9} \times 36\pi$$ $$= 4\pi$$ |

## INSCRIBED POLYGONS AND CIRCLES

A polygon is inscribed in a circle if all of its vertices are points on the circle. If a rectangle is inscribed in a circle, the diagonals of the rectangle are diameters of the circle, as shown in Figure 6.19.

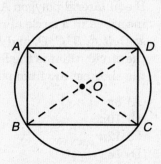

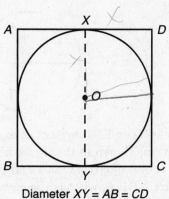

Diagonals *AC* and *BD* are diameters of circle *O*.

Diameter *XY = AB = CD*

**Figure 6.19**  *An Inscribed Rectangle*

**Figure 6.20**  *A Circle Inscribed in a Square*

A circle is inscribed in a polygon if the circle intersects each side of the polgyon in exactly one point. If a circle is inscribed in a square, a side of the square has the same length as a diameter of the circle (see Figure 6.20).

## LESSON 6-6  TUNE-UP EXERCISES

**1**

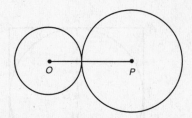

Circles *O* and *P* intersect at exactly one point, as shown in the figure above. If the radius of circle *O* is 4 and the radius of circle *P* is 6, what is the circumference of any circle that has *OP* as a diameter?

(A) 4π
(B) 8π
(C) 12π
(D) 16π
(E) 64π

**2** What is the area of a circle with a circumference of 10π?

(A) √10π
(B) 5π
(C) 25π
(D) 100π
(E) 100π²

**3**

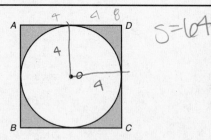

In the figure above, a circle *O* of radius 4 is inscribed in square *ABCD*. What is the area of the shaded region?

(A) 16 − 4π
(B) 32 − 4π
(C) 32 − 8π
(D) 64 − 8π
(E) 64 − 16π

**4** If equilateral polygon *ABCDE* is inscribed in a circle of radius 20 inches so that *A*, *B*, *C*, *D*, and *E* are points on the circle, what is the length in inches of the shortest arc from point *A* to point *C*?

(A) 8π
(B) 10π
(C) 12π
(D) 16π
(E) 24π

**5** What is the circumference of a circle in which a 5 by 12 rectangle is inscribed?

(A) 13π
(B) 17π
(C) 26π
(D) 34π
(E) 60π

**6** What is the perimeter of a square that has the same area as a circle with a circumference of π?

(A) 4
(B) 2π
(C) 4π
(D) 2√π
(E) 4√π

**7**

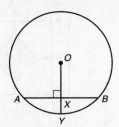

In the figure above, if the radius length of circle $O$ is 10 and $AB = 16$, what is the length of segment $XY$?

(A) 2
(B) 3
(C) 4
(D) 6
(E) It cannot be determined from the information given.

**8**  If a bicycle wheel has traveled $\frac{f}{\pi}$ feet after $n$ complete revolutions, what is the length in feet of the diameter of the bicycle wheel?

(A) $\frac{f}{n\pi^2}$

(B) $\frac{\pi^2}{fn}$

(C) $\frac{nf}{\pi^2}$

(D) $nf$

(E) $\frac{f}{n}$

**9**

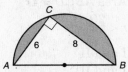

In the figure above, arc $ACB$ is a semi-circle of which $AB$ is a diameter. If $AC = 6$ and $BC = 8$, what is the area of the shaded region?

(A) $100\pi - 48$
(B) $50\pi - 24$
(C) $25\pi - 48$
(D) $12.5\pi - 48$
(E) $12.5\pi - 24$

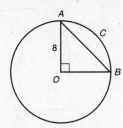

Questions 10 and 11 refer to the diagram above.

**10**  What is the perimeter of the figure bounded by arc $ACB$, radius $OA$, and radius $OB$?

(A) $32 + 8\pi$
(B) $16 + 4\pi$
(C) $16 + 8\pi$
(D) $8\sqrt{2} + 4\pi$
(E) $24\pi$

**11**  What is the area of the region of the circle bounded by chord $AB$ and arc $ACB$?

(A) $16\pi - 8\sqrt{2}$
(B) $8\pi - 16$
(C) $16\pi - 8$
(D) $16\pi - 32$
(E) $32\pi - 16$

**12**

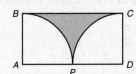

In rectangle $ABCD$ above, arcs $BP$ and $CP$ are quarter circles with centers at points $A$ and $D$, respectively. If the area of each quarter circle is $\pi$, what is the area of the shaded region?

(A) $4 - \frac{\pi}{2}$
(B) $4 - \pi$
(C) $8 - \pi$
(D) $8 - 2\pi$
(E) $8$

**13**

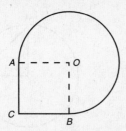

In the figure above, *OACB* is a square with area $4x^2$. If *OA* and *OB* are radii of a sector of a circle *O*, what is the perimeter, in terms of *x*, of the unbroken figure?

(A) $x(4 + 3\pi)$
(B) $x(3 + 4\pi)$
(C) $x(6 + 4\pi)$
(D) $4(x + 2\pi)$
(E) $7\pi x$

**14**

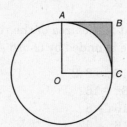

In the figure above, *OABC* is a square. If the area of circle *O* is $2\pi$, what is the area of the shaded region?

(A) $\dfrac{\pi}{2} - 1$

(B) $2 - \dfrac{\pi}{2}$

(C) $\pi - 2$

(D) $\dfrac{\pi - 1}{2}$

(E) $\dfrac{2\pi - 1}{4}$

**15** If the circumference of a circle of radius *r* inches is equal to the perimeter of a square with a side length of *s* inches, $\dfrac{r}{s} =$

(A) $\dfrac{4}{\pi}$

(B) $\dfrac{2}{\pi}$

(C) $\sqrt{\dfrac{2}{\pi}}$

(D) $\dfrac{\sqrt{2}}{\pi}$

(E) $\dfrac{1}{2\pi}$

**16**

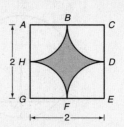

In the figure above, the vertices of square *ACEG* are the centers of four quarter circles of equal area. What is the best approximation for the area of the shaded region? (Use $\pi = 3.14$.)

(A) 0.64
(B) 0.79
(C) 0.86
(D) 1.57
(E) 2.36

**17**

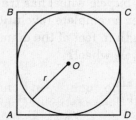

In the figure above, if circle *O* is inscribed in square *ABCD* in such a way that each side of the square is tangent to the circle, which of the following statements must be true?

I. $AB \times CD < \pi \times r \times r$
II. Area $ABCD = 4r^2$
III. $r < \dfrac{2(CD)}{\pi}$

(A) I and II
(B) I and III
(C) II and III
(D) II only
(E) III only

**18**

In the figure above, arc *PBQ* is one-quarter of a circle with center at *O*, and *OABC* is a rectangle. If *AOC* is an isosceles right triangle with $AC = 8$, what is the perimeter of the figure that encloses the shaded region?

(A) $24 - 4\pi$

(B) $24 - 4\sqrt{2} + 4\pi$

(C) $16 - 4\sqrt{2} + 4\pi$

(D) $16 + 4\pi$

(E) $16\sqrt{2} + 8\pi$

---

### Quantitative Comparison

Each question consists of two quantities in boxes, one in Column A and one in Column B. You are to compare the two quantities and on the answer sheet fill in

A  if the quantity in Column A is greater;

B  if the quantity in Column B is greater;

C  if the two quantities are equal;

D  if the relationship cannot be determined from the information given

AN E RESPONSE WILL NOT BE SCORED

---

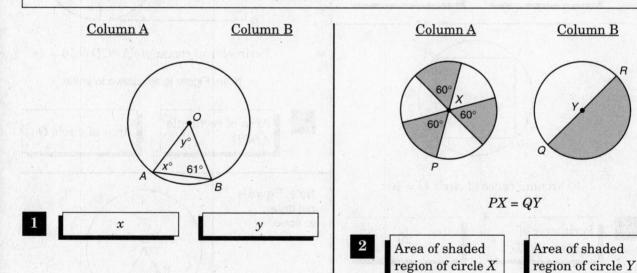

| Column A | Column B |
|---|---|

**1**    $x$          $y$

|  Column A  |  Column B  |
|---|---|

$PX = QY$

**2**  Area of shaded region of circle *X*     Area of shaded region of circle *Y*

Column A      Column B

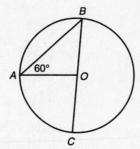

**3**    $(AB)^2$      $OA \times OC$

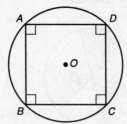

Tangent circles $A$, $B$, and $C$ each have an area of $9\pi$.

**4**    Perimeter of $\triangle ABC$      Circumference of circle $A$

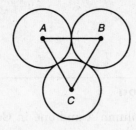

Circumference of circle $O = 10\pi$

**5**    Perimeter of square $ABCD$      20

---

Column A      Column B

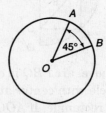 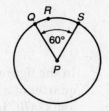

The number of square inches in the area of sector $AOB$ of circle $O$ is $18\pi$.      The number of inches in the length of arc $QRS$ of circle $P$ is $5\pi$.

Note: Figures are not drawn to scale.

**6**    Number of inches in the length of $OA$      Number of inches in the length of $PQ$

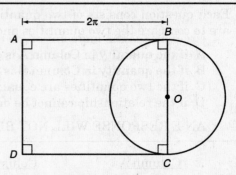

Perimeter of rectangle $ABCD = 16 + 4\pi$

Note: Figure is not drawn to scale.

**7**    Area of rectangle $ABCD$      Area of circle $O$

---

Note: Figure is not drawn to scale.

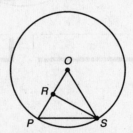

Circle $O$ with $RS > OS$

**8**    $OP$      $PS$

## Grid In

**1**

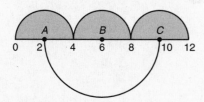

In the figure above, the sum of the areas of the three shaded semicircles with centers at $A$, $B$, and $C$ is $X$, and the area of the larger semicircle below the line is $Y$. If $Y - X = k\pi$, what is the value of $k$?

**2**

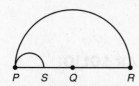

In the figure above, each arc is a semicircle. If $S$ is the midpoint of $PQ$ and $Q$ is the midpoint of $PR$, what is the ratio of the area of semicircle $PS$ to the area of semicircle $PR$?

**3** Through how many degrees does the minute hand of a clock move from 1:25 P.M. to 1:37 P.M. of the same day?

**4**

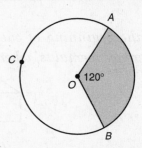

In the figure above, if the area of the shaded region of circle $O$ is $\dfrac{12}{\pi}$, what is the length of arc $ACB$?

# Solid Figures

## OVERVIEW

*At the beginning of each SAT I math section you will find the following reference formulas, where V stands for volume:*

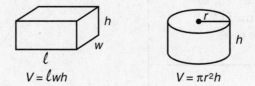

$$V = \ell wh \qquad V = \pi r^2 h$$

## VOLUME AND SURFACE AREA OF A RECTANGULAR SOLID

- The **volume** of a solid figure represents the number of cubes each with an edge length of 1 unit that can be placed in the space that the figure encloses (see Figure 6.21).
- The **surface area** of a rectangular solid is the sum of the areas of each of its six surfaces.

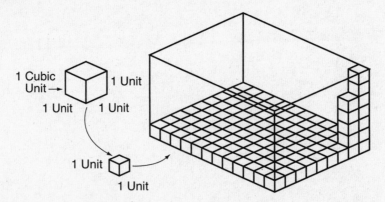

**Figure 6.21**  *Volume of a Rectangle Solid*

| Formulas for a Rectangular Solid | Examples |
|---|---|
| • Volume<br><br>$\text{Volume}_{\text{box}} = \text{length} \times \text{width} \times \text{height}$<br><br><br><br>• Surface area<br><br>$\text{Area}_{\text{surface}} = 2[(\ell \times w) + (\ell \times h) + (w \times h)]$ | If $\ell = 6$, $w = 3$, and $h = 2$, then<br><br>• $V = \ell w h$<br>  $= 6 \times 3 \times 2$<br>  $= 36$<br><br>• $A = 2[(\ell \times w) + (\ell \times h) + (w \times h)]$<br>  $= 2[(6 \times 3) + (6 \times 2) + (3 \times 2)]$<br>  $= 2(18 + 12 + 6)$<br>  $= (36)2$<br>  $= 72$ |

## VOLUME AND SURFACE AREA OF A CUBE

- A **cube** is a rectangular solid in which each of its six square faces has the same edge length. Its volume is the cube of the edge length of any face.
- The surface area of a cube is 6 times the area of any square face of the cube.

| Formulas for a Cube | Examples |
|---|---|
| • Volume<br><br>$\text{Volume}_{\text{cube}} = \text{edge} \times \text{edge} \times \text{edge} = e^3$<br><br><br><br>• Surface area<br><br>$\text{Area}_{\text{surface}} = 6(\text{edge})^2$ | If $e = 4$, then<br><br>• $V = e^3$<br>  $= 4 \times 4 \times 4$<br>  $= 64$<br><br>• $A = 6e^2$<br>  $= 6(4)^2$<br>  $= 6(16)$<br>  $= 96$ |

## VOLUME OF A CYLINDER

The volume of a **cylinder** is the area of its circular base times its height.

| Volume Formula for a Cylinder | Example |
|---|---|
| $\text{Volume}_{\text{cylinder}} = \pi(radius)^2 \times height$ <br><br> | If $r = 3$ and $h = 5$, then <br><br> $V = \pi\, r^2 h$ <br> $= \pi(3)^2(5)$ <br> $= 45\pi$ |

# LESSON 6-7 TUNE-UP EXERCISES

## Multiple Choice

**1** What is the volume of a cube whose surface area is 96?

(A) $16\sqrt{2}$
(B) 32
(C) 64
(D) 125
(E) 216

**2** The length, width, and height of a rectangular solid are in the ratio of $3 : 2 : 1$. If the volume of the box is 48, what is the total surface area of the box?

(A) 27
(B) 32
(C) 44
(D) 64
(E) 88

**3** If $X$ is the center point of a face of a cube with a volume of 8, and $Y$ is the center point of the opposite face of this cube, what is the distance from $X$ to $Y$?

(A) $\sqrt{2}$
(B) 2
(C) $2\sqrt{2}$
(D) 4
(E) 6

**4** A cube whose volume is $\frac{1}{8}$ cubic foot is placed on top of a cube whose volume is 1 cubic foot. The two cubes are then placed on top of a third cube, whose volume is 8 cubic feet. What is the height, in *inches*, of the stacked cubes?

(A) 30
(B) 40
(C) 42
(D) 44
(E) 64

**5** Note: Figure is not drawn to scale.

The height of the solid cone above is 12 inches, and the area of the circular base is $64\pi$ square inches. What is the area, in square inches, of the base of the cone formed when a plane parallel to the base cuts through the original cone 9 inches above the vertex of the cone?

(A) $9\pi$
(B) $16\pi$
(C) $25\pi$
(D) $36\pi$
(E) $49\pi$

**6** If the height of a cylinder is doubled, by what number must the radius of the base be multiplied so that the resulting cylinder has the same volume as the original cylinder?

(A) 4
(B) 2
(C) $\dfrac{1}{\sqrt{2}}$
(D) $\dfrac{1}{2}$
(E) $\dfrac{1}{4}$

**7** A rectangular fish tank has a base 2 feet wide and 3 feet long. When the tank is partially filled with water, a solid cube with an edge length of 1 foot is placed in the tank. If no overflow of water from the tank is assumed, by how many *inches* will the level of the water in the tank rise when the cube becomes completely submerged?

(A) $\frac{1}{6}$

(B) $\frac{1}{2}$

(C) 2

(D) 3

(E) 4

**8** The volume of a cylinder of radius $r$ is $\frac{1}{4}$ of the volume of a rectangular box with a square base of side length $x$. If the cylinder and the box have equal heights, what is $r$ in terms of $x$?

(A) $\frac{x^2}{2\pi}$

(B) $\frac{x}{2\sqrt{\pi}}$

(C) $\frac{\sqrt{2x}}{\pi}$

(D) $\frac{\pi}{2\sqrt{x}}$

(E) $\sqrt{2\pi x}$

**9** The height of sand in a cylinder-shaped can drops 3 inches when 1 cubic foot of sand is poured out. What is the diameter, in *inches*, of the cylinder?

(A) $\frac{2}{\sqrt{\pi}}$

(B) $\frac{4}{\sqrt{\pi}}$

(C) $\frac{16}{\pi}$

(D) $\frac{32}{\sqrt{\pi}}$

(E) $\frac{48}{\sqrt{\pi}}$

**10** The height $h$ of a cylinder equals the circumference of the cylinder. In terms of $h$, what is the volume of the cylinder?

(A) $\frac{h^3}{4\pi}$

(B) $\frac{h^2}{2\pi}$

(C) $\frac{h^3}{2}$

(D) $h^2 + 4\pi$

(E) $\pi h^3$

**11**

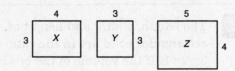

For which of the following combinations of rectangular faces $X$, $Y$, and $Z$ in the figures above can a rectangular solid be formed?

   I. Two of face $X$, two of face $Y$, and two of face $Z$

  II. Four of face $X$ and two of face $Y$

 III. Two of face $Y$ and four of face $Z$

(A) I only

(B) II only

(C) I and III only

(D) II and III only

(E) None

**12** A cylinder with radius $r$ and height $h$ is closed on the top and bottom. Which of the following expressions represents the total surface area of this cylinder ?

(A) $2\pi r(r + h)$

(B) $\pi r(r + 2h)$

(C) $\pi r(2r + h)$

(D) $\pi r^2 + 2h$

(E) $2\pi r^2 + h$

**13**

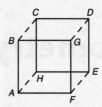

In the figure above, if the edge length of the cube is 4, what is the shortest distance from *A* to *D*?

(A) $4\sqrt{2}$
(B) $4\sqrt{3}$
(C) $8\sqrt{3}$
(D) 8
(E) $4\sqrt{2} + 4$

---

## Grid In

**1** The dimensions of a rectangular box are integers greater than 1. If the area of one side of this box is 12 and the area of another side is 15, what is the volume of the box?

**2** What is the minimum length of $\frac{1}{4}$-inch wide tape needed to cover completely a cube whose volume is 8 cubic inches?

**3** By what percent does the volume of a cube increase when the length of each of its sides is doubled?

# LESSON

# 6-8

# Coordinate Geometry

## OVERVIEW

*A **coordinate plane** is represented by a grid of square boxes that is divided into four **quadrants** by a horizontal number line, called the x-axis, and a vertical number line, called the y-axis, that intersect at their 0 points. Each point in the coordinate plane is located by using an ordered pair of numbers, called **coordinates**, of the form (x,y). For example, point P(3,5) is located 3 units to the right of the origin and 5 units above the origin.*

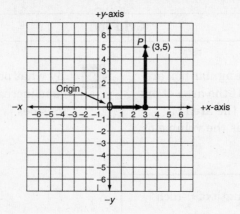

*If you know the coordinates of points A and B, you can use a formula to find:*

- *the distance from A to B; or*
- *the "slant" or slope of the line that passes through A and B; or*
- *the midpoint of line segment AB.*

## GRAPHING ORDERED PAIRS

As shown in Figure 6.22, the signs of the *x*- and *y*-coordinates of a point determine the quadrant in which the point lies.

- A point lies to the right of the *y*-axis when *x* is *positive*, and lies to the left of the *y*-axis when *x* is *negative*.
- A point lies *above* the *x*-axis when *y* is *positive*, and lies *below* the *x*-axis when *y* is *negative*.

| Point | Sign of Coordinates | Quadrant |
|-------|---------------------|----------|
| $A(2,3)$ | $(+, +)$ | I |
| $B(-4,5)$ | $(-, +)$ | II |
| $C(-3,-6)$ | $(-. -)$ | III |
| $D(3,-3)$ | $(+, -)$ | IV |

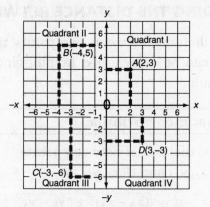

**Figure 6.22**   *Locating Points*

## FINDING LENGTHS OF HORIZONTAL AND VERTICAL SEGMENTS

Points that have the same $x$-coordinate lie on the same vertical line, and points that have same $y$-coordinate lie on the same horizontal line. To find the length of a horizontal or vertical segment, take the positive difference between the two unequal coordinates.

**EXAMPLE:**       In the figure below, what is the perimeter of $\triangle ABC$?

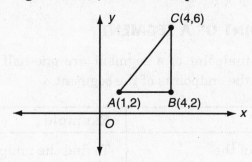

**SOLUTION:**

- The length of horizontal segment $AB$ is $4 - 1 = 3$.
- The length of vertical segment $BC$ is $6 - 2 = 4$.
- Since $\angle ABC$ is formed by horizontal and vertical line segments, it is a right angle. Use the Pythagorean relationship to find the length of $AC$:

$$AC = \sqrt{(AB)^2 + (BC)^2}$$

$$= \sqrt{(3)^2 + (4)^2}$$

$$= \sqrt{25}$$

$$= 5$$

Hence, the perimeter of $\triangle ABC$ is $3 + 4 + 5 = 12$.

## FINDING THE DISTANCE BETWEEN TWO POINTS

The distance between two points that do not lie on the same vertical or horizontal line can be determined by plugging the coordinates of the points into the distance formula.

| Distance Formula | Example |
|---|---|
| The distance $d$ between points $A(x_A, y_A)$ and $B(x_B, y_B)$ is $$d = \sqrt{(x_B - x_A)^2 + (y_B - y_A)^2}$$ | To find the distance $d$ between points $(4, -1)$ and $(7, 5)$, let $(x_A, y_A) = (4, -1)$ and $(x_B, y_B) = (7, 5)$: $$\begin{aligned} d &= \sqrt{(7-4)^2 + (5-(-1))^2} \\ &= \sqrt{3^2 + 6^2} \\ &= \sqrt{45} \\ &= \sqrt{9} \cdot \sqrt{5} \\ &= 3\sqrt{5} \end{aligned}$$ |

## FINDING THE MIDPOINT OF A SEGMENT

The coordinates of the midpoint of a segment are one-half the sums of the corresponding coordinates of the endpoints of the segment.

| Midpoint Formula | Example |
|---|---|
| The coordinates $(\bar{x}, \bar{y})$ of the midpoint of the segment whose endpoints are $A(x_A, y_A)$ and $B(x_B, y_B)$ are $$\bar{x} = \frac{x_A + x_B}{2} \text{ and } \bar{y} = \frac{y_A + y_B}{2}$$ | To find the midpoint of a segment whose endpoints are $(4, -1)$ and $(7, 5)$, let $(x_A, y_A) = (4, -1)$ and $(x_B, y_B) = (7, 5)$. Since $$\bar{x} = \frac{4 + 7}{2} = \frac{11}{2}$$ and $$\bar{y} = \frac{(-1) + 5}{2} = 2,$$ the midpoint is $\left(\frac{11}{2}, 2\right)$. |

## FINDING AREAS USING COORDINATES

If a side of a triangle or parallelogram lies on an axis or is parallel to an axis, you can calculate the area of the figure by finding the lengths of the base and altitude and then using the appropriate area formula.

**EXAMPLE:** What is the area of the shaded region in the figure below?

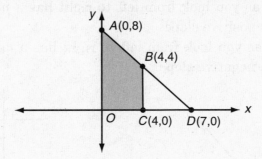

**SOLUTION:**

$$\text{area of } OABC = \text{area of } \triangle AOD - \text{area of } \triangle BCD$$

$$= \frac{1}{2}(OD)(OA) \quad -\frac{1}{2}(CD)(BC)$$

$$= \frac{1}{2}(7)(8) \quad -\frac{1}{2}(3)(4)$$

$$= 28 \quad\quad -6$$

$$= 22$$

## FINDING THE SLOPE OF A LINE

**Slope** is a number that represents the steepness of a line. To find the slope of a line, write the change in the $y$-coordinates of any two points on the line over the corresponding change in the $x$-coordinates of the same two points. Then simplify the fraction that results.

| Slope Formula | Example |
|---|---|
| The slope of a nonvertical line that contains points $A(x_A, y_A)$ and $B(x_B, y_B)$ is given by the formula $$\text{slope} = \frac{\text{vertical change}}{\text{horizontal change}} = \frac{y_B - y_A}{x_B - x_A}$$ | To find the slope of the line that contains points $(4,-1)$ and $(7,5)$, let $(x_A, y_A) = (4,-1)$ and $(x_B, y_B) = (7,5)$. Then: $$\text{slope} = \frac{y_B - y_A}{x_B - x_A} = \frac{5-(-1)}{7-4}$$ $$= \frac{5+1}{3}$$ $$= 2$$ |

## SOME FACTS ABOUT SLOPE

- If you know the slope and the coordinates of one point of a line, you can find the coordinates of other points on the same line. For example, if the slope of a line is 2, then for each 1-unit increase in the $x$-coordinate of a point on this line, the corresponding $y$-coordinate must increase by 2. Hence, if $(4,-1)$ is a point on a line with slope 2, then point $(4 + 1,-1 + 2) = (5,1)$ is on the same line.
- A line that *rises* as you look from left to right has a *positive* slope. Line $p$ in Figure 6.23 has a positive slope.
- A line that *falls* as you look from left to right has a *negative* slope. Line $n$ in Figure 6.24 has a negative slope.

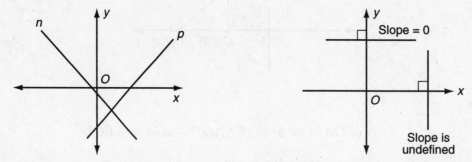

**Figure 6.23** *Lines with Positive and Negative Slopes*   **Figure 6.24** *Lines with 0 and Undefined Slope*

- A horizontal line has a slope of 0, and the slope of a vertical line is not defined. See Figure 6.24.
- The slope of a line does not change when different points on the same line are used to calculate the slope. If points $A$, $B$, and $C$ lie on the same nonvertical line, then

$$\text{slope of } AB = \text{slope of } BC = \text{slope of } AC$$

- Parallel lines have equal slopes.

# LESSON 6-8 TUNE-UP EXERCISES

## Multiple Choice

**1** The length of the line segment whose endpoints are (3,−1) and (6,5) is

(A) 3
(B) 5
(C) $3\sqrt{5}$
(D) $5\sqrt{3}$
(E) $\sqrt{97}$

**2** What is the area of a rectangle whose vertices are (−2,5), (8,5), (8,−2), and (−2,−2)?

(A) 45
(B) 50
(C) 55
(D) 60
(E) 70

**3** What is the area of a parallelogram whose vertices are (−4,−2), (−2,6), (10,6), and (8,−2)?

(A) 32
(B) 48
(C) 72
(D) 96
(E) 104

**4** What is the area of a triangle whose vertices are (−4,0), (2,4), and (4,0)?

(A) 8
(B) 12
(C) 16
(D) 32
(E) 64

**5** If A(−3,0) and C(5,2) are the endpoints of diagonal AC of rectangle ABCD, what is the perimeter of rectangle ABCD?

(A) 24
(B) 20
(C) 16
(D) 14
(E) 10

**6**

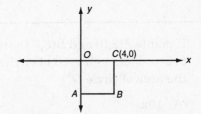

In the figure above, if A, B, C, and O are the vertices of a square and the coordinates of A are (k,p), what are the values of k and p?

(A) k = −4 and p = 0
(B) k = 0 and p = −4
(C) k = −2 and p = 0
(D) k = 0 and p = −2
(E) k = 2 and p = −2

**7** Which point lies on the same line that contains the points whose coordinates are (1,3) and (2,5)?

(A) (0,−1)
(B) (0,1)
(C) (−1,0)
(D) (3,2)
(E) (0,4)

**8**

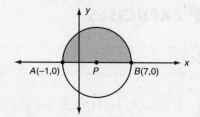

In the figure above, if $AB$ is a diameter of circle $P$, what is the perimeter of the shaded region?

(A) $4\pi + 8$
(B) $8\pi + 4$
(C) $8\pi + 8$
(D) $16\pi + 4$
(E) $16\pi + 8$

**9** If points $A(2,0)$ and $B(8,-4)$ are the endpoints of diameter $AB$ of circle $O$, what is the area of circle $O$?

(A) $10\pi$
(B) $13\pi$
(C) $24\pi$
(D) $26\pi$
(E) $52\pi$

**10** The point whose coordinates are $(4,-2)$ lies on a line whose slope is $\frac{3}{2}$. Which of the following are the coordinates of another point on this line?

(A) $(1,0)$
(B) $(2,1)$
(C) $(6,1)$
(D) $(7,0)$
(E) $(1,4)$

**11** Line segment $AB$ is a diameter of circle $O$. If the coordinates of $O$ are $(-2,1)$ and the coordinates of point $A$ are $(1,2)$, what are the coordinates of point $B$?

(A) $(0,5)$
(B) $(-3,4)$
(C) $(-5,0)$
(D) $(1,-3)$
(E) $(-3,-1)$

**12** If point $E(5,h)$ is on the line that contains $A(0,1)$ and $B(-2,-1)$, what is the value of $h$?

(A) $-1$
(B) $0$
(C) $1$
(D) $3$
(E) $6$

**13** If the line whose equation is $y = x + 2k$ passes through point $(1,-3)$, then $k =$

(A) $-2$
(B) $-1$
(C) $1$
(D) $2$
(E) $4$

**14** A circle that has its center at the origin and passes through $(-8,-6)$ will also pass through which point?

(A) $(1,10)$
(B) $(4,9)$
(C) $(7,7)$
(D) $(9,\sqrt{19})$
(E) $(\sqrt{37},8)$

**15**

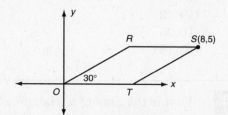

In the figure above, $ORST$ is a parallelogram with $OR = OT$. What is the perimeter of parallelogram $ORST$?

(A) $20$
(B) $32$
(C) $20\sqrt{3}$
(D) $40$
(E) $64$

**16**

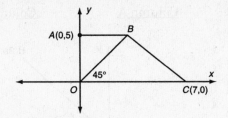

In the figure above, what is the area of quadrilateral $OABC$?

(A) 15
(B) 20
(C) 25
(D) 30
(E) 40

**17**

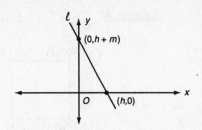

In the figure above, if the slope of line $\ell$ is $m$, what is $m$ in terms of $h$?

(A) $\dfrac{h}{1+h}$

(B) $\dfrac{-h}{1+h}$

(C) $\dfrac{h}{1-h}$

(D) $1+h$

(E) $1-h$

---

## Quantitative Comparison

Each question consists of two quantities in boxes, one in Column A and one in Column B. You are to compare the two quantities and on the answer sheet fill in

    A  if the quantity in Column A is greater;
    B  if the quantity in Column B is greater;
    C  if the two quantities are equal;
    D  if the relationship cannot be determined from the information given

**AN E RESPONSE WILL NOT BE SCORED**

---

| <u>Column A</u> | <u>Column B</u> | | <u>Column A</u> | <u>Column B</u> |
|---|---|---|---|---|

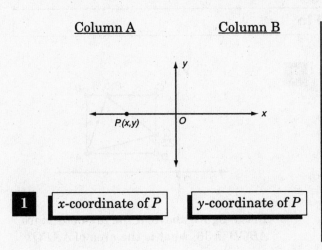

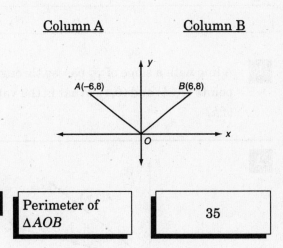

**1** | $x$-coordinate of $P$ | $y$-coordinate of $P$

**2** | Perimeter of $\triangle AOB$ | 35

Column A      Column B      Column A      Column B

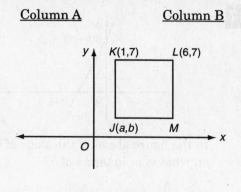

*JKLM* is a square.

**3**    $a$          $b$

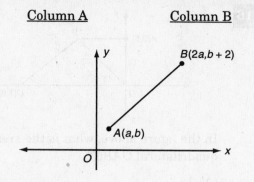

**5**   Distance from $A$ to $B$      $a + 2$

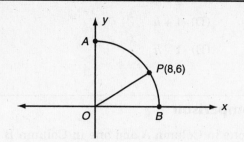

Arc *APB* with center *O*

**4**   Length of arc *APB*      $5\pi$

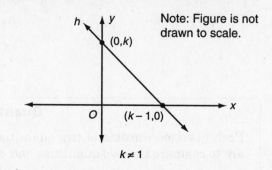

Note: Figure is not drawn to scale.

$k \neq 1$

**6**   Slope of line $h$      $-1$

---

## Grid In

**1**   A line with a slope of $\frac{3}{14}$ passes through points $(7,3k)$ and $(0,k)$. What is the value of $k$?

**2**

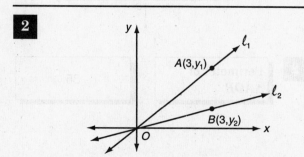

In the figure above, the slope of line $\ell_1$ is $\frac{5}{6}$ and the slope of line $\ell_2$ is $\frac{1}{3}$. What is the distance from point $A$ to point $B$?

**3**

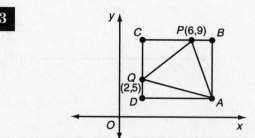

In the figure above, if the area of square *ABCD* is 36, what is the area of $\triangle APQ$?

# ANSWERS TO CHAPTER 6 TUNE-UP EXERCISES

**LESSON 6-1** *(Multiple Choice)*

1. **B** Since vertical angles are equal, the angle opposite the 40° angle also measures 40°. The measure of an exterior angle of a triangle is equal to the sum of the measures of the two nonadjacent interior angles of the triangle. Hence:
$$x° = 80° + 40° = 120°$$
and
$$y° = 40° + 70° = 100°$$
so
$$x + y = 120 + 110 = 230$$

2. **B** Since vertical angles are equal, the angle opposite the angle marked $3x°$ also measures $3x°$. Also, the angle opposite the angle marked $2x°$ also measures $2x°$. The sum of the measures of the angles about a point is 360°. Hence:
$$3x + 2x + x + 3x + 2x + y = 360$$
or $11x + y = 360$. Since vertical angles are equal, $y = x$. Thus:
$$11x + x = 360$$
$$12x = 360$$
$$x = \frac{360}{12} = 30$$
Hence, $y = x = 30$.

3. **E** Since $x + y + z$ forms a straight angle, $x + y + z = 180$. You are given that $\frac{y}{x} = 5$ and $\frac{z}{x} = 4$, so $y = 5x$ and $z = 4x$. Hence:
$$x + 5x + 4x = 180$$
$$10x = 180$$
$$x = \frac{180}{10} = 18$$

4. **B** First find the measures of the two angles that lie above $\ell_2$ and that have the same vertex as angle $x$. The vertical angle opposite the 58° angle also measures 58°. The acute angle above line $\ell_2$ that is adjacent to the 58° angle measures 37° since acute angles formed by parallel lines have equal measures. Since the sum of the measures of the angles that form a straight line is 180°,
$$37° + 58° + x° = 180°$$
so $x = 180 - 95 = 85$.

5. **C** If the angles of a triangle are in the ratio of 3 : 4 : 5, let $3x$, $4x$, and $5x$ represent the measures of the three angles. Since the sum of the measures of the angles of a triangle is 180,
$$3x + 4x + 5x = 180$$
$$12x = 180$$
$$x = \frac{180}{12} = 15$$
Hence, the measure of the smallest angle of the triangle is $3x = 3(15) = 45$.

6. **B** Since $AB \perp BC$, $\angle ABD = 90$. Vertical angles have equal measures, so $\angle DAB = 32$. Since the measures of the acute angles of a right triangle must add up to 90,
$$(x + 14) + (x + 32) = 90$$
$$2x + 46 = 90$$
$$x = \frac{44}{2} = 22$$

7. **A** Since $\ell_1 \parallel \ell_2$, $w + x = 180$ and $y = z$. Adding corresponding sides of these two equations gives
$$w + x + y = 180 + z$$
or $x + y = 180 - w + z$.

8. **B** The measures of the three interior angles of the triangle are $(180 - x)$, $(180 - y)$, and $z$. Since the sum of the measures of the angles of a triangle is 180,
$$(180 - x) + (180 - y) + z = 180$$
$$360 - x - y + z = 180$$
$$360 + z = 180 + x + y - 360$$
$$= x + y - 180$$

9. **D** Angles $x$ and $y$ form a straight line, so $x + y = 180$. Since $\ell_1 \parallel \ell_2$, obtuse angle $y$ equals obtuse angle $3x$. Substituting $3x$ for $y$ in $x + y = 180$ gives $4x = 180$, so $x = \frac{180}{4} = 45$. Hence:
$$y = 3x = 3(45) = 135$$

10. **C** Since $\ell_1 \parallel \ell_2$, the acute angle formed by lines $\ell_2$ and $\ell_3$ equals acute angle $x$. Also, since $\ell_3 \parallel \ell_4$, the obtuse angle whose measure is $y + y$ is supplementary to acute angle $x$, so $x + 2y = 180$. Solving this equation for $y$ gives $2y = 180 - x$ or
$$y = \frac{180}{2} - \frac{x}{2} = 90 - \frac{x}{2}$$

11. **C** Since the measures of vertical angles are equal, the angle opposite the $(3x + 10)°$ angle also measures $(3x + 10)°$. Since the angles marked $x°$, $(3x + 10)°$, and 90° (by the right-angle mark) form a straight line:

*(Lesson 6-1 Multiple Choice continued)*

$$x° + (3x + 10)° + 90° = 180°$$
$$4x + 100 = 180$$
$$x = \frac{80}{4} = 20$$

12. **D**

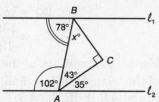

Suppose $\ell_1 \parallel \ell_2$. Then the obtuse angle formed at vertex $A$ is supplementary to the acute 78° angle formed at vertex $B$, so it measures 180° − 78° or 102°. Hence, the measure of $\angle BAC$ is 180° − 102° − 35° or 43°. Since the acute angles of a right triangle are supplementary,
$$x = 90 - 43 = 47$$

13. **A** Since $\ell \parallel m$, the acute angles marked $3y$ and $2y + 25$ must have the same measure, so $3y = 2y + 25$ and $y = 25$. Then
$$2y + 25 = 2(25) + 25 = 75$$
Since vertical angles have the same measures, $x + 15 = 75$, so $x = 60$.

14. **D** Since the sum of the measures of the angles about a point is 360, the sum of the measures of the marked angles and the unmarked angles at the four vertices is $4 \times 360$ or 1440. The unmarked angles are the interior angles of the two triangles, so their sum is $2 \times 180$ or 360. Hence, the sum of the measures of the marked angles is $1440 - 360$ or 1080.

15. **B** The measure of $\angle ACE$ is $x + 2x$ or $3x$ since it is an exterior angle of $\triangle ABC$. Also, $3y = 3x + y$ since the angle marked $3y°$ is an exterior angle of the triangle in which $3x°$ and $y°$ are the measures of the two nonadjacent interior angles. Hence, $2y = 3x$, so $y = \frac{3}{2}x$.

16. **C** Since $\ell_1 \parallel \ell_2$, the obtuse angle that measures $a + a$ or $2a°$ is supplementary to the acute angle that measures $b + b$ or $2b°$. Hence, $2a + 2b = 180$, so $a + b = 90$. The angle that measures $x + x$ or $2x°$ is an exterior angle of the triangle in which the two nonadjacent angles are marked $a°$ and $b°$. Hence, $2x = a + b = 90$, so $x = \frac{90}{2} = 45$.

17. **D** Since the angles marked $7y°$ and $5y°$ form a straight line:
$$7y + 5y = 180$$
$$12y = 180$$
$$y = \frac{180}{12} = 15$$
An exterior angle of a triangle is equal to the sum of the measures of the two non-adjacent angles. Hence, $7y = 3y + x$ or $x = 4y$. Since $y = 15$, then $x = 4(15) = 60$.

18. **D** Since acute angle $ABD$ is supplementary to obtuse angle $CDB$:
$$(y + 2y + y) + 5y = 180$$
$$9y = 180$$
$$y = \frac{180}{9} = 20$$

19. **E** In the smaller triangle, $a + b + 140 = 180$, so $a + b = 40$. In $\triangle RST$, $x + 2a + 2b = 180$. Dividing each member of this equation by 2 gives $\frac{x}{2} + a + b = 90$. Since $a + b = 40$, then
$$\frac{x}{2} + 40 = 90$$
$$\frac{x}{2} = 50$$
$$x = 100$$

20. **E** Determine whether each Roman numeral statement is always true.
- I. Since the measures of vertical angles are equal, $a = d$. Since we cannot tell whether $b = c$, we do not know whether statement I, $a + b = d + c$, is always true.
- II. Since angles $a$, $b$, and $c$ form a straight line, $a + b + c = 180$. Angles $b$ and $e$ are vertical angles, so $b = e$. Substituting $e$ for $b$ in $a + b + c = 180$ gives $a + c + e = 180$, so statement II is always true.
- III. Angles $b$ and $e$ are vertical angles, as are $f$ and $c$. Since $b = e$ and $f = c$, then $b + f = c + e$, so statement III is always true.

Only Roman numeral statements II and III are always true.

*(Quantitative Comparison)*

1. **C** Call the vertical angles 1 and 2. Then $a + b + 1 = 180 = c + d + 2$. The measures of angles 1 and 2 are equal, so they can be canceled from the two sides of the equation. Since $a + b = c + d$, Column A = Column B.

*(Lesson 6-1  Quantitative Comparison continued)*

2. **C**

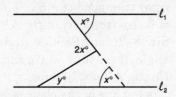

Extend the side of angle $x$ so that it intersects $\ell_2$, forming another acute angle that equals $x$ since $\ell_1 \parallel \ell_2$. Since an exterior angle of a triangle equals the sum of the two nonadjacent interior angles, $2x = x + y$ or $x = y$. Hence, Column A = Column B.

3. **C**  In right triangle $ACB$, the measures of the acute angles must add up to 90, so $x + \angle A = 90$. In right triangle $ADC$, $y + \angle A = 90$. Hence, $x + \angle A = y + \angle A$, so $x = y$. Hence, Column A = Column B.

4. **A**  Since the sum of the measures of the angles about a point is 360,
$$b + 3b + 2a + 2b + a + 3a = 360$$
$$6b + 6a = 360$$
$$b + a = \frac{360}{6} = 60$$
Since $60 > 50$, Column A > Column B.

5. **B**  Since an obtuse angle and an acute angle formed by parallel lines are supplementary,
$$a = 180 - 130 = 50$$
and
$$b = 180 - 150 = 30$$
Vertical angles have the same measure, so the angle opposite the angle marked $c°$ also measures $c°$. This angle, together with the angles marked $a°$ and $b°$, forms a straight line. Hence, $a + c + b = 180$ or $50 + c + 30 = 180$, so
$$c = 180 - 80 = 100$$
Since $a + b = 80$ and $c = 100$, Column B > Column A.

6. **C**  The vertical angles at vertices $A$ and $B$ in the interior of $\triangle ABE$ measure $a$ and $b$, respectively. Since the measures of the angles of a triangle add up to 180, $a + b + \angle E = 180$. The vertical angles at vertices $C$ and $D$ in the interior of $\triangle ECD$ measure $x$ and $y$, respectively. Hence:
$$a + b + \angle E = x + y + \angle E$$
or $a + b = x + y$, so Column A = Column B.

*(Grid In)*

1. **20**  The measures of vertical angles are equal, so $\angle EFC = 60$. In right triangle $CEF$, the measures of the acute angles add up to 90, so $\angle ECF + 60 = 90$ or
$$\angle ECF = 90 - 60 = 30$$
Since the measures of acute angles formed by parallel lines are equal, $y + \angle ECF = 50$. Hence, $y + 30 = 50$, so $y = 20$.

2. **35**  In $\triangle ABC$,
$$\angle ACB = 180 - 25 - 45 = 110$$
Since angles $ACB$ and $DCE$ form a straight line,
$$\angle DCE = 180 - 110 = 70$$
Angle $BED$ is an exterior angle of $\triangle ECD$. Hence:
$$3x = 70 + x$$
$$2x = 70$$
$$x = \frac{70}{2} = 35$$

3. **40**  Since $OP$ and $OQ$ are rotating in opposite directions, for the least value of $t$ for which these segments coincide, $4t$ plus $5t$ represents one complete revolution or $360°$. Thus:
$$4t + 5t = 360$$
$$9t = 360$$
$$t = \frac{360}{9} = 40 \text{ seconds}$$

4. **3**  When $t = 1$ hour $= 60$ seconds, $OP$ completes $\frac{60}{4}$ or 15 revolutions while $OQ$ completes $\frac{60}{5}$ or 12 revolutions. Hence, $OP$ completes $15 - 12$ or 3 more revolutions than $OQ$.

## LESSON 6-2  *(Multiple Choice)*

1. **A**  Since $\triangle RST$ is a right triangle,
$$(RS)^2 + (\sqrt{3})^2 = (\sqrt{7})^2$$
$$(RS)^2 + \phantom{(}3\phantom{)} = 7$$
$$RS = \sqrt{4} = 2$$

2. **E**  You are given that the measures of $\angle A$ and $\angle B$ are in the ratio of 2 to 3. Let $2x$ represent the measure of $\angle A$, and $3x$ represent the measure of $\angle B$. Since the measures of the acute angles of a right triangle add up to 90,
$$2x + 3x = 90$$
$$5x = 90$$
$$x = \frac{90}{5} = 18$$

*(Lesson 6-2   Multiple Choice continued)*

Hence, $\angle B = 3x = 3(18) = 54$. In $\triangle CDB$, $CD = BD$ so $\angle DCB = \angle B = 54$. Since the sum of the measures of the three angles of a triangle is 180,

$$x = 180 - 54 - 54 = 72$$

3. **D**  In $\triangle JKL$:

$$\angle L = 180 - 50 - 65 = 65$$

Since $\angle K = 65$ and $\angle L = 65$, then $\angle K = \angle L$, so the sides opposites these angles, $JK$ and $JL$, must be equal in length. Hence:

$$JK = JL$$
$$3x - 2 = x + 10$$
$$2x = 12$$
$$x = \frac{12}{2} = 6$$

4. **B**  If two angles of a triangle each measure 60°, then the third angle also measures 60° because $180 - 60 - 60 = 60$. Since an equiangular triangle is also equilateral, the length of each side of the acute triangle in the figure is 10. Since the hypotenuse of the right triangle in the figure is 10, the lengths of the sides of the right triangle form a $(6, 8, 10)$ Pythagorean triple, where $x = 6$.

5. **D**  Since you are told that $\triangle ABC$ is equilateral, each angle of the triangle measures 60°. Hence:

$$\angle B = x + 2x + x = 4x = 60$$

so $x = \frac{60}{4} = 15$. The angle that measures $y°$ is an exterior angle of $\triangle BEC$, so

$$y = x + \angle C = 15 + 60 = 75$$

6. **D**  If $AC = BC$, then $\angle A = \angle B = 2x + y$. Since the angle that measures $3y - x$ degrees is an exterior angle of $\triangle ABC$,

$$3y - x = (2x + y) + (2x + y)$$
$$= 4x + 2y$$
$$3y - 2y = 4x - x$$
$$y = 3x$$

7. **C**  Since $\angle D = 108$, the measures of angles $DAC$ and $DCA$ must add up to 72 since $180 - 108 = 72$. Since $DA$ bisects $\angle BAC$ and $DC$ bisects $\angle BCA$, the sum of the measures of $\angle BAC$ and $\angle BCA$ is $2 \times 72$ or 144. Hence, $x = 180 - 144 = 36$.

8. **A**  Since the sum of the measures of the angles that have $C$ as a vertex is 180, $\angle ECD = 60$, so $\angle E = 30$. Hence, $CD$, the

side opposite the 30° angle in a 30°-60° right triangle, is $\frac{1}{2} \times 8$ or 4 so $BC = 10 - 4 = 6$. In right triangle $ABC$, $\angle A = 60$. Since $BC$ is the side opposite the 60° angle, it must be $\sqrt{3}$ times as long as $AB$. Hence, $AB = \frac{6}{\sqrt{3}}$ because $\frac{6}{\sqrt{3}} \times \sqrt{3} = 6$. Since $\frac{6}{\sqrt{3}}$ is not one of the choices, simplify:

$$\frac{6}{\sqrt{3}} = \frac{6}{\sqrt{3}} \times \frac{\sqrt{3}}{\sqrt{3}} = \frac{6\sqrt{3}}{3} = 2\sqrt{3}$$

9. **C**  The lengths of the sides of right triangle $AEB$ form a $(5, 12, 13)$ Pythagorean triple, where $BE = 12$. Since $BC \parallel AD$, the distance between these segments must always be the same, so $CF = BE = 12$. In a 45°-45° right triangle, the hypotenuse is $\sqrt{2}$ times the length of either leg. Hence:

$$CD = CF \times \sqrt{2} = 12\sqrt{2}$$

10. **C**  Let the length of $WS$ be any convenient number. If $WS = 1$, then $TS = 1 \times \sqrt{3} = \sqrt{3}$. In right triangle $RST$, $RS$ is the side opposite the 60° angle, so

$$RS = \sqrt{3} \times TS = \sqrt{3} \times \sqrt{3} = 3$$

Hence, $RW = RS - WS = 3 - 1 = 2$. Since $RW = 2$ and $WS = 1$, the ratio of $RW$ to $WS$ is 2 to 1.

*(Quantitative Comparison)*

1. **B**  Form two right triangles by drawing line segment $AD$. Since the lengths of the sides of right triangle $ABD$ form a $(3, 4, 5)$ Pythagorean triple, hypotenuse $AD = 5$. The hypotenuse is always the longest side of a right triangle, so, in right triangle $ACD$, $AD$, or 5, is greater than $CD$. Hence, Column B > Column A.

2. **C**  Angle $ACB$ measures $180 - 110$ or 70°, $\angle ABC$ measures $180 - 140$ or 40°, and $\angle CAB$ measures $180 - 70 - 40$ or 70°. Since angles $ACB$ and $CAB$ have equal measures, the sides opposite these angles have the same lengths, so $AB = BC$. Hence, Column A = Column B.

3. **A**  In an isosceles right triangle, the length of the hypotenuse is $\sqrt{2}$ times the length of either leg, so $x = 5\sqrt{2}$. To compare $5\sqrt{2}$, the value of $x$ in Column A, with 7 in Column B, eliminate the radical by squaring the quantities in both columns. This gives $25(2)$ or 50 in Column A and $7^2$ or 49 in Column B. Hence, Column A > Column B.

*(Lesson 6-2   Quantitative Comparison continued)*

**4. A**  In right triangle $ACB$, angle $A$ measures $90 - 30$ or $60°$, so, $x$, the side opposite $\angle A$ is one-half the hypotenuse times $\sqrt{3}$:

$$x = \frac{1}{2}(20)(\sqrt{3}) = 10\sqrt{3}$$

To compare $10\sqrt{3}$, the value of $x$, in Column A with 16 in Column B, eliminate the radical by squaring both columns. This gives $100(3)$ or 300 in Column A and $16^2$ or 256 in Column B. Hence, Column A > Column B.

**5. A**  Since $\triangle RST$ is equilateral, it is also equiangular, so $\angle R = 60$ and $h$, the side opposite the $60°$ angle in the right triangle, equals $\frac{1}{2} \times 8 \times \sqrt{3}$ or $4\sqrt{3}$. To compare $4\sqrt{3}$, the value of $h$, in Column A with 6 in Column B, eliminate the radical by squaring both columns. This gives $16(3)$ or 48 in Column A and $6^2$ or 36 in Column B. Hence, Column A > Column B.

**6. C**  In the right triangle, the side opposite the angle marked $x°$ is one-half of the length of the hypotenuse. Hence, $x = 30$, so $y = 90 - 30 = 60$. In Column A replace $y$ with 60, and in Column B replace $x + 30$ with $30 + 30$ or 60. Hence, Column A = Column B.

**7. B**  Since $\frac{PR}{PQ} = 1$, then $PR = PQ$; also, since $\frac{RQ}{PR} = 1$, then $RQ = PR$. Hence, $\triangle PQR$ is equilateral since $PR = PQ = RQ$. An equilateral triangle is also equiangular, so each angle of $\triangle PQR$ measures $60°$. Since the degree measure of $\angle P$ is 60, Column B > Column A.

**8. C**  Since $x + y + z = 180$ and $z = x + y$,

$$z + z = 180$$
$$2z = 180$$
$$z = \frac{180}{2} = 90$$

Hence, side $h$ is the hypotenuse of a right triangle in which the lengths of the other two sides are $k$ and $p$. Using the Pythagorean theorem gives $h^2 = k^2 + p^2$, so $h = \sqrt{k^2 + p^2}$. Hence, Column A = Column B.

**9. C**  Angle $ACB$ measures $30 + 60$ or $90°$. Since $BC$, the side opposite $\angle A$, is $\sqrt{3}$ times the length of the shorter leg of the right triangle, $\angle A$ measures $60°$. Triangle

$DCA$ is equiangular, so it is also equilateral. Hence:

$$DA = DC = CA = 1$$

Since $\angle B = \angle DCB = 30$, then $BD = DC = 1$, so $AD = AB$. Hence, Column A = Column B.

**10. B**  Since $AD = DE = EC = 1$, then $AC = 1 + 1 + 1 = 3$. You are also given that $BC = 3$, so $AC = BC$. Right triangle $ACB$ is isosceles, so $\angle A = \angle ABC$. Thus, $a = x + y + \angle ABE$. Since the degree measure of $\angle ABE$ is some positive number, deleting it from the right side of the equation $a = x + y + \angle ABE$ makes $a > x + y$. Hence, Column B > Column A.

*(Grid In)*

**1.  1.5**  Angle $B$ measures $15 + 30 + 15$ or $60°$, so the sum of the measures of angles $A$ and $C$ is 120. Since $AB = BC = 10$, then $\angle A = \angle C = 60$, so $\triangle ABC$ is equiangular. A triangle that is equiangular is also equilateral, so $AC = 10$. Angles $BDE$ and $BED$ each measure $60 + 15$ or $75°$ since they are exterior angles of triangles $ADB$ and $CEB$. Hence triangles $ADB$ and $CEB$ have the same shape and size, so $AD = CE$. Since you are given that $DE = 7$, then $AD + CE = 3$, so $AD = 1.5$.

**2.  95**

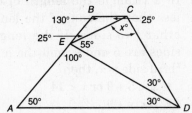

The measures of acute angle $A$ and obtuse angle $B$ must add up to 180 since the angles are formed by parallel lines. Since $\angle A = 50$, then $\angle B = 130$. You are told that $BE = BC$, so $\angle BEC = \angle BCE$. In $\triangle BEC$, the measures of angles $BEC$ and $BCE$ must add up to 50, so $\angle BEC = \angle BCE = 25$. Since the three adjacent angles about point $E$ form a straight angle, $25 + 55 + \angle AED = 180$, so $\angle AED = 100$. In $\triangle AED$,

$$\angle ADE = 180 - 100 - 50 = 30$$

Since you are also told that $ED$ bisects $\angle ADC$, $\angle CDE = \angle ADE = 30$. In $\triangle CED$,

$$x = 180 - 55 - 30 = 95$$

**LESSON 6-3** *(Multiple Choice)*

1. **E** In obtuse triangle $RST$, $RT = TS$. If the obtuse angle were opposite one of the equal sides, there would be another obtuse angle opposite the other equal side, an impossible case. Hence, the obtuse angle must be opposite $RS$.

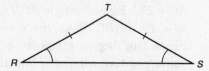

Determine whether each Roman numeral statement is always true.

- I. Since $RT = TS$, the angles opposite these sides are equal, so $\angle R = \angle S$. Hence, statement I is always true.
- II. The obtuse angle must be the angle opposite the unequal side. Since $\angle T$ is opposite side $RS$, $\angle T$ is obtuse. Hence, statement II is always true.
- III. Since $RS$ is opposite obtuse angle $T$, it is the longest side of the triangle. Hence, $RS > TS$ and statement III is always true.

Roman numeral statements I, II, and III are always true.

2. **D** In a triangle the length of any side is less than the sum of the lengths of the other two sides. If the lengths of two sides are 5 and 9, and the length of the third side is $x$, then
- $x < 5 + 9$ or $x < 14$
- $5 < x + 9$
- $9 < x + 5$ or $4 < x$

Since $x < 14$ and $4 < x$, $4 < x < 14$.

3. **B** You are given that, in $\triangle ABC$, $BC > AB$ and $AC < AB$. Hence, $BC > AB > AC$. The measures of the angles opposite these sides are ordered in the same way. Hence, $\angle A$ (opposite $BC$) $> \angle C$ (opposite $AB$) $> \angle B$ (opposite $AC$)

4. **A** In $\triangle ABC$, if $AB = BD$, then $\angle A = x$. Determine whether each Roman numeral statement is always true.
- I. You are not given any information that tells you whether $x > z$, $x < z$, or $x = z$. Statement I may or may not be true.

- II. Since an exterior angle of a triangle is greater than either nonadjacent interior angle of the triangle, $y > \angle A$. Since $\angle A = x$, then $y > x$, so statement II is always true.
- III. Angle $x$ is an exterior angle of $\triangle CBD$, so $x > \angle C$. Since $\angle A = x$, then $\angle A > \angle C$, so $BC$ (the side opposite $\angle A$) $> AB$ (the side opposite $\angle C$. Since it is not true that $AB > BC$, statement III is false. Hence, only Roman numeral statement II is always true.

5. **A** Since $n$ is an integer and $3 < n < 8$, $n$ may be equal to 4, 5, 6, and 7. Test whether a triangle can be formed for each possible value of $n$.
- If $n = 4$, then $4 < 3 + 8$ and $3 < 4 + 8$, but 8 is not less than $3 + 4$, so $n \neq 4$.
- If $n = 5$, then $5 < 3 + 8$ and $3 < 5 + 8$, but 8 is not less than $3 + 5$, so $n \neq 5$.
- If $n = 6$, then $6 < 3 + 8$, $3 < 6 + 8$, *and* $8 < 3 + 6$, so $n$ may be equal to 6.
- If $n = 7$, then $7 < 3 + 8$, $3 < 7 + 8$, *and* $8 < 3 + 7$, so $n$ may be equal to 7.

Hence, two different triangles are possible.

6. **C** Draw $\triangle RST$ so that $RS \perp TS$, $\angle T$ measures 40°, and $W$ is a point on side $RT$ such that $\angle RWS$ measures 100°.

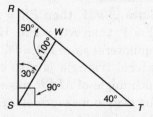

- In $\triangle RWS$, $\angle RSW = 180 - 50 - 100 = 30$. Since $\angle RSW$ is the smallest angle of $\triangle RWS$, then $RW < RS$ and $RW < SW$. Hence, eliminate choices (A) and (E).
- Since $\angle T < \angle R$, $RS < ST$. Since $RW < RS$ and $RS < ST$, then $RW < ST$, so you can also eliminate choice (B).
- Since $\angle TSW = 90 - 30 = 60$, then, $\angle T < \angle TSW$. Hence, in $\triangle TSW$, $SW < TW$. Since $RW < SW$ and $SW < TW$, then $RW < TW$ so you can also eliminate choice (D). Hence, $RW$ is the shortest segment.

*(Lesson 6-3 Quantitative Comparison)*

1. **B** In $\triangle RST$, $\angle S = 180 - 55 - 60 = 65$. Since $\angle S$ is the largest angle of the triangle, the longest side is $RT$. Since $RT > RS$, Column B > Column A.

2. **B** Angle $BDA$ is an exterior angle of $\triangle BDC$, so $\angle ADB > \angle C$. Since $\triangle ABC$ is equilateral, it is also equiangular, so $\angle C = \angle B = \angle A$. Hence, $\angle ADB > \angle A$, so $AB > BD$. Since $AB = BC = AC$, substitute $AC$ for $AB$ in $AB > BD$. Then $AC > BD$, so Column B > Column A.

3. **A** The lengths of the sides of right triangle $ABC$ form a (6, 8, 10) right triangle, where $BC = 8$. Since $BC > AB$, $\angle A > \angle C$, so Column A > Column B.

4. **B** Since an acute angle and an obtuse angle formed by parallel lines are supplementary, the exterior angle of $\triangle ABC$ at vertex $C$ measures $180 - x°$. The measure of an exterior angle of a triangle is greater than the measure of either nonadjacent interior angle, so $180 - x > y$. Adding $x$ and subtracting $y$ on both sides of this inequality gives $180 - y > x$. Hence, Column B > Column A.

5. **A** Since the length of any side of a triangle must be less than the sum of the lengths of the other two sides, $7 < 5 + x$ or $2 < x$. Since $x > 2$ Column A > Column B.

6. **D** Since the measure of only one of the three angles of $\triangle ABC$ is given, pick easy numbers for the measure of $\angle A$ or angle $\angle C$.
   - If $\angle A = 60$, then, in $\triangle ADB$, $\angle A > \angle ABD$, so $BD > AD$. In $\triangle BDC$, $\angle C = 35$, so $\angle DBC > \angle C$ and $DC > BD$. Since $DC > BD$ and $BD > AD$, then $DC > AD$.
   - If $\angle C = 60$, then, in $\triangle BDC$, $\angle C > \angle DBC$, so $BD > DC$. In $\triangle ADB$, $\angle A = 35$, so $\angle ABD > \angle A$ and $AD > BD$. Since $AD > BD$ and $BD > DC$, then $AD > DC$.

   Since two different answers are possible, the correct choice is (D).

7. **B** Since the lengths of the three sides of $\triangle ABC$ are unequal, the measures of the three angles of the triangle are all different. The shortest side of the triangle is $BC$, so the smallest angle measure is $a°$. Assume $a \geq 60$. Then the measures of angles $B$ and $C$ must sum to 120 or a smaller number. Since angles $B$ and $C$ have unequal measures, at least one of these two angles must be less than 60, an impossibility since $a$ is the smallest angle measure of the triangle. Since it cannot be the case that $a \geq 60$, then $60 > a$, so Column B > Column A.

*(Grid In)*

1. **37** If the lengths of two sides of an isosceles triangle are 7 and 15, then the third side must be 7 or 15. Since 15 is not less than $7 + 7$, the third side cannot be 7. Hence, the lengths of the three sides of the triangle must be 7, 15, and 15. The perimeter of this triangle is $7 + 15 + 15$ or 37.

2. **3** You are told that the perimeter of a triangle with integer side lengths is 21 and that the length of one side is 8. Hence, the sum of the lengths of the other two sides must be 13. Test possible combinations of integer side lengths to find one in which the lengths of two of the sides add up to 13 and the third side is 8.
   - Suppose the lengths of the three sides are 1, 12, and 8. This combination is not possible since 12 is not less than $1 + 8$.
   - Suppose the lengths of the three sides are 2, 11, and 8. This combination is not possible since 11 is not less than $2 + 8$.
   - Suppose the lengths of the three sides are 3, 10, and 8. This combination is possible since $3 < 10 + 8$, $10 < 3 + 8$, and $8 < 3 + 10$.

   Hence, the shortest possible length of a side is 3.

### LESSON 6-4 *(Multiple Choice)*

1. **C**

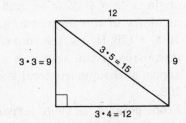

*(Lesson 6-4   Multiple Choice continued)*

The diagonal of a 9 by 12 rectangle is the hypotenuse of a 9-12-*15* right triangle. Since the two diagonals of a rectangle have the same length, the sum of the lengths of the diagonals of a 9 by 12 rectangle is 15 + 15 or 30.

**2. D**

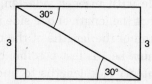

The diagonal of a square is the hypotenuse of a 45°-45° right triangle, so the length of the diagonal is √2 times the length of a side. Since the length of a side of the square is √2, the length of a diagonal is $\sqrt{2} \times \sqrt{2}$ or 2.

**3. E**

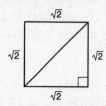

If a diagonal of a rectangle forms a 30° angle with each of the longer sides of the rectangle, the shorter side is the leg of a right triangle that is opposite the 30° angle. Since the length of the shorter side is 3, the length of the diagonal, or hypotenuse of the right triangle, is 2 times 3 or 6.

**4. C**  The sum of the measures of the four angles of a quadrilateral is 360. If the degree measures of the angles of a quadrilateral are $4x, 7x, 9x$, and $10x$, then

$$4x + 7x + 9x + 10x = 360$$
$$30x = 360$$
$$x = \frac{360}{30} = 12$$

The degree measure of the smallest angle is $4x = 4(12) = 48$, and the degree measure of the largest angle is $10x = 10(2) = 120$. Hence, the sum of the degree measures of the smallest and largest angles of the quadrilateral is 48 + 120 or 168.

**5. E**  From point *A*, draw a perpendicular to side *CD*. Call the point of intersection *E*.

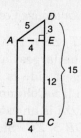

Since *ABCE* is a rectangle, *AE* = *BC* = 4. The lengths of the sides of right triangle *AED* form a (*3-4-5*) Pythagorean triple in which *DE* = 3, so

$$AB = EC = 15 - 3 = 12$$

**6. E**  Since the right angles and the vertical angles at *E* are equal, triangles *AEC* and *BED* are similar. The lengths of corresponding sides of similar triangles are in proportion. Let *x* represent the length of *AE*. Then

$$\frac{AC}{DB} = \frac{AE}{BE}$$
$$\frac{3}{4} = \frac{x}{14-x}$$
$$4x = 3(14 - x)$$
$$= 42 - 3x$$
$$7x = 42$$
$$x = \frac{42}{7} = 6$$

Hence, *AE* = 6.

**7. D**   At each vertex of a polygon, an interior angle is supplementary to an exterior angle drawn at that vertex. In an equiangular polygon, all the exterior angles have the same measures. Dividing 360, the sum of the degree measures of the exterior angles of a polygon, by the measure of an exterior angle gives the number of sides of that polygon. Test each answer choice until you find an exterior angle that does *not* divide evenly into 360.

- (A): The measure of an exterior angle is $180 - 165$ or 15 and $\frac{360}{15} = 24$, so the polygon has 24 sides.
- (B): The measure of an exterior angle is $180 - 162$ or 18 and $\frac{360}{18} = 20$, so the polygon has 20 sides.
- (C): The measure of an exterior angle is $180 - 140$ or 40 and $\frac{360}{40} = 9$, so the polygon has 9 sides.

*(Lesson 6-4   Multiple Choice continued)*

- (D): The measure of an exterior angle is $180 - 125$ or 55, but $\frac{360}{55}$ does not give an integer answer.

Hence, a polygon in which each interior angle measures 125 is not possible.

8. **A** Since consecutive angles of a parallelogram are supplementary,
$$3x + 2x = 180$$
$$5x = 180$$
$$x = \frac{180}{5} = 36$$
Since opposite angles of a parallelogram are equal,
$$y = 3x = 3(36) = 108$$

9. **D** Since side $AC$ of the rotated square cannot have the same length as a diagonal of the original square that contains $AC$ as a side, Figure II is the only figure that can *not* represent a rotation of square $ABCD$ about point $A$.

10. **A** If $S = 540t$ and $S = (n - 2)180$, then
$$(n - 2)180 = 540t$$
Dividing both sides of the equation by 180 gives
$$n - 2 = \frac{540t}{180} = 3t$$
Since $n - 2 = 3t$, then $n = 3t + 2$.

11. **D** In any polygon the sum of the degree measures of the exterior angles is 360. If the sum of the degree measures of the interior angles of a polygon is 4 times the sum of the degree measures of the exterior angles, then
$$S = (n - 2)180 = 4 \times 360$$
or $(n - 2)180 = 1440$. Dividing both sides of the equation by 180 gives
$$(n - 2) = \frac{1440}{180} = 8$$
Since $n - 2 = 8$, then $n = 8 + 2$ or 10.

12. **E** You are given that, in quadrilateral $ABCD$, $\angle A + \angle C$ is 2 times $\angle B + \angle D$, and $\angle A = 40$, so
$$40 + \angle C = 2(\angle B + \angle D)$$
Since the sum of the degree measures of the four angles of a quadrilateral is 360,
$$40 + \angle C + (\angle B + \angle D) = 360$$
Substituting $2(\angle B + \angle D)$ for $40 + \angle C$ gives
$$2(\angle B + \angle D) + (\angle B + \angle D) = 360$$

Since the last equation contains two unknowns, the measures of angles $B$ and $D$, it is not possible to find the measure of $\angle B$ (or $\angle D$).

13. **C** Determine whether each Roman numeral statement is true, given that $ABCD$ is a parallelogram and $AB > BD$.
- I. Since opposite sides of a parallelogram have the same length, $CD = AB$. Substituting $CD$ for $AB$ in $AB > BD$ gives $CD > BD$. Hence, statement I is false.
- II. Since $AB > BD$, the measures of the angles opposite these sides have the same size relationship. Thus, $\angle ADB > \angle A$. Since opposite angles of a parallelogram are equal, $\angle A = \angle C$, so $\angle ADB > \angle C$. Hence, statement II is true.
- III. Acute angles formed by parallel lines are equal, so $\angle CBD = \angle ADB$. Since $\angle ADB > \angle A$, then $\angle CBD > \angle A$, so statement III is true.

Only Roman numeral statements II and III must be true.

14. **D** Right triangles $CDE$ and $BAE$ are similar, so the lengths of corresponding sides are in proportion. Since $CD = 1$ and $AB = 2$, each side of $\triangle BAE$ is 2 times the length of the corresponding side of $\triangle CDE$. Since $AD = 6$, it must be the case that $AE = 4$ and $DE = 2$. Since $BC = CE + BE$, use the Pythagorean theorem to find the lengths of $CE$ and $BE$.
- In right triangle $CDE$,
$$(CE)^2 = 1^2 + 2^2 = 1 + 4 = 5$$
so $CE = \sqrt{5}$.
- In right triangle $BAE$,
$$(BE)^2 = 4^2 + 2^2 = 16 + 4 = 20$$
so $BE = \sqrt{20} = \sqrt{4} \cdot \sqrt{5} = 25$.

Hence,
$$BC = CE + BE = \sqrt{5} + 2\sqrt{5} = 3\sqrt{5}$$

15. **C** The sum of the degree measures of the three adjacent angles at each of the marked vertices is equivalent to the sum of the measures of four straight angles, which equals $4 \times 180$ or 720°. Since the sum of 720° includes the sum of the measures of the four angles of the inscribed quadrilateral, which is 360, the sum of the degree measures of only the marked angles is $720 - 360$ or 360.

*(Lesson 6-4   Quantitative Comparison)*

1. **A**  In a square, the diagonal is the hypotenuse of a 45°-45° right triangle, so its length is $\sqrt{2}$ times the length of a side of the square. Hence, in Column A, the length of a diagonal is $10\sqrt{2}$. In Column B, the diagonal of a 6 by 8 rectangle forms a 6-8-*10* right triangle in which the length of the diagonal is 10. Since $10\sqrt{2} > 10$, Column A > Column B.

2. **B**  Since opposite sides of a parallelogram are parallel and equal in length, $\angle CBD = a$ and $BC = AD = 7$. Hence, in $\triangle BCD$, $BC > CD$. Since the measures of the angles opposite these sides have the same size relationship, $b > a$. Hence, Column B > Column A.

3. **B**  Since the smaller right triangle is similar to the larger right triangle, the lengths of corresponding sides of these two triangles are in proportion. Thus:

$$\frac{\text{side smaller triangle}}{\text{corresponding side larger triangle}} = \frac{x}{y} = \frac{9}{9+18}$$

Since $\frac{x}{y} = \frac{9}{27} = \frac{1}{3}$ and $\frac{1}{2} > \frac{1}{3}$, Column B > Column A.

4. **C**  In $\triangle DCE$, $y + z + \angle C = 180$. Since $ABCD$ is a parallelogram, consecutive angles are supplementary, so $x + \angle C = 180$. Replacing 180 in the equation $y + z + \angle C = 180$ with $x + \angle C$ gives

$$y + z + \angle C = x + \angle C$$

so $y + z = x$.
Hence, Column A = Column B

5. **B**  When the diagonal opposite the 60° angle of a rhombus is drawn, it becomes the base of an isosceles triangle in which the equal legs are adjacent sides of the rhombus.

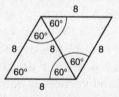

The vertex angle measures 60°, so the sum of the measures of the other two angles must be 120°. Since the measures of the base angles must be equal, each base angle measures 60°, so the triangle is equiangular. Since an equiangular triangle is also equilateral, the length of

the diagonal in Column A is 8. In Column B, the length of a diagonal of a square of side length 8 is 8 times $\sqrt{2}$ or $8\sqrt{2}$. Since $8\sqrt{2} > 8$, Column B > Column A.

6. **C**  In Column A, angles $w$, $x$, and $y$ are exterior angles of the triangle, so $w + x + y = 360$. In Column B, $a + b + c + d = 360$ since the sum of the measures of the angles of a quadrilateral is 360. Hence,

$$w + x + y = a + b + c + d = 360$$

so Column A = Column B.

7. **B**  In Column A, since the length of a diagonal is $\sqrt{2}$ times the length of a side, $x = \sqrt{2}$. In Column B, each angle of an equilateral triangle measures 60, so

$$y = \frac{1}{2}(2)(\sqrt{3}) = \sqrt{3}$$

Since $\sqrt{3} > \sqrt{2}$, Column B > Column A.

## LESSON 6-5   *(Multiple Choice)*

1. **B**  If the area of a square is 25, then the length of a side of that square is $\sqrt{25}$ or 5. The perimeter of the square is $4 \times 5$ or 20.

2. **D**  The height of parallelogram $ABCD$ is the length of perpendicular segment $BE$. The length of base $AD$ is $1 + 5$ or 6. In right triangle $AEB$:

$$(BE)^2 + 1^2 = 2^2$$

or $(BE)^2 + 1 = 4$, so $BE = \sqrt{3}$. Hence:

$$\text{area parallelogram } ABCD = BE \times AD$$
$$= \sqrt{3} \times 6 \text{ or } 6\sqrt{3}$$

3. **D**  If the perimeter of the rectangle is 44, then 2 times the length plus 2 times the width is 44. Hence:

$$2(2x + 1) + 2(x - 3) = 44$$
$$4x + 2 + 2x - 6 = 44$$
$$6x - 4 = 44$$
$$6x = 48$$
$$x = \frac{48}{6} = 8$$

4. **A**  Since the area of the rectangle is 65,

$$(x + 4)(x - 4) = 65$$
$$x^2 - 16 = 65$$
$$x^2 = 81$$
$$x = \sqrt{81} = 9$$

5. **B**  The area of a square is one-half the product of the lengths of its equal diagonals. If the length of a diagonal of a square is $\sqrt{2}$, the area of the square is $\frac{1}{2}(\sqrt{2})(\sqrt{2})$ or $\frac{1}{2}(2)$, which equals 1.

*(Lesson 6-5   Multiple Choice continued)*

6. **C**  To figure out the area of quadrilateral *ABCD*, add the areas of right triangles *DAB* and *DBC*.
   - The area of right triangle *DAB* is $\frac{1}{2} \times 3 \times 4$ or 6.
   - The lengths of the sides of right triangle *DAB* form a (3, 4, *5*) Pythagorean triple in which hypotenuse *BD* = 5. The lengths of the sides of right triangle *DBC* form a (5, *12*, 13) Pythagorean triple in which *BC* = 12. Hence, the area of right triangle *DBC* is $\frac{1}{2} \times 5 \times 12$ or 30.

   The area of quadrilateral *ABCD* is 6 + 30 or 36.

7. **C**  Since the area of rectangle *GDFK* = *KF* × *DF*, you need to find the lengths of line segments *DF* and *KF*.
   - The area of square *ABCD* is 121, so
     $$AB = BC = \sqrt{121} = 11$$
   Since the area of *FKHC* is 6 and the area of *EKHB* is 16, the area of rectangle *FEBC* is 6 + 16 or 22. Also, the area of rectangle *FEBC* = *BC* × *FC* = 22, where *BC* = 11, so *FC* = 2. Hence:
   $$DF = DC - FC = 11 - 2 = 9$$
   - Since the area of rectangle *FKHC* = *FC* × *KF* = 6 and *FC* = 2, then *KF* = 3.

   Area of rectangle *GDFK* = *KF* × *DF* = 3 × 9 or 27.

8. **A**  Since the perimeter of the triangle is 18, the length of the third side is 18 − 7 − 4 or 7. Draw a segment perpendicular to the shortest side from the opposite vertex. In an isosceles triangle, the perpendicular drawn to the base bisects the base.

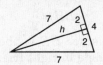

   Hence, if *h* represents the height, then
   $$h^2 + 2^2 = 7^2$$
   $$h^2 = 49 - 4$$
   $$= 45$$
   $$h = \sqrt{45} = \sqrt{9} \cdot \sqrt{5} = 3\sqrt{5}$$
   Since the base is 4, the area of the triangle is $\frac{1}{2} \times 4 \times 3\sqrt{5} = 6\sqrt{5}$.

9. **D**  If the area of square *ABCD* is 64, then the length of each side is $\sqrt{64}$ or 8. Since *BC* = 8, the length of each side of equilateral triangle *BEC* is 8. Hence:

$$\text{area equilateral } \triangle BEC = \frac{(\text{side})^2}{4} \times \sqrt{3}$$
$$= \frac{(8)^2}{4} \times \sqrt{3}$$
$$= \frac{64}{4} \times \sqrt{3}$$
$$= 16\sqrt{3}$$

10. **B**  If *x* represents the lengths of the equal legs of an isosceles triangle whose area is 8, then

$$\frac{1}{2}(x)(x) = 8$$
$$x^2 = 16$$
$$x = \sqrt{16} = 4$$

In an isosceles (45°-45°) right triangle, the length of the hypotenuse is $\sqrt{2}$ times the length of a leg. Since the length of each leg is 4, the length of the hypotenuse is $4\sqrt{2}$. The perimeter of the triangle is $4 + 4 + 4\sqrt{2}$ or $8 + 4\sqrt{2}$.

11. **C**  Since the lengths of the sides of △*ABC* are consecutive integers, let *x*, *x* + 1, and *x* + 2 represent the lengths of the three sides. Since the perimeter of △*ABC* has the same perimeter as an equilateral triangle with a side length of 9,

$$x + (x + 1) + (x + 2) = 9 + 9 + 9$$
$$3x + 3 = 27$$
$$3x = 24$$

Then *x*, the length of the shortest side of △*ABC*, is $\frac{24}{3}$ or 8.

12. **B**  Draw *DE* ⊥ *AB*.

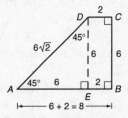

Since quadrilateral *DEBC* is a rectangle, *EB* = *DC* = 2 and *DE* = *CD* = 6. In isosceles right triangle △*AED*, *AE* = *DE* = 6, so
$$AB = AE + EB = 6 + 2 = 8$$
Also, hypotenuse *AD* = $6\sqrt{2}$. Hence:
$$\text{perimeter } ABCD = AD + DC + BC + AB$$
$$= 6\sqrt{2} + 2 + 6 + 8$$
$$= 16 + 6\sqrt{2}$$

*(Lesson 6-5   Multiple Choice continued)*

**13. C**

$$\underset{\text{quadrilateral } ABCD}{\text{area}} = \underset{\text{rectangle } DEBC}{\text{area}} + \underset{\triangle AED}{\text{area}}$$

$$= (2 \times 6) \qquad + \left(\frac{1}{2} \times 6 \times 6\right)$$

$$= 12 \qquad\qquad + 18$$

$$= 30$$

**14. A**   Since the ratio of $AD$ to $DC$ is 3 to 2, $DC$ is $\frac{2}{3+2}$ or $\frac{2}{5}$ of $AC$. The base of $\triangle BDC$ is $\frac{2}{5}$ of the base of $\triangle ABC$, and the heights of triangles $ABC$ and $DBC$ are the same. Hence, the area of $\triangle BDC$ is $\frac{2}{5}$ of the area of $\triangle ABC$ so

$$\text{area } \triangle BDC = \frac{2}{5} \times 40 = 16$$

**15. B**   If the lengths of the adjacent sides of a rectangle are $x$ and $y$, where $x > y$, then the perimeter of the rectangle is $2x + 2y$. The perimeter is 8 times as great as the shorter side of the rectangle, so

$$2x + 2y = 8y$$
$$2x = 6y$$
$$\frac{2}{6} = \frac{y}{x}$$

Since $\frac{2}{6} = \frac{1}{3}$, the ratio of $y$ to $x$ is $1 : 3$.

**16. B**   Since the longer side of a rectangle whose perimeter is 24 is 2 times the length of the shorter side,

$$2x + x + 2x + x = 24$$

where $x$ is the length of the shorter side. Hence, $6x = 24$ or $x = \frac{24}{6} = 4$. Since the shorter side is 4, the longer side is $2 \times 4$ or 8, so the area of the rectangle is $4 \times 8$ or 32. If a square has the same area as this rectangle, then the length of a side of this square is

$$\sqrt{32} = \sqrt{16} \cdot \sqrt{2} = 4\sqrt{2}$$

The perimeter of the square is $4 \times 4\sqrt{2}$ or $16\sqrt{2}$.

**17. E**   The area of right triangle $BEC$ is 8, so $8 = \frac{1}{2} \times BE \times EC$ or $16 = BE \times EC$. You are given that $AB = BE = EC$, so

$$AB = BE = EC = 4$$

In isosceles right triangle $BEC$, hypotenuse $BC = 4\sqrt{2}$. Since opposite sides of a parallelogram have the same length,

$$CD = AB = 4$$

and

$$AD = BC = 4\sqrt{2}$$

Then

perimeter $ABECD = AB + BE + EC + CD + AD$

$$= 4 + 4 + 4 + 4 + 4\sqrt{2}$$
$$= 16 + 4\sqrt{2}$$

**18. C**   You are given that the area of square $PQRS$ is 3, so the length of each side of the square is $\sqrt{3}$. You also know that $\triangle ABC$ is equilateral, so $\angle A = \angle B = \angle C = 60$. Triangle $APQ$ is a 30°-60° right triangle in which the length of the longer leg is $\sqrt{3}$. Since the length of the longer leg is always $\sqrt{3}$ times the length of the shorter leg, $AP = 1$ since $PQ = 1 \times \sqrt{3} = \sqrt{3}$. Similarly, in right triangle $RSC$, $SC = AP = 1$. Hence:

$$AC = AP + PS + SC$$
$$= 1 + \sqrt{3} + 1$$
$$= 2 + \sqrt{3}$$

Since $\triangle ABC$ is equilateral, its perimeter is 3 times the length of any side. Hence, the perimeter of $\triangle ABC$ is $3(2 + \sqrt{3})$ or $6 + 3\sqrt{3}$.

**19. B**   You are given that $AB = BE = 8$ and $\angle A = 60$. Since $\triangle ABE$ is isosceles, $\angle E = 60$, so

$$\angle B = 180 - 60 - 60 = 60$$

Since $\triangle ABE$ is equiangular, it is also equilateral, $AE = 8$ and $ED = 10 - 8 = 2$.

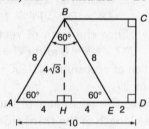

From $B$ draw a perpendicular to $AD$, intersecting it at point $H$. Since $BH$ is the side opposite the 60° angle in right triangle $BHA$,

$$BH = \frac{1}{2} \times 8 \times \sqrt{3} = 4\sqrt{3}$$

Also, since $HE$ is the side opposite the 30° angle in right triangle $BHE$,

$$HE = \frac{1}{2} \times 8 = 4$$

so $HD = 4 + 2 = 6$.

$$\underset{\text{quadrilateral } BCDE}{\text{area}} = \underset{\text{rectangle } BHDC}{\text{area}} - \underset{\text{rt. } \triangle BHE}{\text{area}}$$

$$= BH \times HD \qquad - \frac{1}{2} \times BH \times HE$$

$$= 4\sqrt{3} \times 6 \qquad - \frac{1}{2} \times 4\sqrt{3} \times 4$$

$$= 24\sqrt{3} \qquad\qquad - 8\sqrt{3}$$

$$= 16\sqrt{3}$$

*(Lesson 6-5  Multiple Choice continued)*

20. **E**  Suppose the length of a side of the original square is 10; then the area of that square is $10 \times 10$ or 100. Since one pair of opposite sides of the square are increased in length by 20%, and 20% of 10 is 2, the length of that pair of opposite sides of the new rectangle is $10 + 2$ or 12. Since the other pair of sides are increased in length by 50%, and 50% of 10 is 5, the length of the other pair of opposite sides of the new rectangle is $10 + 5$ or 15. The area of the new rectangle formed is $12 \times 15$ or 180, which is 80% greater than 100, the area of the original square.

*(Quantitative Comparison)*

1. **D**  An altitude dropped from $B$ to side $AD$ is opposite the 60° angle in a right triangle, so its length is equal to $\frac{1}{2} \times 4 \times \sqrt{3}$ or $2\sqrt{3}$. However, since you do not know the length of $AD$, you can't complete the area calculation.

    Since it is impossible to tell how the area of parallelogram $ABCD$ in Column A compares with 15 in Column B, the correct choice is (D).

2. **A**

    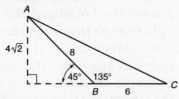

    In Column A, when the altitude from point $A$ is drawn, it intersects side $BC$ extended and becomes a leg of a 45°-45° right triangle. Hence, the length of the altitude is $\frac{1}{2} \times 8 \times \sqrt{2}$ or $4\sqrt{2}$.

    $$\text{area } \triangle ABC = \frac{1}{2} \times BC \times \text{height}$$
    $$= \frac{1}{2} \times 6 \times 4\sqrt{2}$$
    $$= 12\sqrt{2}$$

    In rectangle $QRST$ in Column B, the lengths of the sides of right triangle $RST$ form a (3-4-5) Pythagorean triple where $TS = 3$, so the area of rectangle $QRST$ is $3 \times 4$ or 12. Since $12\sqrt{2} > 12$, Column A > Column B.

3. **A**  In an isosceles right triangle, the length of either leg is $\frac{1}{2}$ times the length of the hypotenuse times $\sqrt{2}$. Hence, $BC = \frac{5}{2}\sqrt{2}$,

so the perimeter of square $ABCD$ in Column A is 4 times the length of a side or

$$4\left(\frac{5}{2}\sqrt{2}\right) = 10\sqrt{2}$$

The perimeter of rectangle $RSTW$ in Column B is 2 times the length plus 2 times the width. The perimeter of the rectangle is 10 since

$$2\left(\frac{7}{2}\right) + 2\left(\frac{3}{2}\right) = 7 + 3 = 10$$

Since $10\sqrt{2} > 10$, Column A > Column B.

4. **A**  Since $ABCD$ is a rectangle, its opposite sides have the same length. Hence, $AB = CD = 6$ and $AD = BC = 11$, so $AE = 11 - 3 = 8$. Since the lengths of the legs of right triangle $FDE$ form a (3-4-5) Pythagorean triple, $EF = 5$. Hence, the perimeter of the shaded region is $AB + BC + CF + FE + AE$ or

$$6 + 11 + 2 + 5 + 8 = 32$$

Since 32 in Column A > 30 in Column B, Column A > Column B.

5. **C**  In Column A,

    $$\text{area } \triangle CDE = \frac{1}{2}(x)(x) = \frac{1}{2}x^2 = 0.50x^2$$

    Since $\triangle CED$ is an isosceles right triangle, the length of hypotenuse $CD$ is $x\sqrt{2}$, so
    $$\text{area square } ABCE = (x\sqrt{2})(x\sqrt{2}) = 2x^2$$
    In Column B, 0.25 times the area of square $ABCD$ is $(0.25)(2x^2)$ or $0.50x^2$. Hence, Column A = Column B.

6. **C**  If the width of a rectangle is 8 and its perimeter is 24, then $24 = (2 \times 8) + (2 \times \text{length})$ or $8 = 2 \times \text{length}$, so the length of the rectangle is 4 and in Column A the area of the rectangle is $4 \times 8$ or 32.
    The area of a square is one-half the product of the lengths of its equal diagonals. The area of a square with diagonal 8 in Column B is $\frac{1}{2} \times 8 \times 8$ or 32. Hence, Column A = Column B.

7. **B**  In Column A:
    $$\text{area rectangle } PQRS = x(x + 2)$$
    $$= x^2 + 2x$$
    In Column B:
    $$\text{area square } ABCD = (x + 1)(x + 1)$$
    $$= x^2 + x \cdot 1 + 1 \cdot x + 1$$
    $$= x^2 + 2x + 1$$
    For all positive values of $x$, $x^2 + 2x + 1$ exceeds $x^2 + 2x$ by 1, so Column B > Column A.

*(Lesson 6-5   Grid In)*

1. **42**  First find the sum of the areas of the four walls:

$$12 \times 8 = 96$$
$$12 \times 8 = 96$$
$$16 \times 8 = 128$$
$$16 \times 8 = 128$$
$$\overline{\text{Sum of areas} = 448}$$

Since 1 gallon of paint provides coverage of an area of at most 150 square feet and $\frac{448}{150} = 2.9...$, a minimum of 3 gallons of paint is needed. The paint costs $14 per gallon, so the minimum cost of the paint needed is $3 \times \$14$ or $42.

2. **1/9**  Since the figure is a square,

$$x = 4x - 1$$
$$3x = 1$$
$$x = \frac{1}{3}$$

The area of the square is

$$x^2 = \left(\frac{1}{3}\right)^2 = \frac{1}{9}$$

Grid in as 1/9.

3. **27/8**  The area of quadrilateral $QBEF$ equals the area of right triangle $PEF$ minus the area of right triangle $PBQ$.

- The area of square $ABCD$ is 9, so the length of each side of the square is $\sqrt{9}$ or 3. Since $P$ and $Q$ are midpoints, $PB = QB = \frac{3}{2}$, so

$$\text{area rt. } \Delta PBQ = \frac{1}{2} \times \frac{3}{2} \times \frac{3}{2} \text{ or } \frac{9}{8}$$

- Since $PB = QB$, right triangle $PBQ$ is isosceles, so $\angle EPF = 45°$. Hence, right triangle $PEF$ is also isosceles, so $EF = PE = 3$, and

$$\text{area rt. } \Delta PEF = \frac{1}{2} \times 3 \times 3 = \frac{9}{2}$$

$$\text{area } QBEF = \text{area rt. } \Delta PEF - \text{area rt. } \Delta PBQ$$

$$= \quad \frac{9}{2} \quad - \quad \frac{9}{8}$$

$$= \quad \frac{36}{8} \quad - \quad \frac{9}{8} = \frac{27}{8}$$

Grid in as 27/8.

**LESSON 6-6**  *(Multiple Choice)*

1. **B**  Since the radius of circle $O$ is 2 and the radius of circle $P$ is 6, $OP = 2 + 6$ or 8. Hence, the circumference of any circle that has $OP$ as a diameter is $8\pi$.

2. **C**  If the circumference of a circle is $10\pi$, its diameter is 10 and its radius is 5. Hence, its area is $\pi(5^2) = 25\pi$.

3. **E**  The area of the shaded region equals the area of the square minus the area of the circle.

- Since a circle of radius 4 is inscribed in square $ABCD$, $AB$ has the same length as a diameter of the circle, so $AB = 4 + 4 = 8$. Hence the area of square $ABCD$ is $8 \times 8$ or 64.
- The area of the inscribed circle is $\pi 4^2$ or $16\pi$.
- The area of the shaded region = $64 - 16\pi$.

4. **D**  Equilateral polygon $ABCDE$ has five sides of equal length. If the polygon is inscribed in a circle, it divides the circle into five arcs of equal length. The circumference of the circle is $2\pi \times 20$ or $40\pi$, so the length of each of the five intercepted arcs is $\frac{1}{5}$ of $40\pi$ or $8\pi$. The length of the shortest arc from point $A$ to point $C$ is the sum of the lengths of arcs $AB$ and $BC$, which is $8\pi + 8\pi$ or $16\pi$.

5. **A**  The length of a diagonal of a 5 by 12 rectangle is 13 since, when a diagonal is drawn, the lengths of the sides of the right triangle that results form a (5-12-*13*) Pythagorean triple. If this rectangle is inscribed in a circle, the diagonals of the rectangle are diameters of the circle, so the length of a diameter of the circle is 13. The circumference of a circle with diameter 13 is $13\pi$.

6. **D**  Since

$$\text{Circumference} = \pi \times \text{Diameter}$$

then, if the circumference of a circle is $\pi$,

$$\text{Diameter} = \frac{\text{Circumference}}{\pi} = \frac{\pi}{\pi} = 1$$

Since the diameter of the circle is 1, the radius of the circle is $\frac{1}{2}$. The area of the circle is $\pi\left(\frac{1}{2}\right)^2$ or $\frac{1}{4}\pi$. If a square has the same area as this circle, then $(\text{side})^2 = \frac{1}{4}\pi$ or

$$\text{side} = \sqrt{\frac{1}{4}\pi} = \frac{1}{2}\sqrt{\pi}$$

The perimeter of the square is 4 times the length of a side or

$$4\left(\frac{1}{2}\sqrt{\pi}\right) = 2\sqrt{\pi}$$

*(Lesson 6-6   Multiple Choice continued)*

**7. C**  To find the length of segment $XY$, first find the length of segment $OX$. Draw radii $OA$ and $OB$.

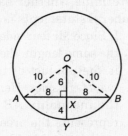

Since all radii of a circle have the same length, $\triangle AOB$ is isosceles, so perpendicular segment $OX$ bisects chord $AB$. In right triangle $AXO$, radius $OA = 10$ and

$$AX = \frac{1}{2}(16) = 8$$

The lengths of the sides of right triangle $AXO$ form a (6-8-10) Pythagorean triple where $OX = 6$. Since radius $OY = 10$,
$$XY = OY - OX = 10 - 6 = 4$$

**8. A**  Let $d$ represent the length in feet of the diameter of the bicycle wheel; then $\pi d$ is the circumference of the wheel. After completing $n$ revolutions, the wheel has traveled $n \times \pi d$ feet. Since it is given that the bicycle wheel has traveled $\frac{f}{\pi}$ feet after $n$ complete revolutions:

$$n\pi d = \frac{f}{\pi}$$

$$\frac{\overset{1}{\cancel{n\pi}}d}{\cancel{n\pi}} = \frac{f}{n\pi \cdot \pi}$$

$$d = \frac{f}{n\pi^2}$$

**9. E**  The area of the shaded region is the area of the semicircle minus the area of the inscribed right triangle. Since $AC = 6$ and $BC = 8$, the lengths of the sides of right triangle $ABC$ form a (6-8-*10*) Pythagorean triple in which diameter $AB = 10$. Hence, the radius of the circle is $\frac{1}{2} \times 10$ or 5.

- The area of right triangle $ABC = \frac{1}{2} \times 6 \times 8 = 24$.
- The area of semicircle $ACB$ is one-half the area of the circle that contains $AB$ as a diameter. Hence:
$$\text{area semicircle} = \frac{1}{2}(\pi 5^2) = \frac{25}{2}\pi = 12.5\pi$$

- The area of the shaded region = $12.5\pi - 24$.

**10. B**  Since all radii of a circle have the same length, $OA = OB = 8$. The circumference of circle $O$ is $2 \times \pi \times 8$ or $16\pi$. Since a 90° central angle intercepts arc $ACB$:

$$\text{length arc } ACB = \frac{90°}{360°} \times 16\pi$$

$$= \frac{1}{4} \times 16\pi$$

$$= 4\pi$$

The perimeter of the figure bounded by $OA$, $OB$, and arc $ACB$ is $8 + 8 + 4\pi$ or $16 + 4\pi$.

**11. D**  The area of the region of the circle bounded by chord $AB$ and arc $ACB$ is the area of sector $AOB$ minus the area of right triangle $AOB$.

- Since the area of circle $O$ is $\pi(8^2)$ or $64\pi$,

$$\text{area sector } AOB = \frac{90°}{360°} \times 64\pi$$

$$= \frac{1}{4} \times 64\pi$$

$$= 16\pi$$

- The area of right triangle $AOB$ is $\frac{1}{2} \times 8 \times 8$ or 32.
- The area of the shaded region is $16\pi - 32$.

**12. D**  The area of the shaded region is the area of the rectangle minus the sum of the areas of the two quarter circles, $BP$ and $CP$.

- Since the area of quarter circle $BP$ is $\pi$,

$$\frac{1}{4} \times \pi \times (AB)^2 = \pi$$

Hence, $(AB)^2 = 4$ so $AB = AP = 2$.

- The area of quarter circle $CP$ is also $\pi$, so $DP = AP = 2$.
- Since $AD = AP + PD = 2 + 2 = 4$, the area of rectangle $ABCD = AB \times AD = 2 \times 4 = 8$.
- The sum of the areas of the two quarter circles is $\pi + \pi$ or $2\pi$.
- The area of the shaded region is $8 - 2\pi$.

**13. A**  The perimeter of the unbroken figure is the sum of the lengths of $AC$, $BC$, and major arc $AB$.

- The area of square $OACB$ is $4x^2$, so
$$AC = BC = \sqrt{4x^2} = 2x$$
- Since $OACB$ is a square, it is equilateral, so radius $OA = AC = 2x$. The

*(Lesson 6-6   Multiple Choice continued)*

circumference of circle $O$ is $2 \times \pi \times 2x$ or $4\pi x$. Angle $AOB$ measures $90°$, so the central angle that intercepts major arc $AB$ is $360 - 90$ or $270°$. Hence:

$$\text{length major arc } AB = \frac{270°}{360°} \times 4\pi x$$

$$= \frac{3}{4} \times 4\pi x$$

$$= 3\pi x$$

- The perimeter of the unbroken figure is $2x + 2x + 3\pi x$ or $4x + 3\pi x$ or, factoring out $x$,
$$\text{perimeter} = x(4 + 3\pi)$$

14. **B**  The area of the shaded region is the area of square $OABC$ minus the area of sector $AOC$.

- Since the area of circle $O$ is given as $2\pi$,
$$\text{area sector } AOC = \frac{90°}{360°} \times 2\pi$$

$$= \frac{1}{4} \times 2\pi$$

$$= \frac{\pi}{2}$$

- The area of circle $O = 2\pi = \pi(OA)^2$, so $2 = (OA)^2$. Then
$$\text{area square } OABC = (OA)^2 = 2$$

- The area of the shaded region $= 2 - \frac{\pi}{2}$.

15. **B**  Since the circumference of a circle with radius $r$ inches is equal to the perimeter of a square with a side length of $s$ inches:

$$2\pi r = 4s$$
$$\pi r = 2s$$
$$\frac{r}{s} = \frac{2}{\pi}$$

16. **C**  The area of the shaded region is the area of the square minus the sum of the areas of the four quarter circles.

- The area of the square is $2 \times 2$ or $4$.
- Since the four quarter circles have equal areas, they have equal radii and the sum of their areas is equivalent to the area of one whole circle with the same radius length. The length of side $GE$ is given as $2$. Then the length of the radius of each quarter circle is $1$ since $B$, $D$, $F$, and $H$ are midpoints of the sides of the square. Hence, the sum of the areas of the four quarter circles is $\pi(1^2)$ or $\pi$.

- The area of the shaded region $= 4 - \pi$. Using $\pi = 3.14$, you find that the best approximation for the area of the shaded region is $4 - 3.14$ or $0.86$.

17. **C**  Determine whether each of the Roman numeral statements is always true.

- I. Since all four sides of a square have the same length, the area of the square is the product of the lengths of any two of its sides. Hence, the left side of the inequality $AB \times CD < \pi \times r \times r$ represents the area of square $ABCD$, and the right side of the inequality represents the area of inscribed circle $O$. Since the area of the square is greater, not less, than the area of the inscribed circle, statement I is not true.

- II. Since a diameter of the circle can be drawn whose endpoints are points at which circle $O$ intersects two sides of the square, the length of a side of the square is $2r$, so the area of the square can be represented as $2r \times 2r$ or $4r^2$. Hence, statement II is true.

- Since the circumference of the circle is less than the perimeter of the square:
$$2\pi r < 4(CD)$$
$$r < \frac{4(CD)}{2\pi}$$
$$r < \frac{2(CD)}{\pi}$$

Hence, statement III is true. Only Roman numeral statements II and III must be true.

18. **D**  The perimeter of the figure that encloses the shaded region is the the length of arc $PBQ$ plus the sum of the lengths of segments $AP$, $CQ$, $AB$, and $BC$.

- Draw radius $OB$. You are told that $OABC$ is a rectangle. Since the diagonals of a rectangle have the same length, $OB = AC = 8$. The circumference of the circle that has its center at $O$ and that contains arc $PBQ$ is $2 \times \pi \times 8$ or $16\pi$. Since a $90°$ central angle intercepts arc $PBQ$:

$$\text{length arc } PBQ = \frac{90°}{360°} \times 16\pi$$

$$= \frac{1}{4} \times 16\pi$$

$$= 4\pi$$

*(Lesson 6-6 Multiple Choice continued)*

• In isosceles right triangle $AOC$,
$$OA = OC = \frac{1}{2} \times 8 \times \sqrt{2} = 4\sqrt{2}$$

Hence:
$$AP = OP - OA = 8 - 4\sqrt{2}$$
and
$$CQ = OQ - OC = 8 - 4\sqrt{2}$$

• In isosceles right triangle $ABC$,
$$AB = BC = \frac{1}{2} \times 8 \times \sqrt{2} = 4\sqrt{2}$$

• The perimeter of the figure that encloses the shaded region

$$= \text{arc } PBQ + \quad AP \quad + \quad CQ \quad + AB + BC$$
$$= \quad 4\pi \quad + (8 - 4\sqrt{2}) + (8 - 4\sqrt{2}) + 4\sqrt{2} + 4\sqrt{2}$$
$$= 4\pi + 16 \quad \text{or} \quad 16 + 4\pi$$

*(Quantitative Comparison)*

1. **A** Since all radii of a circle have the same length, $OA = OB$, so $\angle A = \angle B$, making $x = 61$ in Column A. Hence, in Column B:
$$y = 180 - 61 - 61 = 58$$
Since $x > y$, Column A > Column B.

2. **C** In circle $X$, the sum of the measures of the three central angles is $60° + 60° + 60°$ or $180°$, which is the degree measure of a semicircle. Hence, the sum of the areas of the three shaded sectors in Column A is equivalent to the area of a semicircle. For Column B, the region in circle $Y$, that is shaded is a semicircle. Since you are given that $PX = PY$, the radii of the two circles, $X$ and $Y$, are equal, so the area of a semicircle of one circle must be equal to the area of a semicircle of the other circle. Hence, Column A = Column B.

3. **C** Since all radii of a circle are equal, $OA = OB$. Also, $\angle B = \angle A = 60$, so,
$$\angle C = 180 - 60 - 60 = 60$$
Since $\triangle AOB$ is equiangular, it is also equilateral, so chord $AB$ has the same length as a radius of circle $O$, and $AB = OA = OB = OC$. In Column A, $(AB)^2 = AB \times AB$. Rewriting one factor as $OA$ and the second factor as $OC$ gives $(AB)^2 = OA \times OC$, the expression in Column B. Hence, Column A = Column B.

4. **B** Circles $A$, $B$, and $C$ each have an area of $9\pi$. The radius of each circle is 3 since $9\pi = \pi r^2$ and $r = \sqrt{9} = 3$. Hence, the length of each side of $\triangle ABC$ is $3 + 3$ or 6, so the perimeter of $\triangle ABC$ in Column A is $6 + 6 + 6$ or 18. In Column B, the circumference of circle $A$ is $2 \times \pi \times 3$ or $6\pi$. Since $\pi > 3$, then $6\pi > 18$, so Column B > Column A.

5. **A** Since the circumference of circle $O$ is $10\pi$, the length of a diameter of the circle is 10. Also, since a diagonal of square $ABCD$ is the hypotenuse of an isosceles right triangle in which the legs are the adjacent sides of the square, the length of each side of the square is $\frac{1}{2} \times 10 \times \sqrt{2}$ or $5\sqrt{2}$. Hence, in Column A the perimeter of square $ABCD$ is $4 \times 5\sqrt{2}$ or $20\sqrt{2}$. Since $20\sqrt{2} > 20$, the value in Column B, Column A > Column B.

6. **B** In circle $O$, the area of sector $AOB$ is $\frac{45°}{360°}$ or $\frac{1}{8}$ of the area of circle $O$. Since you are given that the area of sector $AOB$ is $18\pi$, the area of the circle must be $8 \times 18\pi$ or $144\pi$. Then
$$\text{area} = \pi \times (\text{radius})^2 = \pi \times (OA)^2 = 144\pi$$
Hence, in Column A the number of inches in the length of $OA = \sqrt{144} = 12$.
In circle $P$, the length of arc $QRS$ is $\frac{60°}{360°}$ or $\frac{1}{6}$ of the circumference of circle $P$. Since you are given that the length of arc $QRS$ is $5\pi$, the circumference of circle $P$ is $6 \times 5\pi$ or $30\pi$. Then
$$\text{circumference} = 2 \times \pi \times \text{radius}$$
$$= 2 \times \pi \times PQ$$
$$= 30\pi$$
Hence, in Column B the number of inches in the length of $PQ = \frac{30\pi}{2\pi} = 15$. Since $15 > 12$, Column B > Column A.

7. **C** You are given that the perimeter of rectangle $ABCD = 16 + 4\pi$ and that $AB = 2\pi$. Since
$$AB + CD = 2\pi + 2\pi = 4\pi$$
then $AD + BC = 16$, so $AD = BC = 8$. Hence, in Column A:
$$\text{area rectangle } ABCD = AB \times AD$$
$$= 2\pi \times 8$$
$$= 16\pi$$
In Column B, since the length of a diameter of circle $O$ is 8, the radius of the circle is 4.
Hence, its area is $\pi 4^2$ or $16\pi$, and Column A = Column B.

8. **B** Since $RS > OS$, then $\angle O > \angle ORS$. Also, since $\angle ORS$ is an exterior angle of $\triangle PRS$,

*(Lesson 6-6  Quantitative Comparison continued)*

then $\angle ORS > \angle P$, so $\angle O > \angle P$. In $\triangle POS$, $PS$ (the side opposite $\angle O$) is longer than $OS$ (the side opposite $\angle P$). Since $PS > OS$ and $OP = OS$ because all radii of a circle have the same length, $PS > OP$. Hence Column B > Column A.

*(Grid In)*

1. **2**  The length of the radius of each semicircle above the line is 2, so

$$\text{area each semicircle} = \frac{1}{2} \times \pi \times 2^2$$

$$= \frac{1}{2} \times 4\pi$$

$$= 2\pi$$

Hence,

$$X = 2\pi + 2\pi + 2\pi = 6\pi$$

The diameter of the larger semicircle below the line is 8, so its radius is 4. Then

$$\text{area larger semicircle} = \frac{1}{2} \times \pi \times 4^2$$

$$= \frac{1}{2} \times 16\pi$$

so $Y = 8\pi$.
Hence,

$$Y - X = 8\pi - 6\pi = 2\pi = k\pi$$

so $k = 2$.

2. **2/32 (or 1/16)**  Pick an easy number for the length of the diameter of the smaller semicircle. Then find the areas of semicircles *PS* and *PR*.

- If diameter $PS = 4$, the length of the radius of the smaller semicircle is 2, so:

$$\text{area smaller semicircle} = \frac{1}{2} \times \pi \times 2^2$$

$$= \frac{1}{2} \times 4\pi$$

$$= 2\pi$$

- Since $PS = 4$ and $S$ is the midpoint of $PQ$, the length of radius $PQ$ of the larger semicircle is 8. Then:

$$\text{area larger semicircle} = \frac{1}{2} \times \pi \times 8^2$$

$$= \frac{1}{2} \times 64\pi$$

$$= 32\pi$$

- The ratio of the area of semicircle *PS* to the area of semicircle *PR* is $\frac{2\pi}{32\pi}$ or $\frac{2}{32}$ $\left(\text{or } \frac{1}{16}\right)$.

3. **72**  From 1:25 P.M. to 1:37 P.M. of the same day, the minute hand of the clock moves 12 minutes since $37 - 25 = 12$. There are 60 minutes in 1 hour, so 12 minutes represents $\frac{12}{60}$ of a complete rotation. Since there are 360° in a complete rotation, the minute hand moves

$$\frac{12}{\cancel{60}} \times \cancel{360}^{\,6}$$

or $12 \times 6$ or $72°$.

4. **8**

- Find the length of the radius of circle *O*. The central angle of the shaded region is 120°, so the area of the shaded region is $\frac{120°}{360°}$ or $\frac{1}{3}$ of the area of circle *O*. Since the area of the shaded region is $\frac{12}{\pi}$, the area of the circle is $3 \times \frac{12}{\pi}$ or $\frac{36}{\pi}$. Let $r$ represent the length of the radius of circle *O*; then:

$$\pi r^2 = \frac{36}{\pi}$$

$$r^2 = \frac{36}{\pi^2}$$

$$r = \sqrt{\frac{36}{\pi^2}} = \frac{6}{\pi}$$

- Find the circumference of circle *O*. The circumference of circle *O* is $2 \times \pi \times \frac{6}{\pi} = 12$.

- Find the length of arc *ACB*. The degree measure of the central angle that intercepts major arc *ACB* is $360 - 120$ or 240. The length of arc *ACB* is $\frac{240°}{360°}$ times the circumference of the circle. Hence:

$$\text{length arc } ACB = \frac{240°}{360°} \times 12$$

$$= \frac{2}{\cancel{3}} \times \cancel{12}^{\,4}$$

$$= 8$$

## LESSON 6-7  *(Multiple Choice)*

1. **C**  A cube has six surfaces. If the surface area of the cube is 96, then the area of each square surface is $\frac{96}{6}$ or 16, so the edge length is $\sqrt{16}$ or 4. Hence, the volume of the cube is $4 \times 4 \times 4$ or 64.

2. **E**  If the length, width, and height of a rectangular solid are in the ratio of $3 : 2 : 1$, then let $3x$, $2x$, and $x$ represent the

*(Lesson 6-7 Multiple Choice continued)*

dimensions of the solid. You are given that the volume of the box is 48. Hence:

$$(3x)(2x)(x) = 48$$
$$6x^3 = 48$$
$$x^3 = \frac{48}{6} = 8$$

The dimensions of the box are 2, 4, and 6 since $x = 2$, $2x = 2(2) = 4$, and $3x = 3(2) = 6$. The surface area of a rectangular box is 2 times the sum of the products of the three pairs of dimensions:

surface area of box $= 2[(\ell \times w) + (\ell \times h) + (h \times w)]$
$$= 2(2 \times 4 + 2 \times 6 + 4 \times 6)$$
$$= 2(8 + 12 + 24)$$
$$= 2(44)$$
$$= 88$$

3. **B** If the volume of a cube is 8, then the edge length of the cube is 2 since $2 \times 2 \times 2 = 8$. The distance from point $X$ at the center of a face of a cube to point $Y$ at the center of the opposite face of the cube equals the edge length, which is 2.

4. **C** The height of the stacked cubes will be the sum of the edge lengths of the three cubes.

   • If the volume of the first cube is $\frac{1}{8}$ cubic feet, its edge length is $\frac{1}{2}$ of a foot or 6 inches.

   • If the volume of the second cube is 1 cubic foot, its edge length is 1 foot or 12 inches.

   • If the volume of the third cube is 8 cubic feet, its edge length is 2 feet or $2 \times 12$ or 24 inches.

   The height of the stacked cubes, in *inches*, is 6 + 12 + 24 or 42.

5. **D** First find the radius of the circular base of the smaller cone formed when a plane parallel to the base cuts through the original cone.

   • Since the height of the original cone is 12 inches and the height of the smaller cone is 9 inches, the radius of the circular base of the smaller cone is $\frac{9}{12}$ or $\frac{3}{4}$ of the radius of the original cone.

   • You are given that the area of the original cone is $64\pi$, so the radius of its base is $\sqrt{64}$ or 8. Hence, the radius of the circular base of the smaller cone is $\frac{3}{4} \times 8$ or 6.

   Since the radius of the base of the smaller cone is 6 inches, the area, in *square inches*, of the base is $\pi 6^2$ or $36\pi$.

6. **C** The volume $V$ of a cylinder is given by the formula $V = \pi r^2 h$, where $r$ is the radius of the circular base and $h$ is the height. If $h$ is replaced by $2h$ and the radius is multiplied by a constant $k$, so that the volume $V$ does not change, then

$$V = \pi r^2 h = \pi (kr)^2 (2h)$$

   so

$$\pi r^2 h = \pi (k^2 r^2)(2h)$$
$$1 = 2k^2$$
$$\frac{1}{2} = k^2$$
$$\frac{1}{\sqrt{2}} = k$$

7. **C** When a solid cube with an edge length of 1 foot is placed in the fish tank, it displaces a volume of water that is equal to the volume of the cube, which is $1 \times 1 \times 1$ or 1 cubic foot. If $h$ represents the change in the number of feet in the height of the water in the fish tank, then

volume of displaced water $= 2 \times 3 \times h = 1$
$$h = \frac{1}{6} \text{ foot}$$

   Hence, the number of *inches* the level of the water in the tank will rise is $\frac{1}{6} \times 12$ or 2.

8. **B** Let $h$ represent the equal heights of the cylinder and the rectangular box. Since you are given that the volume of the cylinder of radius $r$ is $\frac{1}{4}$ of the volume of the rectangular box with a square base of side length $x$,

$$\overbrace{\pi r^2 h}^{\text{volume of cylinder}} = \frac{1}{4} \overbrace{(x \cdot x \cdot h)}^{\text{volume of box}}$$
$$\pi r^2 h = \frac{x^2}{4} h$$
$$r^2 = \frac{x^2}{4\pi}$$
$$r = \frac{\sqrt{x^2}}{\sqrt{4\pi}} = \frac{x}{2\sqrt{\pi}}$$

9. **E** You are given that the height of the sand in the cylinder-shaped can drops 3 inches or $\frac{1}{4}$ foot when 1 cubic foot of sand is

*(Lesson 6-7  Multiple Choice continued)*

poured out. If $r$ represents the length in feet of the radius of the circular base, then the volume of the sand poured out must be equal to 1. Hence:

$$\text{volume of sand} = \pi r^2\left(\frac{1}{4}\right) = 1$$

$$\pi r^2 = 1 \times 4$$

$$r^2 = \frac{4}{\pi}$$

$$r = \frac{\sqrt{4}}{\sqrt{\pi}} = \frac{2}{\sqrt{\pi}} \text{ feet}$$

$$= \frac{2}{\sqrt{\pi}} \text{ feet} \times 12 \text{ inches/feet}$$

$$= \frac{24}{\sqrt{\pi}} \text{ inches}$$

Hence the radius of the cylinder is $\frac{2}{\sqrt{\pi}}$ feet $\times$ 12 or $\frac{24}{\sqrt{\pi}}$ inches, and the diameter in *inches* is $2 \times \frac{24}{\sqrt{\pi}}$ or $\frac{48}{\sqrt{\pi}}$.

10. **A**  If the height $h$ of a cylinder equals the circumference of the cylinder with radius $r$, then $h = 2\pi r$, so $r = \frac{h}{2\pi}$. Hence:

$$\text{volume of cylinder} = \pi r^2 h$$

$$= \pi\left(\frac{h}{2\pi}\right)^2 h$$

$$= \pi\left(\frac{h^2}{4\pi^2}\right) h$$

$$= \frac{\pi h^3}{4\pi^2}$$

$$= \frac{h^3}{4\pi}$$

11. **B**  For each Roman numeral combination determine whether a rectangular solid can be formed.

- I. If the set of six faces of a rectangular solid includes two of face $X$ and two of face $Y$, then the rectangular solid must be a 3 by 3 by 4 solid. Since face $Z$ is 4 by 5, it cannot be used to complete the rectangular solid. Hence, the six faces in combination I cannot be used to form a rectangular solid.

- II. If the set of six faces of a rectangular solid include four of face $X$, then these four faces can be arranged so that they form an open box with square 3 by 3 bases at opposite ends of the box. Since face $Y$ is a 3 by 3 square, it can be used to complete the rectangular solid. Hence, the six faces in combination II can be used to form a rectangular solid.

- III. If the set of six faces of a rectangular solid includes four of face $Z$, then these four faces can be arranged so that they form an open box with either 4 by 4 squares or 5 by 5 squares at opposite ends of the box. Since face $Y$ is a 3 by 3 square, it cannot be used to complete the rectangular solid. Hence, the six faces in combination III cannot be used to form a rectangular solid.

Only Roman numeral combination II gives faces from which a rectangular solid can be formed.

12. **A**  If a cylinder with radius $r$ and height $h$ is closed on the top and bottom, then the sum of the areas of the circular top and bottom is $\pi r^2 + \pi r^2$ or $2\pi r^2$. The area of the curved surface is the height $h$ times the distance around the curved surface, $2\pi r$. Hence, the total surface area is $2\pi r^2 + 2\pi rh$ or $2\pi r(r + h)$.

13. **B**  The shortest distance from $A$ to $D$ is the length of $AD$. Draw $AD$, which is the hypotenuse of a right triangle whose legs are $AE$ and $ED$. Since the edge length of the cube is given as 4, $ED = AF = 4$. Segment $AE$ is the diagonal of square $AFEH$, so $AE = 4\sqrt{2}$.

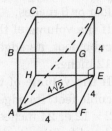

In right triangle $AED$,
$$(AD)^2 = (AE)^2 + (ED)^2$$
$$= (4\sqrt{2})^2 + (4)^2$$
$$= (16 \cdot 2) + 16$$
$$= 32 + 16$$
$$AD = \sqrt{48} = \sqrt{16} \cdot \sqrt{3} = 4\sqrt{3}$$

*(Grid In)*

1. **60**  You are given that all of the dimensions of a rectangular box are integers greater than 1. Since the area of one side of this box is 12, the dimensions of this side must be either 2 by 6 or 3 by 4. The area of another side of the box is given as 15, so the dimensions of this side must be 3

*(Lesson 6-7  Grid In continued)*

by 5. Since the two sides must have at least one dimension in common, the dimensions of the box are 3 by 4 by 5, so its volume is $3 \times 4 \times 5$ or 60.

2. **96**  A cube whose volume is 8 cubic inches has an edge length of 2 inches since $2 \times 2 \times 2 = 8$. Since a cube has six square faces of equal area, the surface area of this cube is $6 \times 2^2$ or $6 \times 4$ or 24. The minimum length $L$ of $\frac{1}{4}$-inch-wide tape needed to completely cover the cube must have the same surface area as the cube. Hence, $L \times \frac{1}{4} = 24$ and $L = 24 \times 4 = 96$.

3. **700**  Pick an easy number for the edge length of the cube. If the edge length is 1, the volume of the cube is $1 \times 1 \times 1$ or 1. If the length of each side of this cube is doubled, a cube with an edge length of 2 results. The volume of the new cube is $2 \times 2 \times 2$ or 8. Hence:

$$\% \text{ increase in volume} = \frac{\text{increase in volume}}{\text{original volume}} \times 100\%$$
$$= \frac{8-1}{1} \times 100\%$$
$$= 700\%$$

## LESSON 6-8  *(Multiple Choice)*

1. **C**  Let $(x_A, y_A) = (3,-1)$ and $(x_B, y_B) = (6,5)$. Use the distance formula to find the distance $d$ between these points:

$$d = \sqrt{(x_B - x_A)^2 + (y_B - y_A)^2}$$
$$= \sqrt{(6-3)^2 + (5-(-1))^2}$$
$$= \sqrt{3^2 + (5+1)^2}$$
$$= \sqrt{9+36}$$
$$= \sqrt{45} = \sqrt{9}\sqrt{5} = 3\sqrt{5}$$

The length of the line segment whose endpoints are $(3,-1)$ and $(6,5)$ is $3\sqrt{5}$.

2. **E**  Sketch the rectangle whose vertices are $(-2,5)$, $(8,5)$, $(8,-2)$, and $(-2,-2)$.

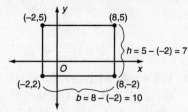

Since
$$b = 8 - (-2) = 8 + 2 = 10$$
and
$$h = 5 - (-2) = 5 + 2 = 7$$

then
$$\text{area of rectangle} = b \times h$$
$$= 10 \times 7$$
$$= 70$$

3. **D**  Sketch the parallelogram whose vertices are $(-4,-2)$, $(-2,6)$, $(10,6)$, and $(8,-2)$.

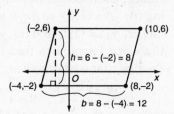

Since
$$b = 8 - (-4) = 8 + 4 = 12$$
and
$$h = 6 - (-2) = 6 + 2 = 8$$
then
$$\text{area of parallelogram} = b \times h$$
$$= 12 \times 8$$
$$= 96$$

4. **C**  Sketch the triangle whose vertices are $(-4,0)$, $(2,4)$, and $(4,0)$.

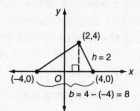

Since
$$b = 4 - (-4) = 4 + 4 = 8$$
and $h = 4$, then
$$\text{area of triangle} = \frac{1}{2} \times b \times h$$
$$= \frac{1}{2} \times 8 \times 4$$
$$= 16$$

5. **B**  If $A(-3,0)$ and $C(5,2)$ are the endpoints of diagonal $AC$ of rectangle $ABCD$, then the other two vertices are $(-3,2)$ and $(5,0)$.

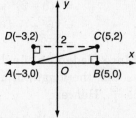

Since the length of one side of the rectangle is 8 and the length of an adjacent side is 2, the perimeter of rectangle $ABCD$ is $2(8 + 2)$ or 20.

*(Lesson 6-8 Multiple Choice continued)*

6. **B** Since point $A$ is on the $y$-axis and below the $x$-axis, its $x$-coordinate is 0 and its $y$-coordinate is negative. You are told that $OABC$ is a square, so $OA = OC = 4$. Hence, the $y$-coordinate of $A$ is $-4$, making $(0,-4) = (k,p)$, so $k = 0$ and $p = -4$.

7. **B** Let $(x_A, y_A) = (1,3)$ and $(x_B, y_B) = (2,5)$. Use the slope formula to find the slope $m$ of the line that contains these points:
$$m = \frac{y_B - y_A}{x_B - x_A} = \frac{5 - 3}{2 - 1} = \frac{2}{1} = 2$$
Calculate the slopes of the lines that contain $(x_A, y_A)$ and each of the points in the set of answer choices until you find the point that makes the slope of a line equal to 2. Choice (B) works since, if $(x_C, y_C) = (0,1)$, then
$$m = \frac{y_A - y_C}{x_A - x_C} = \frac{3 - 1}{1 - 0} = \frac{2}{1} = 2$$
Hence, point $(0,1)$ lies on the same line as points $(1,3)$ and $(2,5)$.

8. **A** The perimeter of the shaded region is the sum of the circumference of semicircle $P$ and the length of diameter $AB$. Since
$$AB = 7 - (-1) = 8$$
the diameter of circle $P$ is 8, so
$$\text{circumference semicircle } P = \frac{1}{2}(\pi D)$$
$$= \frac{1}{2}(8\pi)$$
$$= 4\pi$$
Hence, the perimeter of the shaded region is $4\pi + 8$.

9. **B** If points $A(2,0)$ and $B(8,-4)$ are the endpoints of diameter $AB$ of circle $O$, then let $(x_A, y_A) = (2,0)$ and $(x_B, y_B) = (8,-4)$. Use the distance formula to find the length of diameter $AB$:
$$AB = \sqrt{(x_B - x_A)^2 + (y_B - y_A)^2}$$
$$= \sqrt{(8 - 2)^2 + (-4 - 0)^2}$$
$$= \sqrt{6^2 + (-4)^2}$$
$$= \sqrt{52}$$
Since the length of a diameter of circle $O$ is $\sqrt{52}$, the length of a radius of circle $O$ is $\frac{\sqrt{52}}{2}$. Hence:
$$\text{area circle } O = \pi r^2 = \pi\left(\frac{\sqrt{52}}{2}\right)^2$$
$$= \pi\left(\frac{52}{4}\right)$$
$$= 13\pi$$

10. **C** Test each point in the set of answer choices until you find the point that makes the slope $m$ of the line containing that point and $(4,-2)$ equal to $\frac{3}{2}$. Let $(x_A, y_A) = (4,-2)$. Choice (C) works since, if $(x_B, y_B) = (6,1)$, then
$$m = \frac{y_B - y_A}{x_B - x_A} = \frac{1 - (-2)}{6 - 4} = \frac{1 + 2}{2} = \frac{3}{2}$$
The coordinates of another point on the line are $(6,1)$.

11. **C** The center of a circle is the midpoint of any diameter of the circle. Let $(x_A, y_A) = (1,2)$ and $(x_B, y_B)$ represent the coordinates of point $B$.
*Solution 1*: Since the coordinates of $O$ are $(-2,1)$,
- $-2 = \frac{1 + x_B}{2}$, so $1 + x_B = -4$ or $x_B = -5$.
- $1 = \frac{2 + y_B}{2}$, so $2 + y_B = 2$ or $y_B = 0$.
- The coordinates of point $B = (x_B, y_B) = (-5,0)$.

*Solution 2*: Find the midpoint of each segment whose endpoints are point $A$, and test each point in the set of answer choices until you obtain $O(-2,1)$ as the midpoint of $AB$.

12. **E** Since it is given that point $E(5,h)$ is on the line that contains $A(0,1)$ and $B(-2,-1)$,
$$\text{slope of } EA = \text{slope of } AB$$
$$= \frac{-1 - 1}{-2 - 0} = \frac{-2}{-2} = 1$$
Hence,
$$\text{slope of } EA = \frac{h - 1}{5 - 0} = 1$$
$$h - 1 = 5$$
$$h = 6$$

13. **A** If the line whose equation is $y = x + 2k$ passes through point $(1,-3)$, then substituting $-3$ for $y$ and 1 for $x$ will make the resulting equation a true statement. Hence:
$$-3 = 1 + 2k$$
$$-4 = 2k$$
$$k = \frac{-4}{2} = -2$$

14. **D** If a circle with center at the origin passes through $(-8,-6)$, the length of a radius of the circle is the length of the segment whose endpoints are $A(0,0)$ and $B(-8,-6)$. Use the distance formula to find the length of $AB$:

*(Lesson 6-8  Multiple Choice continued)*

$$AB = \sqrt{(-8-0)^2 + (-6-0)^2}$$
$$= \sqrt{64 \qquad + 36}$$
$$= \sqrt{100}$$
$$= 10$$

Any point that lies on the circle will be the same distance from the origin as point $B$. Find the distance from the origin to each point in the set of answer choices until you obtain a distance of 10. Choice D works since

$$\sqrt{(9-0)^2 + (\sqrt{19}-0)^2} = \sqrt{81+19}$$
$$= \sqrt{100} = 10$$

The circle passes through $(9, \sqrt{19})$ since the distance of this point from the origin is 10.

15. **D**  Drop a perpendicular to the $x$-axis from point $S$. Call the point where it intersects the $x$-axis point $H$. Since point $S$ is 5 units above the $x$-axis, $SH = 5$. Opposite sides of a parallelogram are parallel, so $\angle STH = \angle O = 30°$. In 30°-60° right triangle $SHT$, hypotenuse $ST$ is 2 times the length of $SH$ (the side opposite the 30° angle). Hence, $ST = 2 \times 5 = 10$. Since opposite sides of a parallelogram have the same length, $OR = ST = 10$. Also, since $OR = OT$, then $OT = RS = 10$. Hence, the perimeter of parallelogram $ORST$ is $10 + 10 + 10 + 10$ or 40.

16. **D**  The area of quadrilateral $OABC$ is the sum of the areas of triangles $OAB$ and $OBC$.

- Since $\angle BOC = 45$, then
$$AOB = 90 - 45 = 45$$
Also, since point $A$ is 5 units above the $x$-axis, $OA = OB = 5$, so
$$\text{area rt. } \triangle OAB = \frac{1}{2} \times 5 \times 5$$
$$= 12.5$$

- Since $AB \perp y$-axis, point $B$ is 5 units above the $x$-axis, so the height of $\triangle OBC$ is 5 and the length of base $OC$ is 7. Hence:
$$\text{area } \triangle OBC = \frac{1}{2} \times 7 \times 5$$
$$= 17.5$$

- The area of quadrilateral $OABC$ is $12.5 + 17.5$ or 30.

17. **B**  If the slope of line $\ell$ is $m$, then

$$m = \frac{(h+m)-0}{0-h} = \frac{h+m}{-h}$$

Hence, $-hm = h + m$ or $-h = m + mh$. Factoring out $m$ from the right side of the equation gives $-h = m(1 + h)$ so

$$m = \frac{-h}{1+h}$$

*(Quantitative Comparison)*

1. **B**  Since point $P$ is on the $x$-axis to the *left* of the origin, $x < 0$ and $y > 0$. Thus, $y > x$, so Column B > Column A.

2. **B**  Since
$$OB = OA = \sqrt{(6-0)^2 + (8-0)^2}$$
$$= \sqrt{36+64} = \sqrt{100} = 10$$
and
$$AB = 6 - (-6) = 6 + 6 = 12$$
the perimeter of $\triangle AOB$ in Column A is $10 + 10 + 12$ or 32. Hence, Column B > Column A.

3. **B**  The length of $KL$ is $6 - 1$ or 5. Since $JKLM$ is a square, $JK = KL = 5$. The length of vertical segment $JK$ is 5, so $7 - b = 5$ or $b = 2$ in Column B. Points $J$ and $K$ lie on the same vertical line, so they must have the same $x$-coordinates. Hence, $a = 1$ in Column A, so Column B > Column A.

4. **C**  In arc $APB$,
$$\text{length radius } OP = \sqrt{(8-0)^2 + (6-0)^2}$$
$$= \sqrt{100} = 10$$
Since $\angle AOB$ measures 90°,
$$\text{length arc } APB = \frac{90°}{360°}$$
or $\frac{1}{4}$ of the circumference of circle $O$. In Column A,
$$\text{length arc } APB = \frac{1}{4} \times 2\pi \times 10$$
$$= 5\pi$$
so Column A = Column B.

5. **B**  In Column A,
$$\text{distance from } A \text{ to } B = \sqrt{(2a-a)^2 + (b+2-b)^2}$$
$$= \sqrt{a^2 + 4}$$
To compare $\sqrt{a^2 + 4}$ and $a + 2$, eliminate the radical by squaring both quantities. Since $\left(\sqrt{a^2 + 4}\right)^2 = a^2 + 4$,
$$(a + 2)^2 = (a + 2)(a + 2)$$
$$= a^2 + 4a + 4$$
Since $a > 0$, then $a + 2 > \sqrt{a^2 + 4}$, so Column B > Column A.

6. **B**  In Column A,

*(Lesson 6-8  Quantitative Comparison continued)*

$$\text{slope line } h = \frac{0-k}{(k-1)-0} = -\frac{k}{k-1}$$

where $k > 0$ and $k - 1 > 0$. Since the denominator of $\frac{k}{k-1}$ is less than the numerator, the fraction is greater than 1, so the value of $-\frac{k}{k-1}$ is less than $-1$. Hence, Column B > Column A.

*(Grid In)*

1. **3/4**  Since the line that passes through points $(7,3k)$ and $(0,k)$ has a slope of $\frac{3}{14}$,

$$\frac{3k-k}{7-0} = \frac{3}{14}$$

$$\frac{2k}{7} = \frac{3}{14}$$

$$28k = 21$$

$$k = \frac{21}{28} = \frac{3}{4}$$

2. **3/2**  Since the slope of line $\ell_1$ is $\frac{5}{6}$, then

$$\frac{y_1 - 0}{3 - 0} = \frac{5}{6} \quad \text{or} \quad y_1 = \frac{15}{6} = \frac{5}{2}$$

The slope of line $\ell_2$ is $\frac{1}{3}$, so

$$\frac{y_2 - 0}{3 - 0} = \frac{1}{3} \quad \text{or} \quad y_2 = \frac{3}{3} = 1$$

Since points $A$ and $B$ have the same $x$-coordinates, they lie on the same vertical line, so

$$\text{distance from } A \text{ to } B = y_1 - y_2$$

$$= \frac{5}{2} - 1 \text{ or } \frac{3}{2}$$

3. **16**  The area of $\triangle APQ$ is equal to the area of square $ABCD$ minus the sum of the areas of the three corner right triangles.

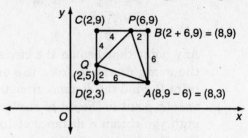

- Find the coordinates of the vertices of square $ABCD$, as shown in the accompanying diagram.
- Find the area of each of the right triangles.

$$\text{Area rt. } \triangle QDA = \text{area rt. } \triangle PBA$$

$$= \frac{1}{2} \times 2 \times 6 = 6$$

$$\text{Area rt. } \triangle QCP = \frac{1}{2} \times 4 \times 4 = 8$$

- Then

$$\text{area } \triangle APQ = 36 - (6 + 6 + 8)$$

$$= 36 - 20 = 16$$

# Special Problem Types

This chapter reviews some special types of problems
that frequently appear on SAT exams.

---

## LESSONS IN THIS CHAPTER

---

---

# LESSON 7-1

# Average Problems

## OVERVIEW

*There are three types of statistics that can be used to describe a set of N numbers:*

- *The **average (arithmetic mean)**, which is calculated by dividing the sum of the numbers in the set by N.*
- *The **median**, which is the middle value when the numbers in the set are listed in size order.*
- *The **mode**, which is the number in the set that occurs most frequently. The term* average *always refers to the arithmetic mean of a set of numbers.*

## FINDING THE AVERAGE (ARITHMETIC MEAN)

To find the average of a set of $n$ numbers, add all the numbers together and then divide the sum by $n$. Thus:

$$\text{Average} = \frac{\text{Sum of the } n \text{ values}}{n}$$

For example, the average of three exam grades of 70, 80, and 72 is 74 since

$$\text{average} = \frac{70+80+72}{3} = \frac{222}{3} = 74$$

A set of algebraic expressions can be averaged in the same way as a set of numbers.

**EXAMPLE:**  If the lengths of the sides of a triangle are represented by $2x + 1$, $x + 7$, and $3x - 11$, what is the average length of a side?

**SOLUTION:**  Since there are three expressions to be averaged, find their sum and divide it by 3:

$$\text{average} = \frac{(2x+1)+(x+7)+(3x-11)}{3}$$

$$= \frac{(2x+x+3x)+(1+7-11)}{3}$$

$$= \frac{6x \quad - \quad 3}{3}$$

$$= \frac{6x}{3} - \frac{3}{3}$$

$$= 2x - 1$$

**318**

## FINDING AN UNKNOWN NUMBER WHEN AN AVERAGE IS GIVEN

If you know the average of a set of $n$ numbers, you can find the sum of the $n$ values by using this relationship:

$$\text{Sum of the } n \text{ values} = \text{Average} \times n$$

**EXAMPLE:** The average of a set of four numbers is 78. If three of the numbers in the set are 71, 74, and 83, what is the fourth number?

**SOLUTION:** If the average of four numbers is 78, the sum of the four numbers is $78 \times 4 = 312$. The sum of the three given numbers is $71 + 74 + 83 = 228$. Since $312 - 228 = 84$, the fourth number is 84.

**EXAMPLE:** The average of $w$, $x$, $y$, and $z$ is 31. If the average of $w$ and $y$ is 24, what is the average of $x$ and $z$?

**SOLUTION 1:**

- Since $\dfrac{w+x+y+z}{4} = 31$, $w + x + y + z = 4 \times 31 = 124$.
- Since $\dfrac{w+y}{2} = 24$, $w + y = 2 \times 24 = 48$.
- Thus, $x + z + 48 = 124$, so $x + z = 76$.
- Since $x + z = 76$, $\dfrac{x+z}{2} = \dfrac{76}{2} = 38$.

Hence, the average of $x$ and $z$ is 38.

**SOLUTION 2:** Since the average of two of the four numbers is 7 ($= 31 - 24$) less than the average of the four numbers, the average of the other two numbers must be 7 more than the average of the four numbers. Hence, the average of $x$ and $z$ is $31 + 7 = 38$.

## FINDING THE WEIGHTED AVERAGE

A **weighted average** is the average of two or more sets of numbers that do not contain the same number of values. To find the weighted average of two or more sets of numbers, proceed as follows:

- Multiply the average of each set by the number of values in that set, and then add the products.
- Divide the sum of the products by the total number of values in all of the sets.

**EXAMPLE:** In a class, 18 students had an average midterm exam grade of 85 and the 12 remaining students had an average midterm exam grade of 90. What is the average midterm exam grade of the entire class?

**SOLUTION:**

$$\text{Weighted average} = \frac{\overbrace{(\text{Average} \times \text{number})}^{\text{Group 1}} + \overbrace{(\text{Average} \times \text{number})}^{\text{Group 2}}}{\text{Total number of students}}$$

$$= \frac{(85 \times 18) + (90 \times 12)}{18 + 12}$$

$$= \frac{1530 + 1080}{30}$$

$$= \frac{2610}{30}$$

$$= 87$$

## FINDING THE MEDIAN

To find the median of a set of numbers, first arrange the numbers in size order.

- If a set contains an odd number of values, the median is the middle value. For example, the median of the set of numbers

$$8, 12, 15, \underbrace{17}_{\text{Median}} 19, 20, 25$$

is 17 since the number of values in the set that are less than 17 is the same as the number of values in the set that are greater than 17.

- If a set contains an even number of values, the median is the average (arithmetic mean) of the two middle values. For example, the set of numbers

$$10, 20, 24, 30, 40, 50$$

contains six values. Since the two middle values in the set are 24 and 30, the median is their average:

$$10, 20, \underbrace{24, 30}, 40, 50$$

$$\text{Median} = \frac{24 + 30}{2} = 27$$

## FINDING THE MODE

The mode of a set of numbers is the number that appears the greatest number of times in the set. For example, the mode of the set

$$7, 2, 6, 3, 2, 6, 7, 3, 9, 6$$

is 6, since 6 appears more times than any other number in the set.

# LESSON 7-1   TUNE-UP EXERCISES

**Multiple Choice**

**1** The average (arithmetic mean) of a set of seven numbers is 81. If one of the numbers is discarded, the average of the remaining numbers is 78. What is the value of the number that was discarded?

(A) 98
(B) 99
(C) 100
(D) 101
(E) 102

**2**

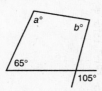

In the figure above, what is the average of $a$ and $b$?

(A) 75
(B) 80
(C) 85
(D) 90
(E) 95

**3** If the average of the two-digit numbers $1N$, $N7$, and $NN$ is 35, then $N =$

(A) 2
(B) 3
(C) 4
(D) 5
(E) 6

**4** What is the area of the circle whose radius is the average of the radii of two circles with areas of $16\pi$ and $100\pi$?

(A) $25\pi$
(B) $36\pi$
(C) $49\pi$
(D) $64\pi$
(E) $81\pi$

**5** If the average of $x$, $3x + 1$, and $5x - 13$ is 47, then $x =$

(A) 15
(B) 17
(C) 18
(D) 20
(E) 21

**6** When $x$ is subtracted from $2y$, the difference is equal to the average of $x$ and $y$. What is the value of $\frac{x}{y}$?

(A) $\frac{1}{2}$

(B) $\frac{2}{3}$

(C) 1

(D) $\frac{3}{2}$

(E) 2

**7** If the average of $x$, $y$, and $z$ is 32 and the average of $y$ and $z$ is 27, what is the average of $x$ and $2x$?

(A) 42
(B) 45
(C) 48
(D) 50
(E) 63

**8** For which set of numbers do the average, median, and mode all have the same value?

(A) 2, 2, 2, 2, 4
(B) 1, 3, 3, 3, 5
(C) 1, 1, 2, 5, 6
(D) 1, 1, 1, 2, 5
(E) 1, 1, 3, 5, 10

**9** A man drove a car at an average rate of speed of 45 miles per hour for the first 3 hours of a 7-hour car trip. If the average rate of speed for the entire trip was 53 miles per hour, what was the average rate of speed in miles per hour for the remaining part of the trip?

(A) 50
(B) 55
(C) 57
(D) 59
(E) 62

**10** For the set of numbers 2, 2, 4, 5, and 12, which statement is true?

(A) Mean = median
(B) Mean > mode
(C) Mean < mode
(D) Mode = median
(E) Mean = mode

**11** Susan received grades of 78, 93, 82, and 76 on four math exams. What is the lowest score she can receive on her next math exam and have an average of at least 85 on the five exams?

(A) 96
(B) 94
(C) 92
(D) 90
(E) 88

**12** What is the average of $(x + y)^2$ and $(x - y)^2$?

(A) $\dfrac{x + y}{2}$
(B) $xy$
(C) $x^2 - y^2$
(D) $\dfrac{xy}{2}$
(E) $x^2 + y^2$

**13** If the average of $x$ and $y$ is $\dfrac{11}{2}$ and the average of $\dfrac{1}{x}$ and $\dfrac{1}{y}$ is $\dfrac{11}{24}$, then $xy =$

(A) 4
(B) 6
(C) 11
(D) 12
(E) 14

**14** If the ratio of $a$ to $b$ is $\dfrac{1}{2}$ and the ratio of $c$ to $b$ is $\dfrac{1}{3}$, what is the average of $a$ and $c$ in terms of $b$?

(A) $\dfrac{b}{6}$
(B) $\dfrac{b}{4}$
(C) $\dfrac{5b}{12}$
(D) $\dfrac{b}{2}$
(E) $\dfrac{5b}{6}$

**15** The average of $a$, $b$, $c$, $d$, and $e$ is 28. If the average of $a$, $c$, and $e$ is 24, what is the average of $b$ and $d$?

(A) 31
(B) 32
(C) 33
(D) 34
(E) 36

**16** If $2a + b = 7$ and $b + 2c = 23$, what is the average of $a$, $b$, and $c$?

(A) 5
(B) 7.5
(C) 10
(D) 12.25
(E) 15

**17** The average final exam grade of 10 students is $x$, and the average final exam grade of 15 students is $y$. If the average final exam grade of the 25 students is $t$, what is $x$ in terms of $y$ and $t$?

(A) $\dfrac{5t - 3y}{2}$

(B) $\dfrac{3t - 5y}{2}$

(C) $\dfrac{3t - 2y}{5}$

(D) $\dfrac{t - 15y}{10}$

(E) $\dfrac{t - 10y}{15}$

**18** The average of $a$, $b$, $c$, and $d$ is $p$. If the average of $a$ and $c$ is $q$, what is the average of $b$ and $d$ in terms of $p$ and $q$?

(A) $2p + q$

(B) $2p - q$

(C) $2q + p$

(D) $2q - p$

(E) $\dfrac{2p + q}{3}$

---

## Quantitative Comparison

Each question consists of two quantities in boxes, one in Column A and one in Column B. You are to compare the two quantities and on the answer sheet fill in

A  if the quantity in Column A is greater;

B  if the quantity in Column B is greater;

C  if the two quantities are equal;

D  if the relationship cannot be determined from the information given

AN E RESPONSE WILL NOT BE SCORED

---

| Column A | Column B |
|---|---|

$$\dfrac{p + q}{6} = r$$

**1**

| The average of $p$ and $q$ | $3r$ |

The average of $3x + 1$ and $x - 7$ is 6.

**2**

| $x$ | 4 |

The average of $a$, $b$, and $c$ is $A$:
$$k > 0$$

**3**

| The average of $a + k$, $b + k$, and $c + k$ | $A + 3k$ |

| Column A | Column B |
|---|---|

$$a - 2b = 7$$
$$a + 2b = 3$$

**4**

| The average of $a$ and $b$ | 3 |

The graph below shows the distribution of temperatures recorded at the same time each day.

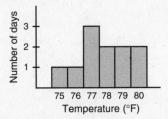

**5**

| The temperature that is the mode | The median temperature |

---

## Grid In

**1**    What is the greatest of four consecutive even integers whose average is 19?

**2**    If the average of $x$, $y$, and $z$ is 12, what is the average of $3x$, $3y$, and $3z$?

**3**    If the average of three consecutive multiples of 4 is 132, what is the LEAST of these multiples?

# Counting Problems

## OVERVIEW

*Some counting problems can be solved simply by compiling a list and then adding up the number of items on the list. Rather than listing the number of ways in which two or more events can happen, it may be faster to multiply together the numbers of ways in which all of these events can occur. Overlapping circles can be used to help solve counting problems that involve sets with members in common.*

## COUNTING BY MAKING A LIST

Some counting problems can be solved by systematically listing and then counting all possible items that fit the conditions of the problem.

**EXAMPLE:** If $1 \leq x \leq 5$, $0 \leq y \leq 3$, and $x$ and $y$ are integers, how many ordered pairs $(x, y)$ can be formed in which the sum of $x$ and $y$ is less than 4?

**SOLUTION:** Make a list by picking 1, 2, and 3 as values for $x$ and matching possible values of $y$ that make their sums equal to 3 or less:

$$(1, 0) \quad (2, 0) \quad (3, 0)$$

$$(1, 1) \quad (2, 1)$$

$$(1, 2)$$

Hence, there are six such ordered pairs.

## USING THE MULTIPLICATION PRINCIPLE OF COUNTING

When it is not practical to count by making a list, you may be able to use the multiplication principle of counting. This principle states that, if one event can happen in $p$ ways and another event can happen in $q$ ways, then both events can happen in $p \times q$ ways. For example:

• If there are three roads that can be traveled by car to get from town $A$ to town $B$, and two roads that can be traveled by car to get from town $B$ to town $C$, then there is a total of $3 \times 2 = 6$ ways in which a car can travel from town $A$ to town $B$, and then to town $C$.

**325**

- A man with four different ties, five different shirts, and three different pairs of slacks can put together $4 \times 5 \times 3 = 60$ different outfits consisting of one tie, one shirt, and one pair of slacks.
- The total number of different three-digit combinations possible for a combination lock is $10 \times 10 \times 10 = 1000$ since each of the three lock settings can accept any one of the ten digits from 0 to 9.

**EXAMPLE:**    Using the numbers 1, 2, 3, 4, 5, 6, and 7, how many three-digit whole numbers greater than 500 can be formed if the same digit cannot be used more than once?

**SOLUTION:**    To find the number of three-digit whole numbers that satisfy the conditions of the problem, multiply together the numbers of possible digits that can be used to fill all of the three positions of the number that is being formed.

- Since the three digit number must be greater than 500, there are only three possible choices for the first digit: 5, 6, and 7.
- After a digit is selected for the first position, any one of the remaining six digits can be used to fill the middle position of the three-digit number. After the middle digit is selected, any one of the five remaining digits can be used to fill the last position.
- Using the multiplication principle of counting gives

$$3 \times 6 \times 5 = 90$$

Thus, 90 three-digit whole numbers greater than 500 can be formed without reusing any of the given digits.

If digits could be repeated, then $3 \times 7 \times 7 = 147$ represents the number of three-digit whole numbers greater than 500 that could be formed from the digits 1, 2, 3, 4, 5, 6, and 7.

## COUNTING USING VENN DIAGRAMS

For any sets $A$ and $B$, let

$n(A)$ = the number of elements in set $A$;

$n(B)$ = the number of elements in set $B$; and

$n(A \cap B)$ = the number of elements that belong to both sets

These relationships of the elements in sets $A$ and $B$ are shown in the **Venn diagram** below.

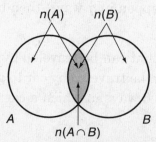

The number of elements in set $A$ *or* in set $B$ or in *both* sets $A$ and $B$ is denoted by $n(A \cup B)$, where

$$n(A \cup B) = n(A) + n(B) - n(A \cap B)$$

**EXAMPLE:** In a class of 30 students, 20 students are studying French, 12 students are studying Spanish, and 7 students are studying both French and Spanish. How many students in this class are studying neither French nor Spanish?

**SOLUTION:** If set $A$ represents students who are studying French and set $B$ represents students who are studying Spanish, then $n(A) = 20$, $n(B) = 12$, and $n(A \cap B) = 7$. Hence,

$$n(A \cup B) = n(A) + n(B) - n(A \cap B)$$

$$= 20 + 12 - 7$$

$$= 25$$

Since 25 of the 30 students in the class are studying either French or Spanish or both French and Spanish, 5 ($= 30 - 25$) students in the class are studying neither French nor Spanish.

**EXAMPLE:** If in the preceeding example Spanish and French are the only languages studied, how many students study exactly one language?

**SOLUTION:**

- $20 - 7 = 13$ students study Spanish but not French.
- $12 - 7 = 5$ students study French but not Spanish.
- $13 + 5 = 18$ students study Spanish or French but not both.

Hence, 18 students study exactly one language.

# LESSON 7-2 TUNE-UP EXERCISES

---

**Multiple Choice**

---

**1** An ice cream parlor makes a sundae using one of six different flavors of ice cream, one of three different flavors of syrup, and one of four different toppings. What is the total number of different sundaes that this ice cream parlor can make?

(A) 72
(B) 36
(C) 30
(D) 26
(E) 13

**2** Five blocks are painted red, blue, green, yellow, and white. In how many different ways can the blocks be arranged in a row if the red block is always first?

(A) 120
(B) 96
(C) 48
(D) 24
(E) 12

**3**

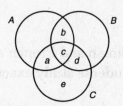

In the figure above, five nonoverlapping regions formed by sets *A*, *B*, and *C* are labeled *a* through *e*. Which region represents the set of elements that belong to sets *B* and *C* but not to set *A*?

(A) *a*
(B) *b*
(C) *c*
(D) *d*
(E) *e*

**4** In how many different ways can five students be seated in three chairs?

(A) 15
(B) 18
(C) 30
(D) 60
(E) 120

**5** A quarter, a dime, a nickel, and a penny are placed in a box. One coin is drawn from the box and then put back before a second coin is drawn. In how many different ways can the two coins be drawn so that the sum of the values of the two coins is at least 25 cents?

(A) 9
(B) 7
(C) 6
(D) 5
(E) 4

**6** The students in a certain physical education class are on either the basketball team or the tennis team, are on both these teams, or are not on either team. If 15 students are on the basketball team, 18 students are on the tennis team, 11 students are on both teams, and 14 students are not on any of these teams, how many students are in the class?

(A) 48
(B) 40
(C) 36
(D) 30
(E) 28

**7** How many four-digit numbers greater than 1000 can be formed from the digits 0, 1, 2, 3, 4, and 5 if the same digit cannot be used more than once?

(A) 1296
(B) 625
(C) 360
(D) 300
(E) 120

**8** From the letters of the word "TRIANGLE," how many three-letter arrangements can be formed if the first letter must be $T$, one of the other letters must be $A$, and no letter can be used more than once in any arrangement?

(A) 6
(B) 10
(C) 12
(D) 42
(E) 84

**9** In a French class, each student belongs to the French Club or to the Spanish Club. In this class, $F$ students are in the French Club, $S$ students are in the Spanish Club, and $N$ of these students are in both the French and the Spanish clubs. What fraction of the class belongs to the French Club but not to the Spanish Club?

(A) $\dfrac{F-N}{F+S}$

(B) $\dfrac{F-N}{F+S+N}$

(C) $\dfrac{F-N}{F+S-N}$

(D) $\dfrac{F}{F+S-2N}$

(E) $\dfrac{F-N}{F+S-2N}$

**10** How many numbers greater than 400 but less than 9999 can be formed using the digits 0, 1, 3, 4, and 5 if no digit may be used more than once in any number?

(A) 24
(B) 48
(C) 96
(D) 120
(E) 180

**11**

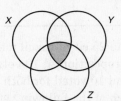

In the figure above, circle $X$ represents the set of all positive odd integers, circle $Y$ represents the set of all numbers whose square roots are integers, and circle $Z$ represents the set of all positive multiples of 5. Which of the following numbers is a member of the set represented by the shaded region?

(A) 25
(B) 49
(C) 75
(D) 81
(E) 100

**12**
$$X = \{1, 2, 4\}$$
$$Y = \{1, 3, 4\}$$

If, in the sets above, $x$ is any number in set $X$ and $y$ is any number in set $Y$, how many different values of $x + y$ are possible?

(A) Five
(B) Six
(C) Seven
(D) Eight
(E) Nine

**13** In a music class of 30 students, 50% study woodwinds, 40% study strings, and 30% study both woodwinds and strings. What percent of the students in the class do NOT study either woodwind or string instruments?

(A) 25%
(B) 40%
(C) 50%
(D) 60%
(E) 75%

**14** Of the 25 city council members, 12 voted for proposition *A*, 19 voted for proposition *B*, and 10 voted for both propositions. Which of the following statements must be true?

   I. Seven members voted for proposition *B* but not for proposition *A*.
  II. Four members voted against both propositions.
 III. Eleven members voted for one proposition and against the other proposition.

(A) I only
(B) II only
(C) I and II only
(D) I and III only
(E) II and III only

**15** In a survey of 63 people, 23 people subscribed to magazine *A*, 21 people subscribed to magazine *B*, and 17 subscribed to magazine *C*. For any two of the magazines, 4 people subscribed to both magazines but not to the third magazine. If 5 people in the survey did not subscribe to any of the three magazines, how many people subscribed to all three magazines?

(A) 9
(B) 7
(C) 5
(D) 2
(E) It cannot be determined from the information given.

## LESSON 7-3

# Probability Problems

## OVERVIEW

*The **probability** of an event is a number from 0 to 1 that indicates the likelihood that the event will occur. The closer a probability value for an event is to 1, the greater the likelihood that this event will occur.*

## EXPRESSING PROBABILITY AS A FRACTION

If an event can happen in $r$ of $n$ equally likely ways, then the probability that this event will occur is $\frac{r}{n}$. For example, if a jar contains one red marble, two yellow marbles, and four green marbles, the probability that a marble picked from the jar at *random* (that is, without looking) will be yellow is

$$\frac{2}{1+2+4} = \frac{2}{7}$$

In the probability fraction above, 2 corresponds to the number of yellow marbles or "favorable outcomes," and 7 represents the total number of marbles or possible outcomes. In general,

$$\text{Probability of an event} = \frac{\text{Number of favorable outcomes}}{\text{Total number of outcomes}}$$

## FINDING A PROBABILITY BY LISTING OUTCOMES

To figure out the number of favorable outcomes for an event, it may be helpful to first list all the different possible outcomes for that probability experiment.

**EXAMPLE:** A penny, a nickel, and a dime are placed in a hat. A coin is drawn, its value is noted, and then the coin is replaced in the same hat. If a second coin is drawn, what is the probability that the sum of the values of the two coins will be more than 10 cents?

**SOLUTION:**

• All possible outcomes can be listed as a set of ordered pairs:

| | | |
|---|---|---|
| (penny, penny) | (nickel, penny) | (dime, penny) |
| (penny, nickel) | (nickel, nickel) | (dime, nickel) |
| (penny, dime) | (nickel, dime) | (dime, dime) |

- Five of the nine possible outcomes are favorable:

(penny, dime) = 11 cents   (dime, penny) = 11 cents

(nickel, dime) = 15 cents   (dime, nickel) = 15 cents

(dime, dime) = 20 cents

- Hence:

$$\text{Probability} = \frac{\text{Number of favorable outcomes}}{\text{Total number of outcomes}} = \frac{5}{9}$$

## MULTIPLYING PROBABILITIES

Multiplying the probability that one event will occur by the probability that a second event will occur gives the probability that both events will occur. For example, suppose a fair coin is tossed 2 times. Since the probability of getting a head on each toss is $\frac{1}{2}$, the probability of tossing two heads is $\frac{1}{2} \times \frac{1}{2}$ or $\frac{1}{4}$. Also, the probability of tossing three heads is $\frac{1}{2} \times \frac{1}{2} \times \frac{1}{2}$ or $\frac{1}{8}$.

## SOME PROBABILITY FACTS

Since the numerator of a probability fraction must be less than or equal to the denominator, the probability of an event must be a number from 0 to 1.

- If an event is certain to happen, its probability is 1.
- If an event can never happen, its probability is 0.
- The probability that an event will *not* happen is 1 minus the probability that the event will happen. For example, if the probability that a basketball team will win tomorrow is $\frac{1}{5}$, then the probability that it will lose is $1 - \frac{1}{5} = \frac{4}{5}$.
- The sum of the probabilities of the events that comprise a probability experiment is 1. For example, if a fair coin is tossed, then

$$P(\text{head}) + P(\text{tail}) = \frac{1}{2} + \frac{1}{2} = 1$$

As another example, consider a jar that contains only white, blue, and orange marbles. If the probability of selecting a white marble is $\frac{1}{3}$ and the probability of selecting a blue marble is $\frac{1}{4}$, then the probability of selecting an orange marble must be the fraction that, when added to $\frac{1}{3}$ and $\frac{1}{4}$, gives 1. Hence, the probability of selecting an orange marble is

$$1 - \left(\frac{1}{3} + \frac{1}{4}\right) = 1 - \left(\frac{4}{12} + \frac{3}{12}\right) = \frac{5}{12}$$

# LESSON 7-3   TUNE-UP EXERCISES

**Multiple Choice**

**1** A bag contains three green marbles, four blue marbles, and two orange marbles. If a marble is picked at random, what is the probability that an orange marble will NOT be picked?

(A) $\frac{1}{4}$

(B) $\frac{1}{3}$

(C) $\frac{4}{11}$

(D) $\frac{1}{2}$

(E) $\frac{7}{9}$

1, 2, 2, 3, 3, 3, 4, 4, 4, 4

**2** What is the probability that a number selected at random from the set of numbers above will be the average of the set?

(A) 0

(B) $\frac{1}{10}$

(C) $\frac{1}{5}$

(D) $\frac{3}{10}$

(E) $\frac{2}{5}$

**3** If the probability that an event will occur is $\frac{x}{4}$ and $x \neq 0$, what is the probability that this event will NOT occur ?

(A) $\frac{1-x}{4}$

(B) $\frac{4-x}{4}$

(C) $\frac{4-x}{x}$

(D) $\frac{4}{x}$

(E) $\frac{x}{x-4}$

**Questions 4 and 5.**

A jar contains only red, blue, and green marbles. The probability of picking a red marble is 0.25, and the probability of picking a blue marble is 0.40.

**4** What is the probability of picking a green marble?

(A) 0.25
(B) 0.35
(C) 0.60
(D) 0.75
(E) 1.00

**5** What is the LEAST number of marbles that could be in the jar?

(A) 4
(B) 5
(C) 10
(D) 20
(E) It cannot be determined from the information given.

**6** A cube whose faces are numbered from 1 to 6 is rolled. Which of the following statements must be true?

  I. The probability of getting a 5 on the top face is $\frac{1}{5}$.
 II. The probability of getting an even number on the top face is $\frac{1}{2}$.
III. The probability of getting a prime number on the top face is $\frac{1}{2}$.

(A) I and II only
(B) I and III only
(C) II only
(D) II and III only
(E) All are true.

**7** If the letters $L, O, G, I,$ and $C$ are randomly arranged to form a five-letter "word," what is the probability that the result will be the word LOGIC?

(A) $\frac{1}{120}$

(B) $\frac{1}{24}$

(C) $\frac{1}{5}$

(D) $\frac{1}{4}$

(E) $\frac{1}{2}$

**8**

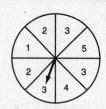

The circle above is divided into eight sectors of equal area. What is the probability that the spinner will land on an even-numbered region in each of two consecutive spins?

(A) $\frac{3}{4}$

(B) $\frac{3}{8}$

(C) $\frac{25}{64}$

(D) $\frac{1}{4}$

(E) $\frac{9}{64}$

**9** Three fair coins are tossed at the same time. What is the probability that all three coins will come up heads *or* all will come up tails?

(A) $\frac{1}{8}$

(B) $\frac{1}{6}$

(C) $\frac{1}{4}$

(D) $\frac{1}{3}$

(E) $\frac{3}{8}$

**10** The faces of a red cube and a yellow cube are numbered from 1 to 6. Both cubes are rolled. What is the probability that the top face of each cube will have the same number?

(A) $\frac{1}{8}$

(B) $\frac{1}{6}$

(C) $\frac{1}{4}$

(D) $\frac{1}{3}$

(E) $\frac{1}{2}$

**11** A jar contains 54 marbles each of which is either blue, green, or white. The probability of selecting a blue marble at random from the jar is $\frac{1}{3}$, and the probability of selecting a green marble at random is $\frac{4}{9}$. How many white marbles does the jar contain?

(A) 6
(B) 8
(C) 9
(D) 12
(E) 18

**12** $$A = \{-2, -1, 0, 1, 2\}$$

If $x$ represents a number picked at random from set $A$ above, what is the probability that $x^2 < 2$?

(A) $\frac{1}{5}$

(B) $\frac{2}{5}$

(C) $\frac{3}{5}$

(D) $\frac{4}{5}$

(E) 1

**13** The probability of guessing the correct answer to a certain test question is $\frac{x}{12}$. If the probability of NOT guessing the correct answer to this question is $\frac{2}{3}$, what is the value of $x$?

(A) 2
(B) 3
(C) 4
(D) 6
(E) 12

**14**
$$A = \{1, 2, 3\}$$
$$B = \{1, 4, 9\}$$

A number is randomly selected from set $A$ above, and then a second number is randomly selected from set $B$. What is the probability that the product of the two numbers selected will be less than 9?

(A) $\frac{1}{3}$

(B) $\frac{5}{9}$

(C) $\frac{2}{3}$

(D) $\frac{7}{9}$

(E) $\frac{5}{6}$

## Quantitative Comparison

Each question consists of two quantities in boxes, one in Column A and one in Column B. You are to compare the two quantities and on the answer sheet fill in

A  if the quantity in Column A is greater;
B  if the quantity in Column B is greater;
C  if the two quantities are equal;
D  if the relationship cannot be determined from the information given

AN E RESPONSE WILL NOT BE SCORED

| Column A | Column B |
|---|---|

One card is picked at random from 20 cards numbered 1 through 20.

**1**

| The probability that the card will have a prime number | The probability that the card will have a number less than 4 or greater than 16 |
|---|---|

The probability that event $E$ will occur is $\frac{1}{p}$ $(p > 1)$.

**2**

| The probability that event $E$ will *not* occur | $p$ |
|---|---|

| Column A | Column B |
|---|---|

From a jar that contains three blue marbles and three green marbles, a marble is picked at random and is not put back into the jar. Another marble is then picked at random from the same jar.

**3**

| The probability that the two marbles will be different colors | $\frac{1}{3}$ |
|---|---|

A fair coin is tossed 10 times.

**4**

| The probability of tossing exactly three heads | The probability of tossing exactly seven tails |
|---|---|

Column A          Column B

$A = \{1, 3\}$ and $B = \{2, 4\}$
$a$ is a number picked from set $A$.
$b$ is a number picked from set $B$.

**5** | The probability that $ab$ is greater than 5 | The probability that $a + b$ equals 5

---

## Grid In

**1** The probability of selecting a green marble at random from a jar that contains only green, white, and yellow marbles is $\frac{1}{4}$. The probability of selecting a white marble at random from the same jar is $\frac{1}{3}$. If this jar contains 10 yellow marbles, what is the total number of marbles in the jar?

**2** Each of five identical cards is numbered on one of its sides with a different integer from 1 to 5. The cards are turned over and then shuffled so that they are not in a predictable order. If two cards are selected at random without replacement, what is the probability that their sum is at least 8?

# Miscellaneous Problem Types

## OVERVIEW

*Some SAT questions do not fall into a particular branch of mathematics. These questions may involve:*

- *performing an unfamiliar arithmetic operation by following a given rule;*
- *finding a missing digit in a correctly worked out arithmetic problem; or*
- *figuring out a particular term in a list of numbers that follow a rule.*

## NEW OPERATIONS THAT USE SPECIAL SYMBOLS

In some SAT I math questions a new arithmetic operation is invented by giving a rule that uses an unfamiliar operation symbol. For example, if for all positive integer values of $a$ and $b$, $\otimes$ is defined by the equation

$$a \otimes b = \left(\frac{a}{b}\right)^2$$

then the value of $6 \otimes 2$ can be obtained by substituting 6 for $a$ and 2 for $b$ in the given formula. Thus:

$$a \otimes b = \left(\frac{a}{b}\right)^2$$

$$6 \otimes 2 = \left(\frac{6}{2}\right)^2 = (3)^2 = 9$$

## MISSING-DIGIT PROBLEMS

In some SAT I math questions a correctly worked out arithmetic problem is presented with one or more digits replaced by letters or symbols. You need to use mathematical reasoning, together with systematic trial and error, to figure out the missing digits.

**EXAMPLE:**

$$\begin{array}{r} A\,B \\ \times \quad B \\ \hline 1\,A\,9 \end{array}$$

In the correctly worked out multiplication problem above, $A$ and $B$ are single digits. What is a possible value of $A$?

**SOLUTION:**

- Note that, since $B$ times $B$ contains the digit 9, the only possible choice for $B$ is 3 or 7.
- Try $B = 3$ by systematically multiplying numbers of the form $A3$ by 3 to see whether any products of the form $1A9$ are possible. Since the product must be a three-digit number, $A$ in this case must be greater than 3. The product on the left below is a three-digit number, but $A$ is 4 in the multiplicand and 2 in the product. Try $A = 5$, and note that the product on the right has the desired form.

$$\begin{array}{r} 43 \\ \times \ \ 3 \\ \hline 129 \end{array} \qquad \begin{array}{r} 53 \\ \times \ \ 3 \\ \hline 159 \end{array}$$

- If no choices for $A$ work with $B = 3$, repeat the process with $B = 7$.

## NUMBER SEQUENCES

A list of numbers that follow a predictable pattern is called a **sequence.** For example, in the sequence

$$3, 6, 12, 24, 48, \ldots$$

each number after the first is obtained by multiplying the number that comes before it by 2. The three trailing dots (. . .) indicate that the pattern continues without ever ending. SAT I test questions may ask you to find or compare the values of particular terms of a sequence.

**EXAMPLE:**  In the sequence below, each term after the first term, $x$, is obtained by doubling the term that comes before it and then subtracting 3. What is the value of $x + z$?

$$x, y, 19, 35, 67, z, \ldots$$

**SOLUTION:**

- Since you need to find the terms that come *before* 19, reverse the given rule. *Add* 3 to 19 to get 22, and then take *one-half* of 22 to get $y = 11$.
- To find $x$, add 3 to 11 (the value of $y$) to get 14, and then take one-half of 14 to get $x = 7$.
- To find $z$, double the preceding term, 67, to get 134. Subtracting 3 from 134 gives $z = 131$.

Hence, $x + z = 7 + 131 = 138$.

## LESSON 7-4  TUNE-UP EXERCISES

> **Multiple Choice**

**1** For all nonnegative numbers $a$, $\boxed{a} = \sqrt{a} + 1$. Which of the following is equal to the sum of $\boxed{36} + \boxed{64}$?

(A) 10
(B) 12
(C) 16
(D) 52
(E) 102

**2** If

$$_b\!\!\overset{a}{\triangle}\!\!_c = \frac{a+b}{c} + \frac{a+c}{b} + \frac{b+c}{a}$$

for all nonzero $a$, $b$, and $c$, then $_2\!\!\overset{1}{\triangle}\!\!_3 =$

(A) 12
(B) 10
(C) 8
(D) 5
(E) 3

Questions 3 and 4 refer to the following definition:

For all positive integers $k$,

let $\boxed{k\!\!\rhd} = 2(k-1)$  if $k$ is even;

let $\boxed{k\!\!\rhd} = \frac{1}{2}(k+1)$  if $k$ is odd.

**3** The product $\boxed{6\!\!\rhd} \times \boxed{11\!\!\rhd} =$

(A) 25
(B) 30
(C) 31
(D) 59
(E) 60

**4** If $\boxed{N\!\!\rhd}$ is a multiple of 4, which of the following can be the value of $N$?

(A) 8
(B) 12
(C) 25
(D) 31
(E) 40

**5**
$$\begin{array}{r} 1\,A\,6 \\ B\,4 \\ +\ \ B\,8 \\ \hline 148 \end{array}$$

In the correctly worked out addition example above, what is the greatest possible value of digit $B$?

(A) 9
(B) 8
(C) 6
(D) 5
(E) 4

**6**      $a_1, a_2, a_3, a_4, a_5, \ldots, a_n$

In the sequence of positive integers above, $a_1 = a_2 = 1$, $a_3 = 2$, $a_4 = 3$, and $a_5 = 5$. If each term after the second is obtained by adding the two terms that come before it and if $a_n = 55$, what is the value of $n$?

(A) 12
(B) 10
(C) 9
(D) 8
(E) 5

**7**
$$\begin{array}{r} 6\,3 \\ \times\ \ 4\,P \\ \hline 3\,1\,P \\ 2\,P\,2 \\ \hline 2\,Q\,3\,P \end{array}$$

In the correctly worked out multiplication problem above, what digit does $Q$ represent?

(A) 8
(B) 7
(C) 6
(D) 5
(E) 4

Questions 8 and 9 refer to the following definition:

Let the operation $\Phi$ be defined for all positive integers $a$ and $b$ by the equation
$$a \Phi b = ab - a$$

**8** For what value of $a$ is $a \Phi 4 = 24$?

(A) 3
(B) 4
(C) 6
(D) $\frac{25}{4}$
(E) 8

**9** Which of the following must be true?

I. $a \Phi b = b \Phi a$
II. $a \Phi (a + 1) = a^2$
III. $\frac{a}{2} \Phi \frac{b}{2} = \frac{a \Phi b}{2}$

(A) I only
(B) II only
(C) III only
(D) I and III only
(E) II and III only

**10**
$$
\begin{array}{r}
C\,5 \\
B\,7 \\
+\ 1\,A\,9 \\
\hline
3\,6\,1
\end{array}
$$

In the correctly worked out addition problem above, which of the following could be digit $A$?

I. 0
II. 4
III. 8

(A) I only
(B) III only
(C) I and II only
(D) II and III only
(E) I, II, and III

Questions 11 and 12 refer to the following definition:

For all positive integers $y$,
$$\triangle y = 2\sqrt{y}.$$

**11** Which of the following equals 8?

(A) $\triangle 4$
(B) $\triangle 8$
(C) $\triangle 16$
(D) $\triangle 32$
(E) $\triangle 64$

**12** $\triangle y \times \triangle y =$

(A) 2
(B) 4
(C) $4\sqrt{y}$
(D) $4y$
(E) $4y^2$

**13** Let $\smile x \smile$ be defined for all real values of $x$ by the equation
$$\smile x \smile = 2x + 1$$
If $3 - \smile x \smile = x$, then $x =$

(A) $\frac{2}{3}$
(B) 1
(C) $\frac{3}{2}$
(D) 2
(E) 3

**14**

$$
\begin{array}{r}
8\ 3\ A \\
+\ D\ B\ B \\
\hline
C\ A\ C\ 2
\end{array}
$$

In the correctly worked out addition problem above, what digit does $D$ represent?

(A) 2
(B) 4
(C) 5
(D) 6
(E) 7

Questions 15 and 16 refer to the following definition:

For all positive integers, let  be defined by the equation

$$\langle p \rangle = p + k$$

where $k$ is the greatest divisor of $p$ and $k < p$.

**15**  =

(A) 56
(B) 42
(C) 35
(D) 32
(E) 30

**16** If $\langle 15 \rangle = n$, then $\langle n \rangle =$

(A) 20
(B) 24
(C) 25
(D) 30
(E) 40

**17**

$$\frac{1}{2}, -\frac{1}{4}, \frac{1}{8}, -\frac{1}{16}, \ldots, a_n$$

In the above sequence, $a_1 = \frac{1}{2}$, $a_2 = -\frac{1}{4}$, $a_3 = \frac{1}{8}$, $a_4 = -\frac{1}{16}$, and $a_n$ represents the $n$th term. Which equation expresses $a_n$ in terms of $n$?

(A) $(-1)^{n+1} \cdot \dfrac{1}{2^n}$

(B) $(-1)^n \cdot \dfrac{1}{2^n}$

(C) $(-1)^{n+1} \cdot \dfrac{1}{2^{-n}}$

(D) $\left(\dfrac{-1}{2}\right)^n$

(E) $\dfrac{-1}{2^n}$

## Quantitative Comparison

Each question consists of two quantities in boxes, one in Column A and one in Column B. You are to compare the two quantities and on the answer sheet fill in

A if the quantity in Column A is greater;
B if the quantity in Column B is greater;
C if the two quantities are equal;
D if the relationship cannot be determined from the information given

AN E RESPONSE WILL NOT BE SCORED

---

Column A      Column B

*a* and *b* are consecutive terms of the sequence

$$-1, 0, 1, -1, 0, 1, -1, 0, 1, \ldots$$

in which −1, 0, and 1 endlessly repeat.

**1**    $a + b$      $a - b$

---

$$
\begin{array}{r}
P \\
Q \\
+\ R \\
\hline
2X
\end{array}
$$

In the correctly completed addition problem above, $P$, $Q$, and $R$ are consecutive nonzero digits.

**2**    $X$      $3$

---

$$
\begin{array}{r}
R\,8 \\
+\ 8\,S \\
\hline
T\,S\,T
\end{array}
$$

In the correctly completed addition problem above, $R$, $S$, and $T$ represent nonzero digits.

**3**    $R$      $S$

---

Column A      Column B

Let  be defined for all real values of *x* by the equation

 $= x^3 - x + 1$

**4**         

---

### Questions 5 and 6.

For all nonzero numbers *x* and *y*, the operation $\otimes$ is defined by the equation

$$x \otimes y = \frac{x+y}{x^2+y^2}$$

**5**    $x \otimes y$      $(-x) \otimes (-y)$

**6**    $x \otimes x$      $\dfrac{1}{x}$

---

$$
\begin{array}{r}
6\,R\,3 \\
\times\qquad 7 \\
\hline
4\,S\,R\,1
\end{array}
$$

In the correctly completed multiplication problem above, $R$ and $S$ represent nonzero digits

**7**    $R$      $S$

| Column A | Column B |
|---|---|

For all positive integers $k$, $<k>$ equals the integer that is closest or equal to $\sqrt{k}$.

**8** | $<k> \times <k>$ | $k$ |

| Column A | Column B |
|---|---|

$$
\begin{array}{r}
Q\,7 \\
8\,3 \\
4\,P \\
+\ P\,Q \\
\hline
2\,P\,1
\end{array}
$$

In the correctly worked out addition problem above, $P$ and $Q$ represent nonzero digits

**9** | $P$ | $Q$ |

---

## Grid In

**1** If the operation $\Omega$ is defined by the equation

$$x \,\Omega\, y = 2x + 3y$$

what is the value of $a$ in the equation $a \,\Omega\, 4 = 1 \,\Omega\, a$?

**2**
$$x, y, 22, 14, 10, \ldots$$

In the sequence above, each term after the first term, $x$, is obtained by halving the term that comes before it and then adding 3 to that number. What is the value of $x - y$?

**3** In a certain sequence of positive integers, if $m$ and $n$ are consecutive terms, the next consecutive term is

$$\dfrac{m+n}{2} \qquad \text{if } m+n \text{ is even; and}$$

$$m+n-3 \qquad \text{if } m+n \text{ is odd}$$

For example, if two consecutive terms are 11 and 17, the next two consecutive terms are 14 and 28. In a sequence that follows the same rule, if the first two terms are 15 and 19, what is the seventh term?

**4** Let the operation $\blacktriangle$ be defined by the equation

$$a \,\blacktriangle\, b = ab - (a + b)$$

If $5 \,\blacktriangle\, b = 1$, what is the value of $b$?

# ANSWERS TO CHAPTER 7 TUNE-UP EXERCISES

## LESSON 7-1  *(Multiple Choice)*

1. **B**  If the average (arithmetic mean) of a set of seven numbers is 81, then the sum of these seven numbers is $7 \times 81$ or 567. Since, if one of the numbers is discarded, the average of the six remaining numbers is 78, the sum of these six numbers is $6 \times 78$ or 468. Since $567 - 468 = 99$, the value of the number that was discarded is 99.

2. **E**  Since vertical angles are equal, the angle of the quadrilateral opposite the 105° angle also measures 105°. Since the sum of the measures of the angles of a quadrilateral is 360°,

$$a + b + 105° + 65° = 360°$$
$$a + b = 360° - 170°$$
$$\frac{a + b}{2} = \frac{190}{2} = 95°$$

3. **C**  If the average of the two-digit numbers $1N$, $N7$, and $NN$ is 35, then the sum of the three numbers is $3 \times 35$ or 105. Since $1N + N7 + NN = 105$, look for a value of $N$ in the set of answer choices such that $N + 7 + N$ is 15 or 25. For choice (C), $N = 4$,

$$14 + 47 + 44 = 105$$

4. **C**  The radii of circles with areas of $16\pi$ and $100\pi$ are 4 and 10, respectively. The average of 4 and 10 is

$$\frac{4 + 10}{2} = \frac{14}{2} = 7$$

The area of a circle of radius 7 is $\pi(7)^2$ or $49\pi$.

5. **B**  If the average of $x$, $3x + 1$, and $5x - 13$ is 47, then

$$\frac{x + (3x + 1) + (5x - 13)}{3} = 47$$
$$9x - 12 = 3(47)$$
$$9x = 141 + 12$$
$$x = \frac{153}{9} = 17$$

6. **C**  When $x$ is subtracted from $2y$ the difference is equal to the average of $x$ and $y$, so $2y - x = \frac{x+y}{2}$ or $2(2y - x) = x + y$. Hence,

$$4y - 2x = x + y$$
$$4y - y = x + 2x$$
$$3y = 3x$$

Dividing both sides of $3y = 3x$ by $3y$ gives $\frac{x}{y} = 1$.

7. **E**  If the average of $x$, $y$, and $z$ is 32, then $\frac{x+y+z}{3} = 32$, so

$$x + y + z = 3 \times 32 = 96$$

Since the average of $y$ and $z$ is 27, then $\frac{y+z}{2} = 27$, so

$$y + z = 2 \times 27 = 54$$

Substituting the value of $y + z$ in the equation $x + y + z = 96$ gives $x + 54 = 96$, so

$$x = 96 - 54 = 42$$

Hence, the average of $x$ and $2x$ equals

$$\frac{x + 2x}{2} = \frac{3x}{2} = \frac{3}{2}(42) = 63$$

8. **B**  Eliminate choice (A) because the mode is 2 but the average (mean) of 2, 2, 2, 2, and 4 must be greater than 2. Eliminate choices (C), (D), and (E) because the mode in each case is 1 but the average (mean) in each of these answer choices must be greater than 1. In choice (B) the mode is 3, the average (mean) of 1, 3, 3, 3, and 5 is

$$\frac{1 + 3 + 3 + 3 + 5}{5} = \frac{15}{5} = 3$$

and the median is 3 since two values (1 and 3) are less than or equal to 3 and two other values (3 and 5) are equal to or greater than 3.

9. **D**  Let $x$ represent the average rate of speed in miles per hour for the second part of the trip, which lasts $7 - 3$ or 4 hours. Since rate multiplied by time equals distance, the man drives $45 \times 3$ or 135 miles during the first part of the trip and $4x$ miles during the second part. Total distance divided by total time gives the average rate of speed for the entire trip, so

$$\frac{135 + 4x}{7} = 53$$
$$135 + 4x = 7 \times 53 = 371$$
$$4x = 371 - 135 = 236$$
$$x = \frac{236}{4} = 59$$

10. **B**  For the set of numbers 2, 2, 4, 5, and 12, the mode is 2, the mean (or average) is

$$\frac{2 + 2 + 4 + 5 + 12}{5} = \frac{25}{5} = 5$$

*(Lesson 7-1 Multiple Choice continued)*

and the median or middle value in the list is 4. Hence, the mean is greater than the mode.

11. **A** If $x$ represents the lowest score Susan can receive on her next math exam and have an average of at least 85 on the five exams, then

$$\frac{78 + 93 + 82 + 76 + x}{5} = 85$$

$$329 + x = 5 \times 85$$
$$= 425$$
$$x = 425 - 329$$
$$= 96$$

12. **E** Since

$$(x + y)^2 = x^2 + 2xy + y^2$$

and

$$(x - y)^2 = x^2 - 2xy + y^2$$

the average of $(x + y)^2$ and $(x - y)^2$ is

$$\frac{(x^2 + 2xy + y^2) + (x^2 - 2xy + y^2)}{2}$$

or

$$\frac{2(x^2 + y^2)}{2} = x^2 + y^2$$

13. **D** Since the average of $x$ and $y$ is $\frac{11}{2}$, then,

$$\frac{x + y}{2} = \frac{11}{2} \quad \text{or} \quad x + y = 11$$

If the average of $\frac{1}{x}$ and $\frac{1}{y}$ is $\frac{11}{24}$, then

$$\frac{1}{2}\left(\frac{1}{x} + \frac{1}{y}\right) = \frac{11}{24} \quad \text{or} \quad \frac{1}{x} + \frac{1}{y} = \frac{11}{12}$$

Since

$$\frac{1}{x} + \frac{1}{y} = \frac{x + y}{xy} \quad \text{and} \quad x + y = 11$$

then $\frac{11}{xy} = \frac{11}{12}$, so $xy = 12$.

14. **C** If the ratio of $a$ to $b$ is $\frac{1}{2}$, then $\frac{a}{b} = \frac{1}{2}$, so $2a = b$ or $a = \frac{b}{2}$. Since the ratio of $c$ to $b$ is $\frac{1}{3}$, then $\frac{c}{b} = \frac{1}{3}$, so $3c = b$ or $c = \frac{b}{3}$. Hence, the average of $a$ and $c$ in terms of $b$ is

$$\frac{1}{2}(a + c) = \frac{1}{2}\left(\frac{b}{2} + \frac{b}{3}\right) = \frac{1}{2}\left(\frac{5b}{6}\right) = \frac{5b}{12}$$

15. **D** If the average of $a$, $b$, $c$, $d$, and $e$ is 28, then

$$\frac{a + b + c + d + e}{5} = 28$$

or

$$a + b + c + d + e = 5 \times 28 = 140$$

If the average of $a$, $c$, and $e$ is 24, then

$$\frac{a + c + e}{3} = 24$$

or

$$a + c + e = 3 \times 24 = 72$$

Substituting 72 for $a + c + e$ in $a + b + c + d + e = 140$ gives

$$b + d + 72 = 140$$
$$b + d = 140 - 72 = 68$$
$$\frac{b + d}{2} = \frac{68}{2} = 34$$

16. **A** Adding corresponding sides of the given equations, $2a + b = 7$ and $b + 2c = 23$, gives $2a + 2b + 2c = 30$ or $a + b + c = 15$. Hence:

$$\frac{a + b + c}{3} = \frac{15}{3} = 5$$

17. **A** If the average final exam grade of 10 students is $x$, then the sum of the final exam grades of the 10 students is $10x$. Similarly, the sum of the final exam grades of 15 students is $15y$. If the average final exam grade of the 25 students is $t$, then

$$\frac{10x + 15y}{25} = t$$
$$10x + 15y = 25t$$
$$2x + 3y = 5t$$
$$2x = 5t - 3y$$
$$x = \frac{5t - 3y}{2}$$

18. **B** The average of $a$, $b$, $c$, and $d$ is $p$, so

$$\frac{a + b + c + d}{4} = p$$

or $a + b + c + d = 4p$. Similarly, since the average of $a$ and $c$ is $q$, then $a + c = 2q$, so $b + d + 2q = 4p$. Since $b + d = 4p - 2q$, then

$$\frac{b + d}{2} = \frac{4p}{2} - \frac{2q}{2} = 2p - q$$

*(Quantitative Comparison)*

1. **C** You are given that $\frac{p + q}{6} = r$, so

$$p + q = 6r \quad \text{and} \quad \frac{p + q}{2} = \frac{6r}{2} = 3r$$

Since the average of $p$ and $q$ equals $3r$, Column A = Column B.

2. **A** If the average of $3x + 1$ and $x - 7$ is 6, then

$$\frac{(3x + 1) + (x - 7)}{2} = 6$$

so

$$4x - 6 = 2(6) = 12$$
$$4x = 18$$
$$x = \frac{18}{4} = 4.5$$

Since $4.5 > 4$, Column A > Column B.

*(Lesson 7-1   Quantitative Comparison continued)*

3. **B**  You are told that the average of $a$, $b$, and $c$ is $A$, so $a + b + c = 3A$. In Column A, the average of $a + k$, $b + k$, and $c + k$ is
$$\frac{(a+k)+(b+k)+(c+k)}{3}$$
or
$$\frac{(a+b+c)+(3k)}{3}$$
Since $a + b + c = 3A$,
$$\frac{(a+b+c)+(3k)}{3} = \frac{3(A+k)}{3} = A + k$$
Since $k > 0$, $A + 3k > A + k$, so Column B > Column A.

4. **B**  Adding corresponding sides of the given equations, $a - 2b = 7$ and $a + 2b = 3$, gives $2a = 10$, so $a = 5$. Substituting 5 for $a$ in the first equation gives $5 - 2b = 7$ or $-2b = 2$, so $b = -1$. The average of $a$ and $b$ is
$$\frac{5+(-1)}{2} = \frac{4}{2} = 2$$
Hence, Column B > Column A.

5. **B**  In the bar graph, the height of each bar gives the number of days the temperature along the horizontal axis was recorded. Since the bar with the greatest height is the mode, 77°F is the temperature that is the mode.

Using the vertical axis for the number of days each temperature was recorded, list each temperature the appropriate number of times: 75, 76, 77, 77, 77, 78, 78, 79, 79, 80, and 80. Since there are 11 recorded temperatures, the median is the middle or sixth temperature in the list. Since the median temperature is 78°F, Column B > Column A.

*(Grid In)*

1. **22**  Let $x$, $x + 2$, $x + 4$, and $x + 6$ represent four consecutive even integers. If their average is 19, then
$$\frac{x+(x+2)+(x+4)+(x+6)}{4} = 19$$
or
$$4x + 12 = 4 \times 19 = 76$$
Then
$$4x = 76 - 12 = 64$$
$$x = \frac{64}{4} = 16$$

Hence, $x + 6$, the greatest of the four consecutive even integers, is $16 + 6$ or 22.

2. **12**  *Solution 1*: Since the average of $x$, $y$, and $z$ is 12, then $x + y + z = 3 \times 12 = 36$. Hence,
$$3x + 3y + 3z = 3(36) = 108$$
The average of $3x$, $3y$, and $3z$ is their sum, 108, divided by 9 since nine variable items are being added (3 $x$'s, 3 $y$'s, and 3 $z$'s): $\frac{108}{9}$ or 12.

*Solution 2*: The average of $x$, $y$, and $z$ is 12, so the average of any constant multiple of $x$, $y$, and $z$ is also 12. That constant factor cancels out when the sum of the original set of numbers multiplied by that constant is divided by the number of values times that same constant.

3. **28**  If the average of three consecutive multiples of 4 is 132, then
$$\frac{4k + 4(k+1) + 4(k+2)}{3} = 132$$
where $k$ is some positive integer. Hence,
$$12k + 12 = 3 \times 32$$
$$12k = 84$$
$$k = \frac{84}{12} = 7$$
The least of the three consecutive multiples is $4k = 4(7) = 28$.

## LESSON 7-2   *(Multiple Choice)*

1. **A**  The total number of different sundaes that the ice cream parlor can make is the number of different flavors of ice cream times the number of different flavors of syrup times the number of different toppings: $6 \times 3 \times 4 = 72$.

2. **D**  Although the red block must occupy the first position in the row, the four remaining blocks can be arranged in any order. Hence, there is 1 choice for the first position, 4 choices for the second position, 3 choices for the third position, 2 choices for the fourth position, and 1 choice for the last position. The total number of arrangements is $1 \times 4 \times 3 \times 2 \times 1$ or 24.

3. **D**  The region in the diagram that represents the set of elements that belong to sets $B$ and $C$ but not to set $A$ is the region in which circles $B$ and $C$ overlap outside of circle $A$. This region is labeled $d$.

4. **D**  Any one of the five students can be seated in the first chair, any of the four remaining

*(Lesson 7-2   Multiple Choice continued)*

students in the second chair, and any of the three remaining students in the third chair. Thus, five students can be seated in three chairs in $5 \times 4 \times 3$ or 60 different ways.

5. **B**   Since the first coin is replaced before another coin is picked, the same coin may be drawn twice. There are seven possible ways in which two coins whose values add up to 25 or more cents can be picked: (quarter, quarter), (quarter, dime), (dime, quarter), (quarter, nickel), (nickel, quarter), (quarter, penny), and (penny, quarter).

6. **C**   The number of students on one but not both of the two teams is the sum of the number of students on the basketball team plus the number on the tennis team minus the number on both teams: $15 + 18 - 11 = 22$. Since 22 students are on the basketball or the tennis team and 14 students are not on either team, there are $22 + 14$ or 36 students in the class.

7. **D**   In forming a four-digit number, 0 cannot be used as the first digit, so only five out of the six possible digits can be used to fill the first decimal position of the number. Since the same digit cannot be used more than once, any of the five remaining digits can be used to fill the second position, any of the four remaining digits can be used to fill the third position, and any of the three remaining digits can be used to fill the last position. Thus, $5 \times 5 \times 4 \times 3$ or 300 four-digit numbers greater than 1000 can be formed from the digits 0, 1, 2, 3, 4, and 5 without repeating any digit.

8. **C**   If the first letter is $T$, as the question states, and the second letter is $A$, then any one of the six other letters can fill the last position of each three-letter arrangement. Hence, there are $1 \times 6 \times 1$ or 6 possible three-letter arrangements. If the first letter is $T$ and the third letter is $A$, then any one of the six other letters can fill the middle position of each three-letter arrangement. This possibility also gives $1 \times 6 \times 1$ or 6 possible three-letter arrangements. Hence, the total number

of three-letter arrangements in which the first letter is $T$ and one of the other letters is $A$ is $6 + 6$ or 12.

9. **C**   If $F$ students in the class are in the French Club, $S$ students are in the Spanish Club, and $N$ of these students are in both the French and the Spanish clubs, then $F - N$ students belong only to the French Club and the total number of students in the class is $F + S - N$. Hence, the fraction of the class that belongs to the French Club but not to the Spanish Club is $\frac{F-N}{F+S-N}$.

10. **D**   When the digits 0, 1, 3, 4, and 5 are used to form a three-digit number greater than 400, the first digit must be either 4 or 5. Hence, the number of *three*-digit numbers greater than 400 that can be formed from the digits 0, 1, 3, 4, and 5, without repeating any digit, is $2 \times 4 \times 3$ or 24. When the same set of digits is used to form a four-digit number, the first digit cannot be 0. Hence, the number of *four*-digit numbers that can be formed from the digits 0, 1, 3, 4, and 5, without repeating any digit, is $4 \times 4 \times 3 \times 2$ or 96. Hence, 120 ($= 24 + 96$) different numbers greater than 400 but less than 9999 can be formed.

11. **A**   The shaded region in the diagram represents the set whose members are elements of sets $X$, $Y$, and $Z$. Look in the answer choices for a number that is odd (set $X$), has a square root that is an integer (set $Y$), and is a multiple of 5 (set $Z$). The only number that has all of these three properties is 25, choice (A).

12. **C**   Given $X = \{1, 2, 4\}$ and $Y = \{1, 3, 4\}$, make a list of all the possible sums $x + y$:

$$1 + 1 = 2 \qquad 2 + 1 = 3 \qquad 4 + 1 = 5$$
$$1 + 3 = 4 \qquad 2 + 3 = 5 \qquad 4 + 3 = 7$$
$$1 + 4 = 5 \qquad 2 + 4 = 6 \qquad 4 + 4 = 8$$

Hence, seven different values of $x + y$ are possible: 2, 3, 4, 5, 6, 7, and 8.

13. **B**   If in a music class of 30 students, 50% study woodwinds, 40% study strings, and 30% study both woodwinds and strings, then 50% of 30 or 15 students study woodwinds, 40% of 30 or 12 students study strings, and 30% of 30 or 9 students study both woodwinds and strings. Hence,

*(Lesson 7-2   Multiple Choice continued)*

15 + 12 − 9 or 18 students in the class study either woodwinds or strings. Thus, 30 − 18 or 12 students do not study either woodwind or string instruments. Since

$$\frac{12}{30} = \frac{2}{5} = 0.40$$

40% of the class do not study either woodwind or string instruments.

14. **E**   Determine whether each Roman numeral statement is true or false, given that, of 25 city council members, 12 voted for proposition *A*, 19 voted for proposition *B*, and 10 voted for both propositions.

- I. Since 19 members voted for proposition *B* and 10 voted for both propositions, 19 − 10 or 9 members voted only for proposition *B*. Hence, statement I is false.

- II. Since 12 + 19 − 10 = 21, then 21 members voted for exactly one of the two propositions. Since 25 members voted, 25 − 21 or 4 members voted against both propositions. Hence, statement II is true.

- III. Since 10 members voted for both propositions, 12 − 10 or 2 members voted for proposition *A* but not for proposition *B*. Also, 19 − 10 = 9 members voted for proposition *B* but not for proposition *A*. Since 2 + 9 or 11 members voted for one proposition and against the other proposition, statement III is true.

Only Roman numeral statements II and III are true.

15. **A**   To count the number of elements in set *A* or *B* or *C*, find the sum of the members of each set, subtract from this sum the sum of the elements common to any two of these sets, and then add the number of elements common to all three sets. If *x* represents the number of people who subscribe to all three magazines, the number of people who subscribe to magazine *A* or *B* or *C* is

23 + 21 + 17 − (4 + 4 + 4) + *x*   or   49 + *x*

Since 5 of the 63 people did not subscribe to any of the three magazines, 58 people subscribed to magazine *A* or *B* or *C*. Hence, 58 = 49 + *x*, so *x* = 9.

**LESSON 7-3**   *(Multiple Choice)*

1. **E**   The total number of marbles in the bag is 3 + 4 + 2 or 9. Of these 9 marbles, 3 + 4 or 7 marbles are not orange. Hence, the probability that an orange marble will NOT be picked is $\frac{7}{9}$.

2. **D**   The average of the set 1, 2, 2, 3, 3, 3, 4, 4, 4, 4 is the sum of the 10 numbers divided by 10 or $\frac{30}{10} = 3$. Since three of the 10 numbers are 3, the probability that a number selected at random from the set of 10 numbers will be the average of the set is $\frac{3}{10}$.

3. **B**   If the probability that an event will occur is $\frac{x}{4}$ and $x \neq 0$, then the probability that this event will NOT occur is

$$1 - \frac{x}{4} = \frac{4}{4} - \frac{x}{4} = \frac{4-x}{4}$$

4. **B**   Since the probability of picking a red marble is 0.25 and the probability of picking a blue marble is 0.40, the probability of picking a green marble is 1 − (0.25 + 0.40) or 0.35.

5. **D**   To find the LEAST number of marbles that could be in the jar, multiply each answer choice by 0.25, 0.40, and 0.35 until an integer is obtained for each product. Since 4 times 0.40 is 1.6, rule out choice (A); 5 times 0.25 is 1.25, so rule out choice (B); 10 times 0.35 is 3.5, so rule out choice (C). Since 20 times 0.25 is 5, 20 times 0.40 is 8, and 20 times 0.35 is 7, choice (D) is correct.

6. **D**   Determine whether each Roman numeral statement is true or false when a cube whose faces are numbered from 1 to 6 is rolled.

- I. Since any one of the six numbers may show up on the top face, the probability of getting a 5 on the top face is $\frac{1}{6}$. Hence, statement I is false.

- II. Since there are three even numbers (2, 4, and 6), the probability of getting an even number on the top face is $\frac{3}{6}$ or $\frac{1}{2}$. Hence, statement II is true.

- III. Since there are three prime numbers from 1 to 6 (2, 3, and 5), the probability of getting a prime number on the top face is $\frac{3}{6}$ or $\frac{1}{2}$. Hence, statement III is true.

Only Roman numeral statements II and III are true.

*(Lesson 7-3   Multiple Choice continued)*

7. **A**   The five letters $L$, $O$, $G$, $I$, and $C$ can be arranged in $5 \times 4 \times 3 \times 2 \times 1$ or 120 different ways. Since only one of these arrangements is "LOGIC," the probability that a random arrangement of the five letters will form the word "LOGIC" is $\frac{1}{120}$.

8. **E**   Since three of the eight sectors are even-numbered, the probability that the spinner will land on an even-numbered region is $\frac{3}{8}$. Hence, the probability that the spinner will land on an even-numbered region in each of two consecutive spins is $\frac{3}{8} \times \frac{3}{8}$ or $\frac{9}{64}$.

9. **C**   If three fair coins are tossed at the same time, the probability that all three will come up heads is $\frac{1}{2} \times \frac{1}{2} \times \frac{1}{2}$ or $\frac{1}{8}$. The probability that all three coins will come up tails is also $\frac{1}{2} \times \frac{1}{2} \times \frac{1}{2}$ or $\frac{1}{8}$. Hence, the probability that all three coins will come up heads *or* all will come up tails is $\frac{1}{8} + \frac{1}{8}$ or $\frac{1}{4}$.

10. **B**   The probability that the red and the yellow cube will each show a given number from 1 to 6 is $\frac{1}{6} \times \frac{1}{6} = \frac{1}{36}$. Hence, the probability that both cubes will show a 1 or a 2 or a 3 or a 4 or a 5 or a 6 is

$$\frac{1}{36} + \frac{1}{36} + \frac{1}{36} + \frac{1}{36} + \frac{1}{36} + \frac{1}{36} = \frac{6}{36} = \frac{1}{6}$$

11. **D**   Since the probability of selecting a blue marble at random from the jar is $\frac{1}{3}$ and the probability of selecting a green marble at random from the jar is $\frac{4}{9}$, the probability of selecting a white marble is

$$1 - \left( \frac{1}{3} + \frac{4}{9} \right) = 1 - \frac{7}{9} = \frac{2}{9}$$

If $x$ marbles of the 54 marbles in the jar are white, then

$$\frac{x}{54} = \frac{2}{9}$$
$$9x = 108$$
$$x = \frac{108}{9} = 12$$

12. **C**   The squares of three of the five numbers in the set $A = \{-2, -1, 0, 1, 2\}$ are less than 2:
$$(-1)^2 = 1 < 2, \quad 0^2 = 0 \le 2, \quad 1^2 = 1 < 2$$
Hence, the probability of picking a number $x$ from set $A$ such that $x^2 < 2$ is $\frac{3}{5}$.

13. **C**   You are given that the probability of guessing the correct answer to a certain test question is $\frac{x}{12}$, so the probability of not guessing the correct answer is $1 - \frac{x}{12}$. Since the probability of NOT guessing the correct answer to this question is given as $\frac{2}{3}$,

$$1 - \frac{x}{12} = \frac{2}{3} \quad \text{or} \quad 1 - \frac{2}{3} = \frac{x}{12}$$

so $\frac{1}{3} = \frac{x}{12}$. Hence, $x = 4$.

14. **B**   If a number is randomly selected from set $A = \{1, 2, 3\}$ and then a second number is randomly selected from set $B = \{1, 4, 9\}$, the set of all possible products is as follows:

| | | |
|---|---|---|
| $1 \times 1 = 1$ | $2 \times 1 = 2$ | $3 \times 1 = 3$ |
| $1 \times 4 = 4$ | $2 \times 4 = 8$ | $3 \times 4 = 12$ |
| $1 \times 9 = 9$ | $2 \times 9 = 18$ | $3 \times 9 = 27$ |

Since five of the nine possible products are less than 9, the probability that the product of the two numbers selected will be less than 9 is $\frac{5}{9}$.

*(Quantitative Comparison)*

1. **A**   In a set of 20 cards numbered 1 through 20, there are eight cards with prime numbers: 2, 3, 5, 7, 11, 13, 17, and 19. Hence, the probability of randomly selecting a card with a prime number is $\frac{8}{20}$. Since there are three cards with numbers less than 4 and four cards with numbers greater than 16, the probability of picking a card at random with a number less than 4 or greater than 16 is $\frac{3}{20} + \frac{4}{20}$ or $\frac{7}{20}$. Since $\frac{8}{20} > \frac{7}{20}$, Column A > Column B.

2. **B**   If the probability that event $E$ will occur is $\frac{1}{p}$, then the probability that event $E$ will NOT occur is $1 - \frac{1}{p}$. Since you are given that $p > 1$, Column B > 1, so Column B > Column A.

3. **A**   If two marbles are picked at random without replacement from a jar that contains three blue marbles and three green marbles, the probability of picking a blue marble followed by a green marble is $\frac{3}{6} \times \frac{3}{5} = \frac{9}{30}$. Similarly, the probability of picking a green marble followed by a blue marble is $\frac{3}{6} \times \frac{3}{5} = \frac{9}{30}$. Hence, the probability of picking two marbles with different colors is

$$\frac{9}{30} + \frac{9}{30} = \frac{18}{30}$$

Since $\frac{1}{3} = \frac{10}{30}$ and $\frac{18}{30} > \frac{10}{30}$, Column A > Column B.

*(Lesson 7-3   Quantitative Comparison continued)*

4. **C**  If a fair coin is tossed 10 times and exactly three heads show up, then seven tails must show up on the other tosses. Since the probability of tossing exactly three heads in 10 tosses is the same as the probability of tossing exactly seven tails in 10 tosses, Column A = Column B.

5. **C**  If $a$ is a number picked from set $A = \{1, 3\}$ and $b$ is a number picked from set $B = \{2, 4\}$, the following products and sums are possible:

| $ab$ | $a+b$ |
|------|-------|
| $1 \times 2 = 2$ | $1 + 2 = 3$ |
| $1 \times 4 = 4$ | $1 + 4 = 5$ |
| $3 \times 2 = 6$ | $3 + 2 = 5$ |
| $3 \times 4 = 12$ | $3 + 4 = 7$ |

In Column A, the probability that $ab$ is greater than 5 is $\frac{2}{4}$. In Column B, the probability that $a + b$ equals 5 is $\frac{2}{4}$. Hence, Column A = Column B.

*(Grid In)*

1. **24**  Since
$$1 - \left(\frac{1}{4} + \frac{1}{3}\right) = 1 - \frac{7}{12}$$
the probability of selecting a yellow marble is $\frac{5}{12}$. If 10 of the $x$ marbles in the jar are yellow, then $\frac{5}{12} = \frac{10}{x}$. Since 10 is two times 5, $x$ must be two times 12 or 24.

2. **4/20**  If two cards are selected at random without replacement from a set of five identical cards each numbered on one side with a different integer from 1 to 5, the set of all possible sums is as follows:

$1+2=3$  $2+1=3$  $3+1=4$  $4+1=5$  $5+1=6$
$1+3=4$  $2+3=5$  $3+2=5$  $4+2=6$  $5+2=7$
$1+4=5$  $2+4=6$  $3+4=7$  $4+3=7$  $5+3=8$
$1+5=6$  $2+5=7$  $3+5=8$  $4+5=9$  $5+4=9$

Since four of the 20 possible sums are 8 or more, the probability of picking two cards whose sum will be at least 8 is $\frac{4}{20}$. Grid in as 4/20.

**LESSON 7-4**  *(Multiple Choice)*

1. **C**  Since $\boxed{a} = \sqrt{a} + 1$,
$$\begin{aligned}
\boxed{36} + \boxed{64} &= (\sqrt{36} + 1) + (\sqrt{64} + 1) \\
&= (6 + 1) + (8 + 1) \\
&= 7 + 9 \\
&= 16
\end{aligned}$$

2. **C**  To evaluate $_2\triangle_3^{\,1}$, let $a = 1$, $b = 2$, and $c = 3$ in the given formula:
$$\begin{aligned}
_b\triangle_c^{\,a} &= \frac{a+b}{c} + \frac{a+c}{b} + \frac{b+c}{a} \\
&= \frac{1+2}{3} + \frac{1+3}{2} + \frac{2+3}{1} \\
&= \frac{3}{3} + \frac{4}{2} + \frac{5}{1} \\
&= 1 + 2 + 5 \\
&= 8
\end{aligned}$$

3. **E**  Since the number 6 is even and the number 11 is odd,
$$\boxed{6\!\!\rhd} = 2(6 - 1) = 2(5) = 10$$
and
$$\boxed{11\!\!\rhd} = \frac{1}{2}(11 + 1) = \frac{1}{2}(12) = 6$$
Hence,
$$\boxed{6\!\!\rhd} \times \boxed{11\!\!\rhd} = 10 \times 6 = 60$$

4. **D**  Evaluate $\boxed{N\!\!\rhd}$ for each answer choice until you obtain a number that is a multiple of 4. For choice (D), if $N = 31$, then
$$\boxed{N\!\!\rhd} = \frac{1}{2}(31 + 1) = \frac{1}{2}(32) = 16$$
Since 16 is a multiple of 4, (D) is the correct choice.

5. **C**  Since 2 is carried over from adding the digits in the first column, in the second column $A + B + B + 1 = 14$, so $A + 2B = 13$. Since $B$ must be an integer, $A$ cannot be 0. Hence, the greatest possible value of digit $B$ occurs when $A = 1$ and $B = 6$.

6. **B**  Following the given rule for obtaining terms, write the terms of the sequence up to and including 55:
$$1, 1, 2, 3, 5, 8, 13, 21, 34, 55$$
Since 10 numbers are listed, $n = 10$.

7. **A**  In the first row of multiplication, $63 \times P = 31P$. Hence, $P$ must be equal to 5 or 6 since the product of each of these and 63 is a number between 300 and 400. Since $63 \times 5 = 315$ and $63 \times 6 = 378$, then $P = 5$, so
$$Q = 3 + P = 3 + 5 = 8$$

8. **E**  You are told that $a \,\Phi\, b = ab - a$. If $a \,\Phi\, 4 = 24$, then $b = 4$ and $ab - a = 24$. Hence, $4a - a = 3a = 24$, so
$$a = \frac{24}{3} = 8$$

*(Lesson 7-4   Multiple Choice continued)*

9. **B** Determine whether each Roman numeral equation is true or false when $a \, \Phi \, b = ab - a$.

- I. Since $b \, \Phi \, a = ab - b$ and $ab - a$ is not necessarily equal to $ab - b$, $a \, \Phi \, b = b \, \Phi \, a$ is not always true. Equation I need not be true.
- II. To evaluate $a \, \Phi \, (a + 1)$, let $b = a + 1$:

$$a \, \Phi \, (a + 1) = a(a + 1) - a$$
$$= a^2 + a \ - a$$
$$= a^2$$

Hence, equation II must be true.

- III. Since

$$\frac{a}{2} \, \Phi \, \frac{b}{2} = \left(\frac{a}{2}\right)\left(\frac{b}{2}\right) - a = \frac{ab}{4} - a$$

and

$$\frac{a \, \Phi \, b}{2} = \frac{ab - a}{2} = \frac{ab}{2} - \frac{a}{2}$$

$\frac{a}{2} \, \Phi \, \frac{b}{2}$ does not equal $\frac{a \Phi b}{2}$. Hence, equation III is false.

Only Roman numeral equation II must be true.

10. **B** Since 2 is carried over from the addition in the first column to the second column and 2 is carried over from the addition in the second column to the third column, $A + B + C + 2 = 26$, so $A + B + C = 24$. Determine whether each Roman numeral value could be possible.

- I. If $A = 0$, then $B + C = 24$, which is not possible since the maximum value of $B + C$ is $9 + 9$ or 18. Hence, value I could not be $A$.
- II. If $A = 4$, then $B + C = 20$, which is not possible since the maximum value of $B + C$ is $9 + 9$ or 18. Hence, value II could not be $A$.
- III. If $A = 8$, then $B + C = 16$, which is possible, so value III could be $A$.

Only Roman numeral value III could be digit $A$.

11. **C** Substitute each of the answer choices for $y$ into $\sqrt{y} = 2\sqrt{y}$ until you get 8. Since

$$\boxed{16} = 2\sqrt{16} = 2(4) = 8$$

(C) is the correct choice.

12. **D**
$$\boxed{y} \times \boxed{y} = (2\sqrt{y}) \times (2\sqrt{y})$$
$$= 4(\sqrt{y} \times \sqrt{y})$$
$$= 4y$$

13. **A** If $\sqrt{x} = 2x + 1$ and $3 - \sqrt{x} = x$, then
$$3 - (2x + 1) = x$$
$$3 - 2x - 1 = x$$
$$2 = 3x$$
$$\frac{2}{3} = x$$

14. **D** In the correctly worked out addition problem, $C$ must be equal to 1 since this is the only possible digit that can be carried over from the addition in column 3 to column 4. From the addition in the first column, there are two possibilities to consider: $A + B = 2$ or $A + B = 12$. If $A + B = 2$, then $A = B = 1$, $A = 0$, and $B = 2$, or $A = 2$ and $B = 0$. All of these combinations of values are impossible because they do not make $C = 1$ in the addition of the second column. Since, therefore, $A + B = 12$, 1 is carried from the addition in the first column to the second column. Hence, $3 + B + 1 = 11$ or $B = 7$, so $A = 5$. Also, since 1 is carried over from the addition in column 3 to column 4, $8 + D + 1 = A = 15$, so $D = 6$.

15. **B** Since the greatest divisor of 28 that is less than 28 is 14,

$$\boxed{28} = 28 + 14 = 42$$

16. **D** The greatest divisor of 15 that is less than 15 is 5, so

$$\boxed{15} = 15 + 5 = 20 = n$$

Since the greatest divisor of 20 that is less than 20 is 10,

$$\boxed{20} = 20 + 10 = 30$$

17. **A** Since

$$\frac{1}{2^1} = \frac{1}{2}, \frac{1}{2^2} = \frac{1}{4}, \frac{1}{2^3} = \frac{1}{8}, \ldots,$$

the $n$th term, without regard to its sign, is $\frac{1}{2^n}$. Each odd-numbered term of the sequence is positive, while each even-numbered term is negative. When $n$ is an odd integer, $(-1)^{n+1}$ is $+1$ because the exponent, $n + 1$, is even. Also, when $n$ is an even integer, $(-1)^{n+1}$ is $-1$ since the exponent, $n + 1$, is odd. Hence, the $n$th term of the sequence is $(-1)^{n+1} \cdot \frac{1}{2^n}$.

*(Quantitative Comparison)*

1. **D** If $a$ and $b$ are consecutive terms of the nonending sequence,

$$-1, 0, 1, -1, 0, 1, -1, 0, 1, \ldots$$

*(Lesson 7-4 Quantitative Comparison continued)*

then, when $a = -1$ and $b = 0$,
$$a + b = -1 \quad \text{and} \quad a - b = -1$$
so Column A = Column B. If, however, $a = 0$ and $b = 1$, then
$$a + b = 1 \quad \text{and} \quad a - b = -1$$
so Column A > Column B.
Since two different answers are possible, (D) is the correct choice.

2. **D** Since $5 + 6 + 7 = 18$, $6 + 7 + 8 = 21$, and $7 + 8 + 9 = 24$, $X$ may be equal to 1 or 4. If $X = 1$, Column B > Column A. If, however, $X = 4$, Column A > Column B. Hence, since two different answers are possible, (D) is the correct choice.

3. **A** After the addition in the second column has been done, the greatest integer that can be carried over to the third column is 1, so $T = 1$. From the first column, if $T = 1$, then $S = 3$. Since 1 is carried over from the addition in the first column to the second column, $R + 8 + 1 = 13$, so $R = 4$. Hence, Column A > Column B.

4. **C** In Column A, if $x = 1$, then
$$\boxed{1} = 1^3 - 1 + 1 = 1 - 1 + 1 = 1$$
In Column B, if $x = -1$, then
$$\boxed{-1} = (-1)^3 - (-1) + 1 = -1 + 1 + 1 = 1$$
Hence, Column B = Column A.

5. **D** The quantity in Column A is $\frac{x+y}{x^2+y^2}$. In Column B,
$$(-x) \otimes (-y) = \frac{-x + (-y)}{(-x)^2 + (-y)^2}$$
$$= \frac{-x - y}{x^2 + y^2}$$
Since $x$ and $y$ are nonzero numbers, $x + y > -x - y$ if $x$ and $y$ are both positive, so Column A > Column B. If, however, $x$ and $y$ are both negative, $-x - y > x + y$, so Column B > Column A.
Since two different answers are possible, (D) is the correct choice.

6. **C** Since
$$x \otimes x = \frac{x+x}{x^2+x^2} = \frac{2x}{2x^2} = \frac{1}{x}$$
Column A = Column B.

7. **B** Since $7 \times 3 = 21$, 2 is carried over from the first column to the second column. Substitute different integer values of $R$ until you find the integer that satisfies

the condition that 7 times $R$ plus 2 gives a number with $R$ in the units digit. The only digit that satisfies this condition is 3: 7 times 3 is 21, and $21 + 2 = 23$. Since $R = 3$, $633 \times 7 = 4431$, so $S = 4$.
Hence, Column B > Column A.

8. **D** When $k$ is a perfect square, $<k> = \sqrt{k}$ and
$$<k> \times <k> = \sqrt{k} \times \sqrt{k} = k$$
so Column A = Column B. When, however, $k$ is not a perfect square, $<k> \times <k>$ will not necessarily equal $k$. For example, if $k = 3$, then $<k> = 2$, so
$$<k> \times <k> = 2 \times 2 \quad \text{or} \quad 4$$
Since two answers are possible, the correct choice is (D).

9. **B** In the first column of the addition problem, $P + Q$ must be 1 or 11 in order for the units digit of the column sum to be 1. Since $P$ and $Q$ are nonzero digits, $P + Q = 11$. After the digits in the first column are added, 2 is carried over to the second column. The addition of the digits in the second column becomes $Q + 12 + P + 2$ or $Q + P + 14$. Since $P + Q = 11$, the second column sum is $11 + 14$ or 25. Since the sum of the four numbers is $2P1$, $2P = 25$, so $P = 5$ and $Q = 6$. Hence, Column B > Column A.

10. **C** In Column A, since the greatest common factor of 10 and 15 is 5,
$$10 \oplus 15 = 10 + 15 + 5 = 30$$
In Column B, since 1 is the greatest common factor of 12 and 17,
$$12 \oplus 17 = 12 + 17 + 1 = 30$$
Hence, Column A = Column B.

*(Grid In)*

1. **10** Since $x \, \Omega \, y = 2x + 3y$, evaluate $a \, \Omega \, 4$ by letting $x = a$ and $y = 4$:
$$a \, \Omega \, 4 = 2a + 3(4) = 2a + 12$$
Evaluate $1 \, \Omega \, a$ by letting $x = 1$ and $y = 4$:
$$1 \, \Omega \, a = 2(1) + 3a = 2 + 3a$$
Since $a \, \Omega \, 4 = 1 \, \Omega \, a$, then $2a + 12 = 2 + 3a$, or $12 - 2 = 3a - 2a$, so $10 = a$.

2. **32** In the sequence $x, y, 22, 14, 10, \ldots$ each term after the first term, $x$, is obtained by halving the term that comes before it and then adding 3 to that number. Hence, to obtain $y$ do the opposite to 22: subtract 3 and then double the result, getting 38. To obtain $x$, subtract 3 from

*(Lesson 7-4 Grid In continued)*

38 and then double the result, getting 70. Thus, $x - y = 70 - 38 = 32$.

3. **54** Since the sum of the first two terms, $15 + 19 = 34$, and 34 is even, the next consecutive term is $\frac{15 + 19}{2} = 17$. The first three terms of this sequence are 15, 19, and 17. To obtain the next term, add 19 and 17, getting 36. Since 36 is even, the next consecutive term is $\frac{19 + 17}{2} = 18$. The first four terms of the sequence are 15, 19, 17, and 18. Since $17 + 18 = 35$ and 35 is odd, the next term of the sequence is $17 + 18 - 3$ or 32. Hence, the first five terms of the sequence are 15, 19, 17, 18, and 32. Since $18 + 32 = 50$ and 50 is even, the next term of the sequence is 25, so the first six terms are 15, 19, 17, 18, 32, and 25. Since $32 + 25 = 57$ and 57 is odd, the next term of the sequence is $32 + 25 - 3 = 54$. Hence, the seventh term of the sequence is 54.

4. **6/4** If $a \blacktriangle b = ab - (a + b)$ and $5 \blacktriangle b = 1$, then
$$5 \blacktriangle b = 5b - (5 + b) = 1$$
$$5b - 5 - b = 1$$
$$4b = 6$$
$$b = \frac{6}{4}$$

Grid in as 6/4.

# PART III

## TAKING PRACTICE TESTS

# Practice Tests 1 and 2

This chapter provides two full-length practice SAT I math tests. Each test consists of three timed math sections that should be taken under testlike conditions, using the sample SAT I answer forms at the beginning of each test.

At the end of each practice test you will find an Answer Key, as well as detailed solutions for all questions. If you think you need additional practice on any type of question you missed or skipped over, you can quickly locate the lesson on which the question was based by referring to the number that follows each answer in the Answer Key.

# Answer Sheet — PRACTICE TEST 1

Start with number 1 for each new section. If a section has fewer questions than answer spaces, leave the extra answer spaces blank.

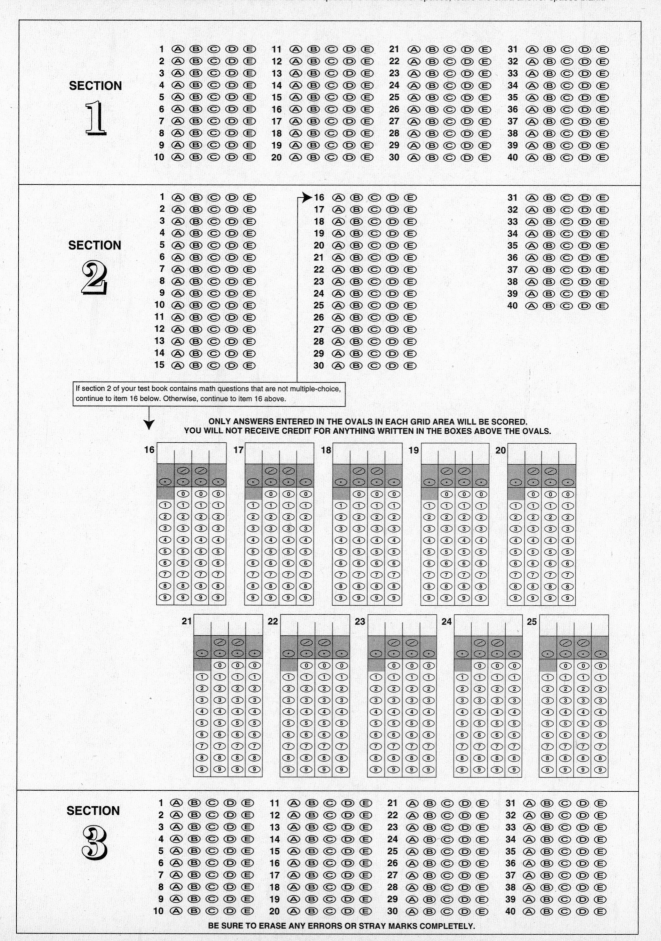

# PRACTICE TEST 1

**Section 1**

**1**

Time—30 Minutes
25 Questions

In this section solve each problem, using any available space on the page for scratchwork. Then decide which is the best of the choices given and fill in the corresponding oval on the answer sheet.

**Notes:**
1. The use of a calculator is permitted. All numbers used are real numbers.
2. Figures that accompany problems in this test are intended to provide information useful in solving the problems. They are drawn as accurately as possible EXCEPT when it is stated in a specific problem that the figure is not drawn to scale. All figures lie in a plane unless otherwise indicated.

**Reference Information**

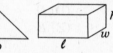

$A = \pi r^2$  $C = 2\pi r$  $A = \ell w$  $A = \frac{1}{2}bh$  $V = \ell wh$  $V = \pi r^2 h$  $c^2 = a^2 + b^2$  Special Right Triangles

The number of degrees of arc in a circle is 360.
The measure in degrees of a straight angle is 180.
The sum of the measures in degrees of the angles of a triangle is 180.

**1** How many positive integers less than 36 are equal to 4 times an *odd* integer ?

(A) Two
(B) Three
(C) Four
(D) Five
(E) Six

**2**

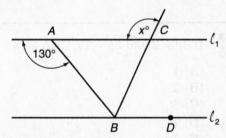

If, in the figure above, $\ell_1 \parallel \ell_2$ and $BC$ bisects $\angle ABD$, then $x =$

(A) 75
(B) 95
(C) 105
(D) 115
(E) 150

GO ON TO THE NEXT PAGE

## Section 1

1

**3** A long-distance telephone call costs $1.80 for the first 3 minutes and $0.40 for each additional minute. If the charge for an $x$-minute long-distance call at this rate was $4.20, then $x =$

(A) 7
(B) 8
(C) 9
(D) 10
(E) 12

**4** What is the LEAST number of squares each of side length 2 inches that is needed to cover completely, without overlap, a larger square with side length 8 inches?

(A) 4
(B) 8
(C) 9
(D) 16
(E) 32

**5** If $3a = 2$ and $2b = 15$, then $ab =$

(A) 1
(B) 2
(C) 5
(D) 7
(E) 12

**6**

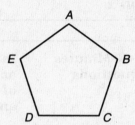

In figure *ABCDE* above, the length of each side is 1 unit. If a point starts at vertex *A* and moves along each side in a clockwise direction, at which vertex will the point be when it has traveled a distance of exactly 715 units?

(A) *A*
(B) *B*
(C) *C*
(D) *D*
(E) *E*

**7**

| Number of Television Sets | Number of Families |
|---|---|
| 0 | 32 |
| 1 | 80 |
| 2 | 160 |
| 3 | 83 |
| More than 3 | 45 |

The table above summarizes the results of a survey in which families reported the number of television sets they have in their homes. What percent of the families surveyed have two or fewer television sets?

(A) 75%
(B) 68%
(C) 60%
(D) 40%
(E) 33%

GO ON TO THE NEXT PAGE

**Section 1**

**1**

**8** If $0 < x < 1$, which of the following expressions must decrease in value as $x$ increases?

   I. $\dfrac{1}{1-x}$

   II. $\dfrac{1}{x^2}$

   III. $1 - \sqrt{x}$

(A) I only
(B) II only
(C) I and II only
(D) I and III only
(E) II and III only

**9** After $\dfrac{1}{8}$ of a ribbon is thrown away, the remaining part is cut into two pieces whose lengths are in the ratio of 4 : 5. If 9 inches of the original ribbon was thrown away, how many inches long is the shorter of the two remaining pieces of ribbon?

(A) 21
(B) 28
(C) 30
(D) 35
(E) 42

**10** If a cube with edge length $\pi$ has the same volume as a cylinder with a height of $\pi$, then the radius of the base of the cylinder is

(A) $\dfrac{1}{\pi}$

(B) $\pi$

(C) $\sqrt{\pi}$

(D) $\dfrac{1}{\sqrt{\pi}}$

(E) $\dfrac{1}{\pi^2}$

**11** If $\dfrac{5}{a} = b$ and $a = 3$, then $a(b + 1) =$

(A) 8

(B) $7\dfrac{1}{3}$

(C) $6\dfrac{2}{3}$

(D) 4

(E) 3

**12** A person spent a total of $720 for dress shirts and sport shirts, each priced at $35 and $20, respectively. If the person purchased two $35 dress shirts for each $20 sport shirt, what is the total number of shirts purchased?

(A) 16
(B) 21
(C) 24
(D) 28
(E) 32

**13** If 10 cubic centimeters of blood contains 1.2 grams of hemoglobin, how many grams of hemoglobin are contained in 35 cubic centimeters of the same blood?

(A) 2.7
(B) 3.0
(C) 3.6
(D) 4.2
(E) 4.8

**GO ON TO THE NEXT PAGE**

**14**

Basketball
20%

Jogging
28%

Swimming
x%

Biking
24%

Fishing
12%

Favorite Sports Activities

The circle graph above summarizes the results of a survey of 2500 students who named their favorite sports activities. If each student in the survey named exactly one activity, what is the total number of students who named either swimming or biking as their favorite sports activity?

(A) 1000
(B) 875
(C) 750
(D) 600
(E) 400

**15** If $x - 3$ is 1 less than $y + 3$, then $x + 2$ exceeds $y$ by what amount?

(A) 4
(B) 5
(C) 6
(D) 7
(E) 9

**16**

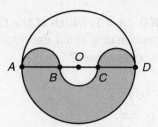

In the figure above, arcs $AB$, $BC$, and $CD$ are semicircles with diameters $AB = BC = CD$. If the diameter of the largest circle with center $O$ is 6, what is the area of the shaded region?

(A) $20\pi$
(B) $24\pi$
(C) $30\pi$
(D) $36\pi$
(E) $40\pi$

**17** For all positive integers $p$, let $p$ be defined by the equation

$$\boxed{\nabla p} = p + \frac{p}{k}$$

where $k$ is the largest prime number that is a factor of $p$, and $k < p$. If $\nabla 110 = x$, what is the value of $\nabla x$?

(A) 120
(B) 133
(C) 144
(D) 148
(E) 160

GO ON TO THE NEXT PAGE

## Section 1

**1**

**18** If a number $N$ is increased by 50% of its value, the result is equal to 35 less 40% of 35. What is the value of $N$?

(A) 10
(B) 14
(C) 18
(D) 21
(E) 24

**19** In $\triangle ABC$, the degree measure of $\angle B$ is twice the degree measure of $\angle A$. If the degree measure of $\angle A$ is subtracted from the degree measure of $\angle C$, the difference is 20. What is the degree measure of the smallest angle of the triangle?

(A) 20
(B) 24
(C) 30
(D) 35
(E) 40

**20** In a certain class, exactly $\frac{2}{3}$ of the students applied for admission to a 4-year college and, of these, $\frac{1}{8}$ applied also to a 2-year junior college. If more than one student applied to both 4-year and 2-year colleges, what is the LEAST number of students who could be in this class?

(A) 12
(B) 24
(C) 28
(D) 30
(E) 36

**21** A group of $p$ people plan to contribute equally to the purchase of a gift that costs $d$ dollars. If $n$ of the $p$ people decide not to contribute, by what amount in dollars does the contribution needed from each of the remaining people increase?

(A) $\dfrac{d}{p-n}$

(B) $\dfrac{pd}{p-n}$

(C) $\dfrac{pd}{n(p-n)}$

(D) $\dfrac{nd}{p(p-n)}$

(E) $\dfrac{d}{n(pd-1)}$

**22** If $p + 1$ is a multiple of 3, what is the greatest multiple of 3 that is LESS than $p + 1$?

(A) $p - 2$
(B) $3p - 1$
(C) $\dfrac{p}{3}$
(D) $p + 4$
(E) $3(p + 1)$

**23** If $h + k + p = 72$, $k = 2h$, and $p = 3k$, then $h =$

(A) 4
(B) 7
(C) 8
(D) 11
(E) 13

**GO ON TO THE NEXT PAGE**

**Section 1**                                                                1

**24**

$$A = \{1, 2, 5\}$$
$$B = \{3, 4\}$$

If a number $a$ is selected at random from set $A$ above and a number $b$ is selected at random from set $B$, which of the following statements must be true?

I. The probability that $a + b$ is greater than 9 is 0.
II. The probability that $a + b$ is a prime number is $\frac{1}{2}$.
III. The probability that $a + b > ab$ is $\frac{1}{2}$.

(A) I only
(B) II only
(C) I and II only
(D) I and III only
(E) II and III only

**25**

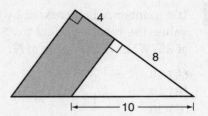

In the figure above, what is the area of the shaded region?

(A) 24
(B) 30
(C) 32
(D) 36
(E) It cannot be determined from the information given.

IF YOU FINISH BEFORE TIME IS CALLED, YOU MAY CHECK YOUR WORK ON THIS SECTION ONLY. DO NOT TURN TO ANY OTHER SECTION IN THE TEST.

**STOP**

## Section 2

**2**

| Time—30 Minutes | **This section contains two types of questions. You have 30 minutes to complete both types. You may use any available space for scratchwork.** |
|---|---|
| 25 Questions | |

### Notes:

1. The use of a calculator is permitted. All numbers used are real numbers.

2. Figures that accompany problems in this test are intended to provide information useful in solving the problems. They are drawn as accurately as possible EXCEPT when it is stated in a specific problem that the figure is not drawn to scale. All figures lie in a plane unless otherwise indicated.

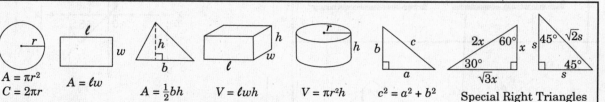

**Reference Information**

$A = \pi r^2$
$C = 2\pi r$

$A = \ell w$

$A = \frac{1}{2}bh$

$V = \ell w h$

$V = \pi r^2 h$

$c^2 = a^2 + b^2$

Special Right Triangles

The number of degrees of arc in a circle is 360.

The measure in degrees of a straight angle is 180.

The sum of the measures in degrees of the angles of a triangle is 180.

### Directions for Quantitative Comparison Questions

Questions 1–15 each consist of two quantities in boxes, one in Column A and one in Column B. You are to compare the two quantities and on the answer sheet fill in oval

 A if the quantity in Column A is greater;
 B if the quantity in Column B is greater;
 C if the two quantities are equal;
 D if the relationship cannot be determined from the information given.

**AN E RESPONSE WILL NOT BE SCORED.**

### Notes:

1. In some questions, information is given about one or both of the quantities to be compared. In such cases, the given information is centered above the two columns and is not boxed.

2. In a given question, a symbol that appears in both columns represents the same thing in Column A as it does in Column B.

3. Letters such as $x$, $n$, and $k$ stand for real numbers.

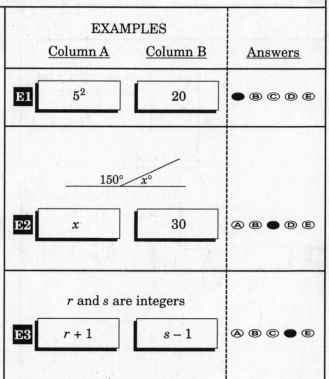

**GO ON TO THE NEXT PAGE**

## Section 2

**2**

Column A | Column B

**1** $4 \div \frac{4}{3}$ | $\frac{1}{\frac{1}{3}}$

**2** $x + y + z$ | $a + b + c + d$

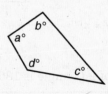

Note: Figure is not drawn to scale.

$b = 2a$

**3** $y$ | $8$

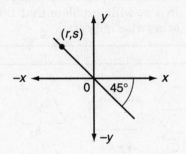

**4** $r^2$ | $s^2$

The area of rectangle $ABCD$ is 10.

**5** Perimeter of $ABCD$ | $14$

13 students are in the Drama Club,
9 students are in the Chess Club,
and 5 students are in both clubs.

**6** The number of students in exactly one of the two clubs | $17$

**GO ON TO THE NEXT PAGE**

## Section 2

**2**

| <u>Column A</u> | <u>Column B</u> |
|---|---|

A coin is weighted so that, when it is tossed, a head is twice as likely to show as a tail.

**7**

| The probability of getting two tails when the coin is tossed two times | The probability of getting three heads when the coin is tossed three times |
|---|---|

*B*

$\frac{1}{3} \times \frac{1}{3} = \frac{1}{9}$    $\frac{2}{3} \times \frac{2}{3} \times \frac{2}{3} = \frac{8}{27}$

$+$ when

$x^2 = y^2$

**8**

| $x^2 + x$ | $y^2 + y$ |
|---|---|

$D$

1, 3, 2, 1, 3, 2, 1, 3, 2, 1, 3, 2, . . .

In the sequence above, the terms 1, 3, 2 repeat in that order without ending.

**9**

| The 57th term of the sequence | The 80th term of the sequence |
|---|---|

*B*

2        3

| <u>Column A</u> | <u>Column B</u> |
|---|---|

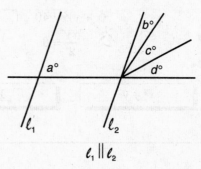

$\ell_1 \parallel \ell_2$

**10**

| $a - b$ | $c + d$ |
|---|---|

$C$

simplify

$$2h + 2k = 8$$
$$7h - 7k = 21$$

**11**

| $h$ | 3 |
|---|---|

$A$

$a = 2b, c = a + b,$
and $a, b, c > 0$

**12**

| The average of $a, b,$ and $c$ | $2b$ |
|---|---|

$C$

$\frac{a+b+c}{3}$

$\frac{2b+b+a+b}{3}$

$\frac{4b+a}{3}$

$\frac{4b+2b}{3}$     $\frac{6b}{3}$   $2b$

**GO ON TO THE NEXT PAGE**

## Section 2

**2**

SUMMARY DIRECTIONS FOR QUANTITATIVE COMPARISON QUESTIONS

<u>Answer:</u>   A  if the quantity in Column A is greater;
B  if the quantity in Column B is greater;
C  if the two quantities are equal;
D  if the relationship cannot be determined from the information given.

| <u>Column A</u> | <u>Column B</u> | | <u>Column A</u> | <u>Column B</u> |

$$(x + y)^2 = 40$$

*Handwritten:* $24$   $x^2 + 2xy + y^2 = 40$   $2\frac{xy}{2} = 6$ . $2xy = 12$

**13**    $x^2 + y^2$  =  $16$

*Handwritten:* A   $x^2 + y^2 = 16$

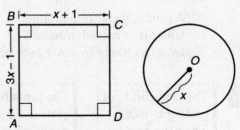

B ⟷ $x+1$ ⟶ C
$3x - 1$
A    D

Note: Figures not drawn to scale.

$$a > b > 1$$

**14**    $\dfrac{1}{b} - \dfrac{1}{a}$     $0$

*Handwritten:* B   Ⓐ

**15**   The area of square *ABCD*    The area of circle *O*

*Handwritten:* A

*Handwritten:*
① $a^2 + b^2 = a^2 + 2ab + b^2$
$a^2 - b^2 = a^2 - 2ab + b^2$
$(a+b)(a-b) = a^2 - b^2$

**GO ON TO THE NEXT PAGE** ⟹

## Section 2

**2**

Directions for Student-Produced Response Questions

Each of the remaining 10 questions (16–25) requires you to solve the problem and enter your answer by marking the ovals in the special grid, as shown in the examples below.

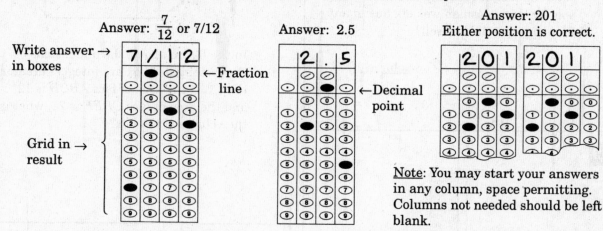

- Mark no more than one oval in any column.

- Because the answer sheet will be machine-scored, **you will receive credit only if the ovals are filled in correctly.**

- Although not required, it is suggested that you write your answer in the boxes at the top of the columns to help you fill in the ovals accurately.

- Some problems may have more than one correct answer. In such cases, grid only one answer.

- No question has a negative answer.

- **Mixed numbers** such as $2\frac{1}{2}$ must be gridded as 2.5 or 5/2. (If $\boxed{2\,|\,1\,/\,2}$ is gridded, it will be interpreted as $\frac{21}{2}$, not $2\frac{1}{2}$.)

- <u>Decimal Accuracy</u>: If you obtain a decimal answer, **enter the most accurate value the grid will accommodate.** For example, if you obtain an answer such as 0.6666 . . ., you should record the result as .666 or .667. **Less accurate values such as .66 or .67 are not acceptable.**

Acceptable ways to grid $\frac{2}{3}$ = .6666 . . .

---

**16** If $x + 2x + 3x + 4x = 1$, then what is the value of $x^2$?

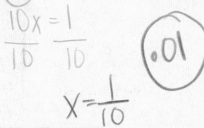

$\dfrac{10x}{10} = \dfrac{1}{10}$

$X = \dfrac{1}{10}$

$\boxed{.01}$

**17** If $2^{x-1} - 1 = 7$, then what is the value of $x$?

$2^{x-1} = 8$

$x = 4$

GO ON TO THE NEXT PAGE ⟶

## Section 2

**2**

**18** During a certain month, a car sales-person sold three economy cars for every two luxury cars. If during that month the total number of economy and luxury cars sold by this salesman was 90, how many economy cars did she sell?

5x    3x:2x
3 3x+2x=90
total=90    x=18×3
54

**19**

(−1,8)    (3,8)

8

In the figure above, what is the perimeter of the shaded rectangle?

24

**20** A committee of 43 people is to be divided into subcommittees so that each person serves on exactly one subcommittee. Each subcommittee must have at least three members but not more than <u>five</u> members. If *M* represents the maximum number of subcommittees that can be formed and *m* represents the least number of subcommittees that can be formed, what is the value of *M* − *m*?

M = 14.5
m = 8.6
5.73

**21**

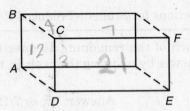

In the figure above, the dimensions of the rectangular box are integers greater than 1. If the area of face *ABCD* is 12 and the area of face *CDEF* is 21, what is the volume of the box?

84

**22** The coordinates of the vertices of a triangle are (1,−2), (9,−2), and (*h,k*). If the area of the triangle is 40, what is a possible value for *k*?

$40=\frac{1}{2}bk$
$40=4k$
k=8

**23** In the games that the Bengals played against the Lions, the Bengals won $\frac{2}{3}$ of the games and lost $\frac{3}{4}$ of the other games. If the teams tied in two games, how many games did the Bengals play against the Lions?

won $\frac{2}{3}$ lost $\frac{1}{4}$
$\frac{8}{12}+\frac{3}{12}+\frac{1}{12}$ tied 2    x² tied 2
$\frac{3}{4}×\frac{1}{3}=\frac{3}{12}=\frac{1}{4}$
x=24

**GO ON TO THE NEXT PAGE**

## Section 2

**2**

**24**

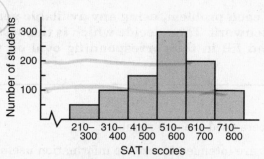

SAT I Scores of Students
at Oceanview High School

If the graph above shows the distribution
of SAT I scores at Oceanview High School,
what percent of the students in this school
scored at least 610? (Omit the percent
sign in your answer).

300

33 ⅓

**25**

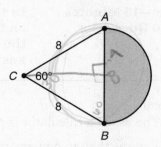

In the figure above, line segment *AB* is a
diameter of semicircle *AB*. What is the
perimeter of the shaded region, rounded
off to the *nearest integer*? [Use 3.14 as an
approximation for π.]

$2\pi r$

$2(3.14)\cancel{A}$

$28.12/2$

$12.56$

$\frac{1}{2}(2\pi r)$

$\pi r + 8$

$4\pi + 8$

$4(3.14) + 8$

$20.56$

21

(21)

┌ ┐

IF YOU FINISH BEFORE TIME IS CALLED, YOU MAY CHECK YOUR WORK ON
THIS SECTION ONLY. DO NOT TURN TO ANY OTHER SECTION IN THE TEST.

**STOP**

## Section 3

**3**

Time—15 Minutes
  10 Questions

In this section solve each problem, using any available space on the page for scratchwork. Then decide which is the best of the choices given and fill in the corresponding oval on the answer sheet.

**Notes:**

1.  The use of a calculator is permitted. All numbers used are real numbers.

2.  Figures that accompany problems in this test are intended to provide information useful in solving the problems. They are drawn as accurately as possible EXCEPT when it is stated in a specific problem that the figure is not drawn to scale. All figures lie in a plane unless otherwise indicated.

**Reference Information**

 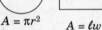    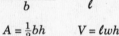   

$A = \pi r^2$
$C = 2\pi r$
$A = \ell w$
$A = \frac{1}{2}bh$
$V = \ell wh$
$V = \pi r^2 h$
$c^2 = a^2 + b^2$
Special Right Triangles

The number of degrees of arc in a circle is 360.
The measure in degrees of a straight angle is 180.
The sum of the measures in degrees of the angles of a triangle is 180.

**1** A bell that rings every $m$ minutes and another bell that rings every $n$ minutes, where $m$ and $n$ are prime numbers, ring at the same time. What is the LEAST number of minutes that must elapse before the two bells again ring at the same time?

(A) $m + n$

(B) $\dfrac{m}{n}$

(C) $mn$

(D) $60mn$

(E) $\dfrac{60}{m + n}$

**2** Pens that usually sell at two for \$3 are on sale at three for \$4. How much money is saved when 18 pens are purchased at the sale price rather than at the usual price?

(A) \$2
(B) \$3
(C) \$4
(D) \$5
(E) \$6

**GO ON TO THE NEXT PAGE**

## Section 3

**3**

5, 10, 14, 16, 20, 23

If each of the six numbers above is increased by $x$, the average of the resulting set of six numbers is 15. The value of $x$ is

(A) $\frac{1}{6}$

(B) $\frac{1}{3}$

(C) $\frac{1}{2}$

(D) $\frac{2}{3}$

(E) $\frac{5}{6}$

**4**

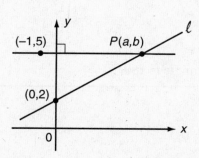

In the figure above, if the slope of line $\ell$ is $\frac{1}{4}$, what are the coordinates $(a,b)$ of point $P$?

(A) $(12,5)$

(B) $(12,8)$

(C) $(8,4)$

(D) $(5,20)$

(E) $\left(\frac{3}{4},5\right)$

**5** If $x$ is an odd integer, for how many values of $x$ is $3 \le 2x \le 56$?

(A) 11

(B) 12

(C) 13

(D) 14

(E) 15

**6**

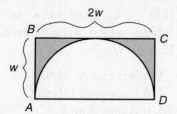

In the figure above, semicircle $AD$ is inscribed in rectangle $ABCD$. What is the area of the shaded region in terms of $w$?

(A) $w^2\left(4 - \frac{\pi}{2}\right)$

(B) $w^2\left(2 - \frac{\pi}{4}\right)$

(C) $w^2\left(2 - \frac{\pi}{2}\right)$

(D) $\pi w^2 - 2$

(E) $2\pi w^2 - 1$

**7**

| $x$ | $y$ |
|---|---|
| 0 | 2 |
| 1 | −1 |

If the ordered pairs of values for $x$ and $y$ in the table above satisfy the equation $ax + by = c$, then $a$ equals which of the following?

(A) $3b$

(B) $b$

(C) $b + 3$

(D) $\frac{b}{3}$

(E) $2b + 1$

**GO ON TO THE NEXT PAGE**

## Section 3

**8** If $3a + b = 24$ and $a$ is a positive even integer, which of the following statements must be true?

  I. $b$ is divisible by 3.
 II. $b$ is an even integer.
III. $a$ is less than 8

  (A) II only
  (B) I and II only
  (C) I and III only
  (D) II and III only
  (E) I, II, and III

**9** After the length and width of a rectangle are each reduced by 40%, the length and width of the new rectangle are each increased by $\frac{1}{3}$ of their new values. In producing the final rectangle, by what percent is the area of the original rectangle reduced?

  (A) 20%
  (B) 24%
  (C) 30%
  (D) 36%
  (E) 40%

**10** If 1 blip + 4 beeps = 1 glitch and 3 blips + 1 beep = 2 glitches, how many beeps equal a glitch?

  (A) 3
  (B) 5
  (C) 8
  (D) 11
  (E) 13

**IF YOU FINISH BEFORE TIME IS CALLED, YOU MAY CHECK YOUR WORK ON THIS SECTION ONLY. DO NOT TURN TO ANY OTHER SECTION IN THE TEST.**

**STOP**

# ANSWER KEY FOR PRACTICE TEST 1

<u>Note</u>: The number inside the brackets that follows each answer is the number of the lesson in Chapter 3, 4, 5, 6, or 7 that relates to the topic or concept that the question tests. In some cases two lessons are relevant, and both numbers are given.

## SECTION 1

| | | | | |
|---|---|---|---|---|
| 1. **C** [3-1] | 6. **A** [3-3] | 11. **A** [4-2] | 16. **A** [6-6] | 21. **D** [5-1] |
| 2. **D** [6-1] | 7. **B** [3-8] | 12. **C** [5-1] | 17. **C** [7-4] | 22. **A** [3-3] |
| 3. **C** [3-5] | 8. **E** [3-5] | 13. **D** [5-4] | 18. **B** [5-2] | 23. **C** [4-2] |
| 4. **D** [6-5] | 9. **B** [5-4] | 14. **A** [3-8] | 19. **E** [5-1, 6-1] | 24. **A** [7-3] |
| 5. **C** [4-2] | 10. **C** [6-7] | 15. **D** [5-1] | 20. **B** [3-5] | 25. **B** [6-5] |

## SECTION 2

| | | | | |
|---|---|---|---|---|
| 1. **C** [3-6] | 6. **B** [7-2] | 11. **A** [4-6] | 16. **.01** [4-1] | 21. **84** [6-7] |
| 2. **B** [6-4] | 7. **B** [7-3] | 12. **C** [7-1] | 17. **54** [5-1] | 22. **8** [6-8] |
| 3. **B** [6-2] | 8. **D** [3-2, 3-4] | 13. **C** [4-6] | 18. **4** [4-1] | 23. **24** [3-7] |
| 4. **C** [6-8] | 9. **B** [7-4] | 14. **A** [4-3] | 19. **24** [6-5] | 24. **33.3** [3-8] |
| 5. **D** [6-5] | 10. **C** [6-1] | 15. **A** [6-5, 6-6] | 20. **5** [7-2] | 25. **21** [6-6] |

## SECTION 3

| | | | | |
|---|---|---|---|---|
| 1. **C** [3-3] | 3. **B** [7-1] | 5. **C** [3-1] | 7. **A** [4-2] | 9. **D** [3-8, 6-5] |
| 2. **B** [3-1] | 4. **A** [6-8] | 6. **C** [6-5, 6-6] | 8. **B** [3-3] | 10. **D** [4-6] |

---

## SELF SCORING CHART

### SECTION 1 (25 Questions)

Number Correct _____ (a)
Number Incorrect _____ (b)
Number Omitted _____ (c)
(a) minus $\frac{1}{4}$(b) = **Raw Score I** _____

### SECTION 2 (25 Questions)

**Part 1 Questions 1–15**

Number Correct _____ (d)
Number Incorrect _____ (e)
(d) minus $\frac{1}{3}$(e) = **Raw Score II** _____

**Part 2 Questions 16–25**

Number Correct **Raw Score III** _____

### SECTION 3 (10 Questions)

Number Correct _____ (f)
Number Incorrect _____ (g)
Number Omitted _____ (h)
(f) minus $\frac{1}{4}$(g) = **Raw Score IV** _____

**Total Score** = Raw Score I + II + III + IV _____

| **Evaluation Chart** | |
|---|---|
| 51–60 | Superior |
| 45–50 | Very Good |
| 40–44 | Good |
| 35–39 | Above Average |
| 30–34 | Average |
| 20–29 | Below Average |
| 0–19 | Below Average; needs intensive study |

# ANSWER EXPLANATIONS FOR PRACTICE TEST 1

## SECTION 1

1. **C** Multiply odd integers beginning with 1 by 4 until the product is more than 36:

$$4 \times 1 = 4 \qquad 4 \times 3 = 12$$
$$4 \times 5 = 20 \qquad 4 \times 7 = 28$$
$$4 \times 9 = 36$$

Hence, there are four positive integers less than 36 (4, 12, 20, and 28) that are equal to 4 times an *odd* integer.

2. **D** Since pairs of obtuse angles formed by parallel lines are equal, the obtuse angles at $A$ and $B$ are equal, so $\angle ABD = 130°$. You are told that $BC$ bisects $\angle ABD$, so

$$\angle DBC = \frac{1}{2}(130°) = 65°$$

An acute and an obtuse angle formed by parallel lines are supplementary. Hence, the obtuse angle at $C$, which is labeled $x°$, is supplementary to $\angle DBC$ and

$$x = 180 - 65 = 115$$

3. **C** If a call lasts $x$ minutes and $x$ is greater than 3, then the charge for the first 3 minutes is \$1.80 and the charge for the next $x - 3$ minutes is $0.40(x - 3)$. Since the total charge was \$4.20,

$$0.40(x - 3) + 1.80 = 4.20$$
$$40(x - 3) + 180 = 420$$
$$40x - 120 + 180 = 420$$
$$40x = 360$$
$$x = \frac{360}{40} = 9$$

4. **D** The area of a square with a side length of 8 inches is $8 \times 8$ or 64 square inches. Since the area of a smaller square with a side length of 2 inches is $2 \times 2$ or 4 square inches, $\frac{64}{4}$ or 16 of the smaller squares are needed to cover completely, without overlap, the larger square.

5. **C** If $3a = 2$ and $2b = 15$, then

$$(3a)(2b) = (2)(15)$$
$$6ab = 30$$
$$ab = \frac{30}{6} = 5$$

6. **A** Since the length of each side of the figure is 1 unit,

$$AB + BC + CD + DE + EA = 5$$

Therefore, each time a point starts and ends at vertex $A$, the point moves 5 units or a multiple of 5 units. Since $\frac{715}{5} = 143$

with no remainder, a point that starts at vertex $A$ and moves a distance of 715 units along each side of the figure will go around the entire figure exactly 143 times, so it will end up at vertex $A$.

7. **B** According to the table, $32 + 80 + 160 + 83 + 45$ or 400 families were surveyed. Since the number of families with two or fewer television sets is $32 + 80 + 160$ or 272, the percent of families surveyed with two or fewer television sets is $\frac{272}{400} \times 100\%$ or 68%.

8. **E** Determine whether each Roman numeral expression decreases when $x$ increases, where $0 < x < 1$.

- I. As $x$ increases, the denominator of the fraction $\frac{1}{1-x}$ gets smaller, so the value of the fraction increases. Hence, expression I does not increase.
- II. As $x$ increases, $x^2$ also increases. Since the denominator of the fraction $\frac{1}{x^2}$ increases, the value of the fraction decreases. Hence, expression II decreases.
- III. As $x$ increases but remains less than 1, $\sqrt{x}$ increases, so the difference $1 - \sqrt{x}$ decreases. Hence, expression III decreases.

Only Roman numeral expressions II and III decrease.

9. **B** Since the length of the discarded ribbon is 9 inches, and this length is $\frac{1}{8}$ of the length of the original ribbon, the original ribbon was $8 \times 9$ or 72 inches long. Hence, the remaining ribbon is $72 - 9$ or 63 inches long. You are given that the 63-inch ribbon is divided into two pieces whose lengths are in the ratio of $4 : 5$. If $x$ represents the length of the shorter ribbon, then $63 - x$ is the length of the longer ribbon. Hence:

$$\frac{\text{shorter ribbon}}{\text{longer ribbon}} = \frac{x}{63-x} = \frac{4}{5}$$

Cross-multiplying gives

$$5x = 4(63 - x)$$
$$= 252 - 4x$$
$$9x = 252$$
$$x = \frac{252}{9} = 28$$

10. **C** If the edge length of a cube is $\pi$, then the volume of the cube is the edge length

raised to the third power or $\pi^3$. If the height of a cylinder is $\pi$ and the radius of its base is represented by $r$, then the volume of the cylinder is the area of the base times its height or $(\pi r^2) \times \pi = \pi^2 r^2$. Since you are told that the volumes of the cube and cylinder are equal,

$$\pi^3 = \pi^2 r^2$$
$$r^2 = \frac{\pi^3}{\pi^2} = \pi$$
$$r = \sqrt{\pi}$$

**11. A** By the distributive law, $a(b + 1) = ab + a$. If $\frac{5}{a} = b$, then $5 = ab$. Since $ab = 5$ and $a = 3$,

$$a(b + 1) = ab + a = 5 + 3 = 8$$

**12. C** If $x$ represents the number of \$20 sport shirts purchased, then $2x$ is the number of \$35 dress shirts purchased. Since a total of \$720 was spent on the shirts, $20(x) + 35(2x) = 720$ or $20x + 70x = 720$, so $90x = 720$. Hence,

$$x = \frac{720}{90} = 8 \quad \text{and} \quad 2x = 2(8) = 16$$

The total number of shirts purchased is $x + 2x$ or $8 + 16 = 24$.

**13. D** If there are 1.2 grams of hemoglobin in 10 cubic centimeters of blood and $x$ represents the number of grams of hemoglobin contained in 35 cubic centimeters of the same blood, then

$$\frac{\text{blood}}{\text{hemoglobin}} = \frac{10}{1.2} = \frac{35}{x}$$
$$10x = 42$$
$$x = \frac{42}{10} = 4.2$$

**14. A** Since the sum of the sectors of a circle graph must add up to 100%,

$$x\% + 20\% + 28\% + 12\% + 24\% = 100\%$$
$$x\% + 84\% = 100\%$$
$$x = 16$$

Since $16\% + 24\% = 40\%$, then $0.40 \times 2500$ or 1000 students named either swimming or biking as their favorite sports activity.

**15. D** If $x - 3$ is 1 less than $y + 3$, then $x - 3 = (y + 3) - 1$, so $x = y + 5$. Since

$$x + 2 = y + 5 + 2 = y + 7$$

then $x + 2$ exceeds $y$ by 7.

**16. A** Since diameter $AD = 6$, $AB = BC = CD = 2$. The areas of semicircles $BC$ and $CD$ are equal, so the area of the shaded region is simply the sum of the areas of semicircles $AD$ and $AB$. The area of semicircle $AD$ is

$\frac{1}{2}(6^2\pi)$ or $18\pi$, and the area of semicircle $AB$ is $\frac{1}{2}(2^2)\pi$ or $2\pi$. Hence, the area of the shaded region is $18\pi + 2\pi$ or $20\pi$.

**17. C** Since $110 = 11 \times 10$, the largest prime factor of 110 is 11. Hence,

$$\sqrt{110} = 110 + \frac{110}{11} = 110 + 10 = 120 = x$$

Since $120 = 3 \times 8 \times 5$, the largest prime factor of 120 is 5. Hence,

$$\sqrt{x} = \sqrt{120} = 120 + \frac{120}{5}$$
$$= 120 + 24 = 144$$

**18. B** A number $N$ increased by 50% of its value is $N + 0.5N$ or $1.5N$. Since you are told that the result is equal to 35 less 40% of 35,

$$1.5N = 35 - (0.40 \times 35)$$
$$= 35 - 14 = 21$$
$$N = \frac{21}{1.5} = 14$$

**19. E** The degree measure of $\angle B$ is twice the degree measure of $\angle A$. Hence, if $x$ is the measure of $\angle A$, then $2x$ is the measure of $\angle B$ and $180 - (x + 2x)$ or $180 - 3x$ is the measure of $\angle C$. When the degree measure of $\angle A$ is subtracted from the degree measure of $\angle C$, the difference is 20, so

$$(180 - 3x) - x = 20$$
$$180 - 4x = 20$$
$$-4x = -160$$
$$x = \frac{-160}{-4} = 40$$

Since the measure of $\angle B$ is $2x = 2(40) = 80$ and the measure of $\angle C$ is $180 - 3x = 180 - 3(40) = 60$, the smallest angle of the triangle measures $40°$.

**20. B** In a certain class, exactly $\frac{2}{3}$ of the students applied for admission to a 4-year college. Of these, $\frac{1}{8}$ or $\frac{1}{8} \times \frac{2}{3} = \frac{2}{24} = \frac{1}{12}$ of the class also applied to a 2-year junior college. Hence, the number of students who could be in this class must be divisible by the least common denominator of $\frac{2}{3}$ and $\frac{1}{12}$, which is 12, or a multiple of the least common denominator. Suppose there are 12 students in the class. Then $\frac{1}{12}$ of 12 or one student applied to both a 4-year and a 2-year college. Since you are told that more than one student applied to both 4-year and 2-year colleges, the LEAST number of students in the class is the next consecutive multiple of 12, which is 24.

21. **D**  If a group of $p$ people plan to contribute equally to the purchase of a gift that costs $d$ dollars, then each person must contribute $\frac{d}{p}$ dollars. If $n$ of the $p$ people decide not to contribute, then each of the $p - n$ people who are left must contribute $\frac{d}{p-n}$ dollars. The difference between the two contribution rates represents the amount of increase for each person who contributes:

$$\frac{d}{p-n} - \frac{d}{p} = \frac{d}{p-n}\left(\frac{p}{p}\right) - \frac{d}{p}\left(\frac{p-n}{p-n}\right)$$
$$= \frac{dp - d(p-n)}{p(p-n)}$$
$$= \frac{dp - dp + nd}{p(p-n)}$$
$$= \frac{nd}{p(p-n)}$$

22. **A**  Multiples of 3 differ by 3. If $p + 1$ is a multiple of 3, then the greatest multiple of 3 that is less than $p + 1$ is $p + 1 - 3$ or $p - 2$.

23. **C**  If $h + k + p = 72$, $k = 2h$, and $p = 3k$ then
$$h + (2h) + 3(2h) = 72$$
$$9h = 72$$
$$h = \frac{72}{9} = 8$$

24. **A**  Determine whether each Roman numeral statement is true or false, given that a number $a$ is selected at random from set $A = \{1, 2, 5\}$ and a number $b$ is selected at random from set $B = \{3, 4\}$.

- I. Since the largest number in set $A$ is 5 and the largest number in set $B$ is 4, $a + b$ cannot be greater than 9. Since the probability of an impossibility is 0, statement I is true.
- II. Six outcomes are possible when $a + b$ is computed:

$$1 + 3 = 4 \qquad 2 + 3 = 5 \qquad 5 + 3 = 8$$
$$1 + 4 = 5 \qquad 2 + 4 = 6 \qquad 5 + 4 = 9$$

Since 5 is a prime number and 5 occurs in two out of the six possible outcomes, the probability that $a + b$ will be a prime number is $\frac{2}{6}$, which is not equal to $\frac{1}{2}$. Hence, statement II is false.
- III. Comparison of $a + b$ with $ab$ for each possible ordered pair of the form $(a, b)$ gives these results:

| $a + b$ | $ab$ |
|---------|------|
| 4 | 3 |
| 5 | 4 |
| 5 | 6 |
| 6 | 8 |
| 8 | 15 |
| 9 | 20 |

For two of the six possible outcomes $a + b > ab$, so the probability that this will happen is $\frac{2}{6}$ or $\frac{1}{3}$. Hence, statement III is false. Only Roman numeral statement I is true.

25. **B**  The area of the shaded region is the difference between the areas of the largest right triangle ($\triangle ACB$) and smallest right triangle ($\triangle ADE$) in the figure below.

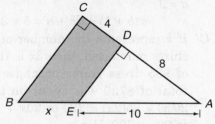

Let $x$ represent the length of $\overline{BE}$. Since triangles $ACB$ and $ADE$ are similar, the lengths of corresponding sides of these triangles have the same ratio. Thus,

$$\frac{8}{4+8} = \frac{10}{x+10}$$
$$8(x + 10) = 120$$
$$8x + 80 = 120$$
$$8x = 40$$
$$x = \frac{40}{8} = 5$$

The hypotenuse of the larger right triangle is $5 + 10$ or 15.

- $\triangle ACB$ is a 9-12-15 right triangle, so its area is $\frac{1}{2}(9 \times 12) = 54$.
- $\triangle ADE$ is a 6-8-10 right triangle, so its area is $\frac{1}{2}(6 \times 8) = 24$.
- The area of the shaded region is $54 - 24$ or 30.

## SECTION 2

1. **C**  In Column A,
$$4 \div \frac{4}{3} = 4 \times \frac{3}{4} = 3$$

In Column B,
$$\frac{1}{\frac{1}{3}} = 1 \div \frac{1}{3} = 1 \times 3 = 3$$

Hence, Column A = Column B.

2. **B**  Since vertical angles are equal, the measures of the three angles of the triangle in

Column A are $x$, $y$, and $z$. Also, since the three angles of a triangle add up to 180°, $x + y + z = 180°$. In Column B, the sum of the four angles of a quadrilateral add up to 360°, so $a + b + c + d = 360°$. Hence, Column B > Column A.

3. **B** In a right triangle, the two acute angles add up 90°. Since you are given that $b = 2a$, then $a = 30°$ and $b = 60°$. In a right triangle whose acute angles measure 30° and 60°, the side opposite the 60° angle is always $\sqrt{3}$ times the side opposite the 30° angle. Hence, $y = 4\sqrt{3}$. Since $\sqrt{3}$ is less than 2, $4\sqrt{3}$ or $y$ is less than 8, so Column B > Column A.

4. **C** Form a right triangle by drawing a perpendicular segment from $(r,s)$ to the $y$-axis.

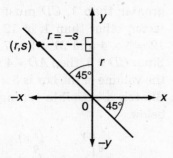

Since the right triangle contains a 45° angle, it is isosceles; therefore, its legs have the same length. Since $(r,s)$ is in quadrant II, $r$ is negative and $s$ is positive, so $r \neq s$ but $r^2 = s^2$. Hence, Column A = Column B.

5. **D** Rectangles of different dimensions can have the same area. If the area of rectangle $ABCD$ is 10, then the dimensions may be 5 by 2 since 5 times 2 is 10. The perimeter of a 5 by 2 rectangle is $5 + 2 + 5 + 2$ or 14. In this case, Column A = Column B. However, since 10 times 1 is also 10, the dimensions of the rectangle could be 10 by 1, in which case the perimeter of the rectangle is $10 + 1 + 10 + 1$ or 22. Then Column B > Column A. Since more than one answer is possible, the correct choice is (D).

6. **B** You are told that 13 students are in the Drama Club, 9 students are in the Chess Club, and 5 students are in both clubs. Hence, there are $13 - 5$ or 8 students in the Drama Club only, and $9 - 5$ or 4 students in the Chess Club only. Since the total number of students in exactly one of the two clubs is $8 + 4$ or 12 students, Column B > Column A.

7. **B** You are given that a coin is weighted so that, when it is tossed, a head is twice as likely to show as a tail. Since the probabilities of all the possible outcomes of a probability experiment must add up to 1, the probability of tossing a head is $\frac{2}{3}$ and the probability of tossing a tail is $\frac{1}{3}$. In Column A, the probability of getting two tails when the coin is tossed two times is $\frac{1}{3} \times \frac{1}{3}$ or $\frac{1}{9}$. In Column B, the probability of getting three heads when the coin is tossed three times is $\frac{2}{3} \times \frac{2}{3} \times \frac{2}{3}$ or $\frac{8}{27}$. Since $\frac{1}{9} = \frac{3}{27}$, and $\frac{8}{27} > \frac{3}{27}$, Column B > Column A.

8. **D** Since $x^2 = y^2$, then $x = \pm\sqrt{y^2} = \pm y$, so $x = y$ or $x = -y$. If $x = y$, Column A = Column B. If, however, $x = -y$ and $x \neq 0$, then $x^2 + x$ could be greater than or less than $y^2 + y$. Since more than one answer is possible, the correct choice is (D).

9. **B** If the number of a term in a sequence is divisible by 3, the term could be the 3rd term, 6th term, 9th term, and so forth. In the nonending sequence

$$1, 3, 2, 1, 3, 2, 1, 3, 2, 1, 3, 2, \ldots$$

the three numbers 1, 3, and 2 repeat in the order given, so 2 is the 3rd term, 6th term, 9th term, and so forth. If the number of the term gives a remainder of 1 when divided by 3, the term could be the 1st term, 4th term, 7th term, and so forth, which is always 1. Similarly, if the number of the term gives a remainder of 2 when divided by 3, the term is always 3. Since $\frac{57}{3}$ is 19 with 0 remainder, the 57th term of the sequence is 2. Also, since $\frac{80}{3}$ is 26 with a remainder of 2, the 80th term of the sequence is 3. Hence, Column B > Column A.

10. **C** Since acute angles formed by parallel lines are equal, $a = b + c + d$, so $a - b = c + d$. Hence, Column A = Column B.

11. **A** Dividing each member of $2h + 2k = 8$ by 2 gives $h + k = 4$, and dividing each member of $7h - 7k = 21$ by 7 gives $h - k = 3$. To eliminate $k$, add corresponding sides of $h + k = 4$ and $h - k = 3$. The result is $2h = 7$, so $h = \frac{7}{2} = 3.5$. Hence, Column A > Column B.

12. **C** Since $a = 2b$ and $c = a + b$,

average of $a$, $b$, and $c = \dfrac{a+b+c}{3} = \dfrac{(2b)+b+(a+b)}{3}$

$$= \dfrac{4b+a}{3}$$

$$= \dfrac{4b+2b}{3}$$

$$= \dfrac{6b}{3}$$

$$= 2b$$

Hence, Column A = Column B.

13. **C** If $\frac{xy}{2} = 6$, then $xy = 12$. Substituting 12 for $xy$ in $(x + y)^2 = x^2 + 2xy + y^2 = 40$ gives

$$x^2 + 2(12) + y^2 = 40$$

or

$$x^2 + y^2 = 16$$

Hence, Column A = Column B.

14. **A** If $a > b > 1$, then $1 > \frac{1}{b} > \frac{1}{a}$. Since $\frac{1}{b} - \frac{1}{a} > 0$, Column A > Column B.

15. **A** In Column A, since $ABCD$ is a square, $AB = BC$. Hence, $3x - 1 = x + 1$ or $2x = 2$, so $x = 1$. Since $BC = x + 1 = 1 + 1 = 2$, the area of square $ABCD$ is $2 \times 2$ or 4. In Column B, the radius of circle $O = x = 1$. Hence the area of circle $O$ is $\pi(1)^2$ or $\pi$. Since $\pi$ is less than 4, Column A > Column B.

16. **.01** If $x + 2x + 3x + 4x = 1$, then $10x = 1$ so

$$x = \frac{1}{10} \quad \text{and} \quad x^2 = \frac{1}{10} \times \frac{1}{10} = \frac{1}{100}$$

Since 1/100 does not fit the grid, grid in .01 instead.

17. **4** If $2^{x-1} - 1 = 7$, then $2^{x-1} = 8 = 2^3$, so $x - 1 = 3$ and $x = 4$.

18. **54** If $2x$ represents the number of luxury cars sold, then $3x$ is the number of economy cars sold. Since the total number of economy and luxury cars sold was 90,

$$2x + 3x = 90$$

$$5x = 90$$

$$x = \frac{90}{5} = 18$$

Hence, the number of economy cars sold was $3x = 3(18) = 54$.

19. **24** One side of the shaded rectangle lies on the $x$-axis, and the side opposite has endpoints $(-1,8)$ and $(3,8)$. Since this side is 8 units above the $x$-axis, the height of the rectangle is 8. The width of this rectangle is the difference between the $x$-coordinates of the points $(-1,8)$ and $(3,8)$ which is $3 - (-1)$ or 4. Hence, the perimeter of the rectangle is $2(4 + 8) = 2(12) = 24$.

20. **5** You are told that each subcommittee formed from a committee of 43 people must have at least three but not more than five members. Since each person serves on exactly one subcommittee, the maximum number of subcommittees is the integer part of $43 \div 3$ or 14. To find the minimum number of subcommittees, divide 43 by 5 to get 8 with a remainder of 3. Hence, the 43 people can be divided into eight subcommittees with five members each plus one subcommittee with three members, for a total of nine subcommittees. Thus, $M - m = 14 - 9 = 5$.

21. **84** Since the area of face $ABCD$ is 12, then $AD \times CD = 12$. The area of face $CDEF$ is 21, so $CD \times DE = 21$. Since the dimensions of the rectangular box are integers greater than 1, $CD$ must be a common factor, other than 1, of 12 and 21. Since $12 = 3 \times 4$ and $21 = 3 \times 7$, then $CD = 3$. Since $CD = 3$, then $AD = 4$ and $DE = 7$, so the volume of the box is $3 \times 4 \times 7$ or 84.

22. **8** Sketch a possible triangle, as shown below.

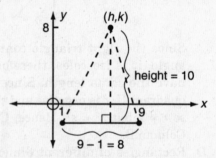

The length of the horizontal base of the triangle is the difference between the $x$-coordinates of the endpoints of the base, which is $9 - 1$ or 8. The area of a triangle is $\frac{1}{2}$ of the product of the base and the height. Since the area of the triangle is given as 40 and $\frac{1}{2} \times 8 \times 10 = 40$, the height of the triangle is 10 units. Since the $y$-coordinate of a point 10 units above the base is $-2 + 10$ or 8, a possible value of $k$ is 8.

23. **24** The Bengals won $\frac{2}{3}$ of the games played, so the number of games that ended in a loss or tie is $\frac{1}{3}$ of the total number of games played. The Bengals lost $\frac{3}{4}$ of the number of games that remained, so they lost $\frac{3}{4}$ of $\frac{1}{3}$ or $\frac{1}{4}$ of the games they played. Since the Bengals won $\frac{2}{3} \left(= \frac{8}{12}\right)$ and lost

$\frac{1}{4}\left(=\frac{3}{12}\right)$ of the games played, $\frac{1}{12}$ of the games played ended in a tie. You are told that the teams tied in two games. Therefore, the total number of games played is 24, since $\frac{1}{12}$ of 24 is 2.

24. **33.3** The total number of students who took the SAT I at Oceanview High School is the sum of the heights of all of the bars on the graph. Thus,

$$50 + 100 + 150 + 300 + 200 + 100$$
or 900 students

took the test. The number of students who scored at least 610 is the sum of the heights of the last two bars, which is 200 + 100 or 300. The percent of the students in this school who scored at least 610 is

$$\frac{300}{900} \times 100\% = 33.3\%$$

25. **21** Since $\angle C = 60$, the sum of the measures of the remaining angles of the triangle is 120. Since $AC = BC = 8$, the other two angles of the triangle have equal degree measures and are each 60°, so $\triangle ABC$ is equiangular. If a triangle is equiangular, it is also equilateral, so $AB = 8$. Since the diameter of semicircle $AB$ is 8, its circumference is $\frac{1}{2}(\pi \times 8)$ or $4\pi$. Hence, the perimeter of the shaded region is $8 + 4\pi$. When 3.14 is used as an approximation for $\pi$,

$$8 + 4\pi = 8 + 4(3.14) = 8 + 12.56 = 20.56$$

To the nearest integer, the perimeter is 21.

# SECTION 3

1. **C** If a bell rings every $m$ minutes and another bell rings every $n$ minutes, then the number of minutes that must elapse before the bells again ring at the same time must be divisible by both $m$ and $n$. Since $m$ and $n$ are prime numbers, the least number of minutes that must elapse before the two bells ring again is the lowest common multiple of $m$ and $n$, which is $m$ times $n$ or $mn$.

2. **B** When 18 pens are purchased at two for $3, the cost is $9 \times \$3$ or $27. When 18 pens are purchased at three for $4, the cost is $6 \times \$4$ or $24. The amount saved is $27 − $24 or $3.

3. **B** The sum of numbers 5, 10, 14, 16, 20, 23

is 88. If each of the six numbers is increased by $x$, their sum is $88 + 6x$. If the average of this new set of six numbers is 15, the sum of these six numbers is $6 \times 15$ or 90. Hence:

$$88 + 6x = 90$$
$$6x = 2$$
$$x = \frac{2}{6} = \frac{1}{3}$$

4. **A** Since point $P$ lies on the same horizontal line as $(-1, 5)$, its $y$-coordinate is also 5. Hence,

$$\text{slope of line } \ell = \frac{\text{change in } y}{\text{change in } x} = \frac{5-2}{a-0} = \frac{1}{4}$$
$$\frac{3}{a} = \frac{1}{4}$$
$$a = 12$$

Since $a = 12$ and $b = 5$, the coordinates of point $P$ are $(12, 5)$.

5. **C** If $3 \le 2x \le 56$, then dividing each member of the inequality gives

$$\frac{3}{2} \le \frac{2x}{2} \le \frac{56}{2} \quad \text{or} \quad \frac{3}{2} \le x \le 28$$

Count the number of odd integers from $\frac{3}{2}$ to 28. There are 13 odd integers in this interval: 3, 5, 7, 9, 11, 13, 15, 17, 19, 21, 23, 25, and 27.

6. **C** Since $AD = BC = 2w$, the diameter of semicircle $AD$ is $2w$, so its radius is $w$. Hence:

$$\underbrace{\text{area}}_{\text{shaded region}} = \underbrace{\text{area}}_{\text{rectangle } ABCD} - \underbrace{\text{area}}_{\text{semicircle } AD}$$

$$= 2w(w) \quad - \quad \frac{1}{2}\pi w^2$$
$$= 2w^2 \quad - \quad \frac{1}{2}\pi w^2$$

Since $2w^2 - \frac{1}{2}\pi w^2$ is not among the answer choices, write it in a different form by factoring out $w^2$ to get $w^2\left(2 - \frac{\pi}{2}\right)$.

7. **A** When $x = 0$ and $y = 2$, then $ax + by = c$ becomes $2b = c$. When $x = 1$ and $y = -1$, then $ax + by = c$ becomes $a - b = c$. Since $2b = c$, then

$$a - b = 2b \text{ or } a = 3b$$

8. **B** Determine whether each Roman numeral statement is true or false when $3a + b = 24$ and $a$ is a positive even integer.

- I. Since $3a + b = 24$, then $b = 24 - 3a = 3(8 - a)$. Since $3(8 - a)$ is divisible by 3, then $b$ is divisible by 3, so statement I is true.

- II. Since $b = 3(8 - a)$ and $a$ is a positive even integer, $8 - a$ is an even integer. Since the product of 3 and an even integer is always an even integer, $b$ is an even integer, so statement II is true.

- III. Since there is no restriction on whether $b$ is positive, negative, or 0, $a$ could be less than 8, greater than 8, or equal to 8. Hence, statement III is false.

Only Roman numeral choices I and II are true.

9. **D** Suppose the length and width of the original rectangle are each 10; then the area of the original rectangle is 10 times 10 or 100. After the length and width of this rectangle are each reduced by 40%, the length and width of the new rectangle are each $10 - (0.40 \times 10)$ or 6. If the length and width of this rectangle are then each increased by $\frac{1}{3}$ of their value, then the length and width of the final rectangle is $6 + \left(\frac{1}{3} \times 6\right)$ or 8. The area of the final rectangle is $8 \times 8$ or 64.

Hence,

$$\begin{aligned} \text{percent of} \\ \text{decrease in area} &= \frac{\text{decrease in area}}{\text{original area}} \times 100\% \\ &= \frac{100-64}{100} \times 100\% \\ &= \frac{36}{100} \times 100\% \\ &= 36\% \end{aligned}$$

10. **D** Multiply the first equation by 3 and then subtract:

$$\begin{array}{rcl} 1\,\text{blip} + 4\,\text{beeps} = 1\,\text{glitch} & \rightarrow & 3\,\text{blips} + 12\,\text{beeps} = 3\,\text{glitches} \\ 3\,\text{blips} + 1\,\text{beep} = 2\,\text{glitches} & \rightarrow & -\,3\,\text{blips} + \phantom{1}1\,\text{beep} \phantom{2} = 2\,\text{glitches} \\ \hline & & 11\,\text{beeps} = 1\,\text{glitch} \end{array}$$

Hence, 11 beeps equal a glitch.

# Answer Sheet — PRACTICE TEST 2

Start with number 1 for each new section. If a section has fewer questions than answer spaces, leave the extra answer spaces blank.

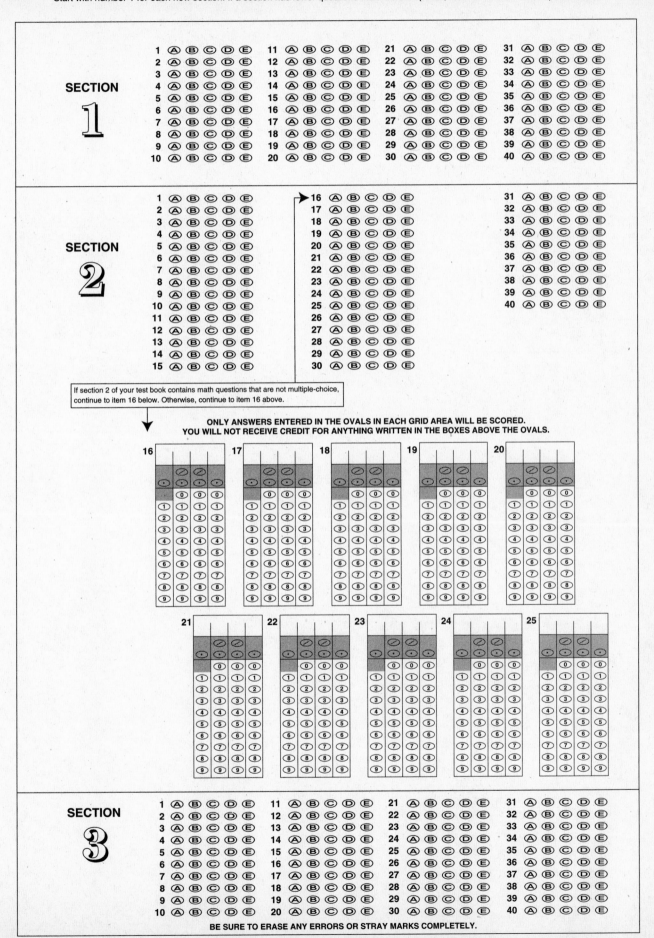

SECTION 1

SECTION 2

If section 2 of your test book contains math questions that are not multiple-choice, continue to item 16 below. Otherwise, continue to item 16 above.

ONLY ANSWERS ENTERED IN THE OVALS IN EACH GRID AREA WILL BE SCORED.
YOU WILL NOT RECEIVE CREDIT FOR ANYTHING WRITTEN IN THE BOXES ABOVE THE OVALS.

SECTION 3

BE SURE TO ERASE ANY ERRORS OR STRAY MARKS COMPLETELY.

# PRACTICE TEST 2

| **Section 1** | **1** |

**Time—30 Minutes**
**25 Questions**

In this section solve each problem, using any available space on the page for scratchwork. Then decide which is the best of the choices given and fill in the corresponding oval on the answer sheet.

**Notes:**

1.  The use of a calculator is permitted. All numbers used are real numbers.

2.  Figures that accompany problems in this test are intended to provide information useful in solving the problems. They are drawn as accurately as possible EXCEPT when it is stated in a specific problem that the figure is not drawn to scale. All figures lie in a plane unless otherwise indicated.

**Reference Information**

 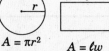   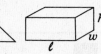 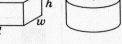   

$A = \pi r^2$
$C = 2\pi r$   $A = \ell w$   $A = \frac{1}{2}bh$   $V = \ell wh$   $V = \pi r^2 h$   $c^2 = a^2 + b^2$   Special Right Triangles

The number of degrees of arc in a circle is 360.
The measure in degrees of a straight angle is 180.
The sum of the measures in degrees of the angles of a triangle is 180.

---

**1** If $(y - 1)(1 - 3) = (5 - 13)$, then $y =$

(A) $-5$

(B) $-\frac{5}{2}$

(C) $\frac{3}{2}$

(D) 5

(E) 6

---

**2** The number of chirps a cricket makes increases with the temperature at a constant rate. If at a temperature of 12° Fahrenheit a cricket chirps 30 times per minute, how many times per minute will the cricket chirp when the temperature is 20° Fahrenheit?

(A) 28
(B) 32
(C) 36
(D) 48
(E) 50

---

**GO ON TO THE NEXT PAGE**

## Section 1

**1**

**3** If one of the angles of a quadrilateral measures 75° and the degree measures of the other angles are in the ratio of $2 : 4 : 9$, what is the degree measure of the largest angle of the quadrilateral?

(A) 171
(B) 168
(C) 135
(D) 105
(E) 63

**4** If the average of $y$, $2y$, and $3y$ is 12, then $y =$

(A) 2
(B) 4
(C) 6
(D) 8
(E) 12

**5** When a whole number $N$ is divided by 2, the remainder is 0. When $N$ is divided by 3, the remainder is 2. Which is a possible value of $N$?

(A) 17
(B) 30
(C) 49
(D) 53
(E) 74

**6** Which of the following expressions is NOT equal to $\frac{1}{3}$?

(A) $\frac{1}{2} - \frac{1}{6}$

(B) $\frac{1}{6} \div 2$

(C) $\frac{4}{9} \times \frac{3}{4}$

(D) $\sqrt{\frac{1}{9}}$

(E) $\dfrac{\frac{1}{12}}{\frac{1}{4}}$

**7**

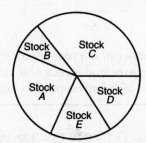

Investment of $4800 in five stocks

The circle graph above shows the distribution of an investment of $4800 among five different stocks. In which stock is the investment closest in value to $1200?

(A) $A$
(B) $B$
(C) $C$
(D) $D$
(E) $E$

GO ON TO THE NEXT PAGE

## Section 1

**1**

**8** A class consists of 15 boys and 12 girls. On Tuesday, all the boys were present and the number of girls who were present represented 40% of the class attendance for that day. How many girls were ABSENT on Tuesday?

(A) Two
(B) Three
(C) Four
(D) Five
(E) Six

**9** If $2^x = 16$ and $2x + y = 16$, then $y =$

(A) 2
(B) 4
(C) 6
(D) 8
(E) 12

**10** A car averages 20 miles per gallon in city driving and 25 miles per gallon in highway driving. If gas costs \$2 per gallon, how much money must be spent on gas for a 500-mile trip if $\frac{3}{5}$ of the distance consists of city driving and the remainder consists entirely of highway driving?

(A) \$36
(B) \$40
(C) \$46
(D) \$48
(E) \$50

**Questions 11–13.**

A yellow marble, a blue marble, a red marble, a white marble, and a green marble are dropped into an empty jar, in the order given. This process is repeated until there are 88 marbles in the jar.

**11** How many red marbles are in the jar?

(A) 17
(B) 18
(C) 19
(D) 20
(E) 21

**12** What is the LEAST number of marbles that must be added to the jar so the jar will contain the same number of marbles of each color?

(A) 0
(B) 1
(C) 2
(D) 3
(E) 4

**13** If a marble is selected at random from the jar with 88 marbles, what is the probability that the marble will be yellow?

(A) $\frac{2}{11}$

(B) $\frac{17}{88}$

(C) $\frac{9}{44}$

(D) $\frac{19}{88}$

(E) It cannot be determined from the information given.

**GO ON TO THE NEXT PAGE**

**Section 1**                                                    **1**

**14**

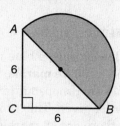

In the figure above, if arc $AB$ is a semi-circle with diameter $AB$, what is the area of the shaded region?

(A) $4\pi$
(B) $9\pi$
(C) $18\pi$
(D) $18\sqrt{2}\pi$
(E) $36\pi$

**15**  If $r = s^{t+1}$ and $s \neq 0$, then $\dfrac{r}{s} =$

(A) $t$
(B) $s^t$
(C) $t + 1$
(D) $s^t + 1$
(E) $\dfrac{1}{s^{t+1}}$

**16**  Which of the following statements must be true when $0 < a < 1$?

I.   $\dfrac{\sqrt{a}}{a} > 1$
II.  $2a < 1$
III. $a^2 - a^3 < 0$

(A) I only
(B) III only
(C) I and III only
(D) II and III only
(E) I, II, and III

**17**  A rectangular container with dimensions of 3 inches by 4 inches by 10 inches is filled to the top with lemonade. The lemonade is then poured into an empty cylindrical pitcher with a diameter of 8 inches. If there is no overflow, what is the height in inches of the lemonade in the pitcher?

(A) $\dfrac{15}{8\pi}$
(B) $\dfrac{15}{2\pi}$
(C) $\dfrac{30}{\pi}$
(D) $15$
(E) $\dfrac{45-\pi}{15}$

**18**  If $\dfrac{a+b}{a} = 3$ and $\dfrac{a+c}{c} = 2$, then $\dfrac{c}{b} =$

(A) $\dfrac{1}{2}$
(B) $\dfrac{2}{3}$
(C) $1$
(D) $\dfrac{3}{2}$
(E) $2$

**19**  In a certain sequence of numbers, each number after the first is 7 less than twice the preceding number. If the third number in the sequence is 23, what is the first number in the sequence?

(A) 9
(B) 11
(C) 13
(D) 15
(E) 17

GO ON TO THE NEXT PAGE

| Section 1 | **1** |
|---|---|

**20** Let

$$\boxed{x} = \frac{x^2}{2} \quad \text{and} \quad \boxed{y} = y + 3$$

If $\boxed{6} \div \boxed{6} = k$, then $\boxed{k} =$

(A) 5
(B) 8
(C) 9
(D) 12
(E) 16

**21**

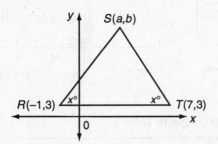

In the figure above, which of the following ordered pairs could be the coordinates $(a,b)$ of point $S$?

(A) (2,9)
(B) (2,11)
(C) (3,9)
(D) (4,8)
(E) (5,10)

**22** If 3 times 1 less than a number $n$ is the same as twice the number increased by 14, what is $n$?

(A) 15
(B) 17
(C) 19
(D) 21
(E) 23

**23** If the length of line segment $AB$ is 9 inches, the total number of points 4 inches from point $A$ and 6 inches from point $B$ is

(A) one
(B) two
(C) three
(D) four
(E) It cannot be determined from the information given

**24**

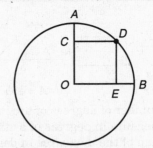

The circle in the figure above has center $O$. If the area of square $OCDE$ is 2, what is the length of arc $ADB$?

(A) $4\pi$
(B) $2\pi$
(C) $\sqrt{2}\pi$
(D) $\pi$
(E) It cannot be determined from the information given

**25** If $w \neq 0$ and the value of $y + 2w$ exceeds $w$ by 75% of $w$, then $\frac{y}{w} =$

(A) $-\frac{3}{2}$
(B) $-\frac{3}{4}$
(C) $-\frac{1}{4}$
(D) $\frac{1}{2}$
(E) $\frac{5}{4}$

**IF YOU FINISH BEFORE TIME IS CALLED, YOU MAY CHECK YOUR WORK ON THIS SECTION ONLY. DO NOT TURN TO ANY OTHER SECTION IN THE TEST.**

**STOP**

## Section 2

**2**

| Time—30 Minutes | This section contains two types of questions. You have 30 |
|---|---|
| 25 Questions | minutes to complete both types. You may use any available |
| | space for scratchwork. |

**Notes:**

1. The use of a calculator is permitted. All numbers used are real numbers.

2. Figures that accompany problems in this test are intended to provide information useful in solving the problems. They are drawn as accurately as possible EXCEPT when it is stated in a specific problem that the figure is not drawn to scale. All figures lie in a plane unless otherwise indicated.

Reference Information

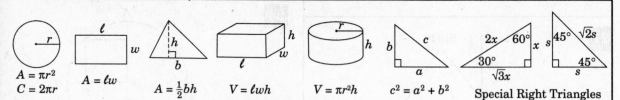

$A = \pi r^2$
$C = 2\pi r$
$A = \ell w$
$A = \frac{1}{2}bh$
$V = \ell wh$
$V = \pi r^2 h$
$c^2 = a^2 + b^2$
Special Right Triangles

The number of degrees of arc in a circle is 360.
The measure in degrees of a straight angle is 180.
The sum of the measures in degrees of the angles of a triangle is 180.

### Directions for Quantitative Comparison Questions

Questions 1–15 each consist of two quantities in boxes, one in Column A and one in Column B. You are to compare the two quantities and on the answer sheet fill in oval

A if the quantity in Column A is greater;
B if the quantity in Column B is greater;
C if the two quantities are equal;
D if the relationship cannot be determined from the information given.

**AN E RESPONSE WILL NOT BE SCORED.**

Notes:

1. In some questions, information is given about one or both of the quantities to be compared. In such cases, the given information is centered above the two columns and is not boxed.

2. In a given question, a symbol that appears in both columns represents the same thing in Column A as it does in Column B.

3. Letters such as $x$, $n$, and $k$ stand for real numbers.

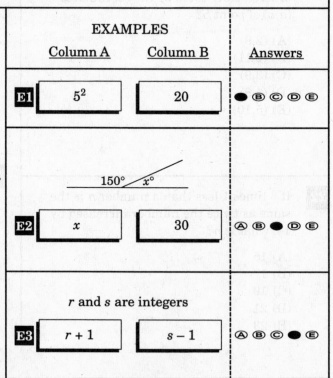

# Section 2

**2**

| Column A | Column B |
|----------|----------|

**1** $3\frac{1}{3}$ hours | 200 minutes

---

$x > 0$

**2** 40% of $\dfrac{x}{2}$ | 75% of $\dfrac{x}{3}$

---

**3** $\sqrt{3}\,(1 + \sqrt{3})$ | $\sqrt{3} + 2$

---

$w = \dfrac{3}{8}$

**4** $7w - 2w + 3w$ | $\dfrac{1}{w}$

---

$x^3 = 7$

**5** $x^6$ | $x^3 + 42$

| Column A | Column B |
|----------|----------|

$y^2 - 2y - 3 = 0$

**6** $y$ | 2

---

$x^2 = 64$

**7** The length of a side of a square whose perimeter is 32 | $x$

---

$0 < a < b < 1$

**8** $\dfrac{1}{1-a}$ | $\dfrac{1}{1-b}$

---

$(p + r)(p - r) = p^2 - 0.0009$
$r > 0$

**9** $r$ | $\dfrac{3}{1000}$

**GO ON TO THE NEXT PAGE**

# Section 2

**2**

SUMMARY DIRECTIONS FOR QUANTITATIVE COMPARISON QUESTIONS

<u>Answer</u>:  A  if the quantity in Column A is greater;
        B  if the quantity in Column B is greater;
        C  if the two quantities are equal;
        D  if the relationship cannot be determined from the information given.

|  <u>Column A</u>  |  <u>Column B</u>  |

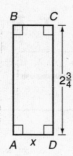

Area of rectangle *ABCD* is 1.

**10**

| $x$ | $\dfrac{1}{3}$ |

---

Three fair coins are tossed at the same time.

**11**

| The probability of getting two heads and one tail | The probability of getting two tails and one head |

---

$$x + y = 3$$
$$z - y = 1$$

**12**

| The average of $x$ and $y$ | The average of $x$ and $z$ |

---

|  <u>Column A</u>  |  <u>Column B</u>  |

Of the 150 hamburgers sold in the student cafeteria, 100 were served with ketchup, 75 were served with onions, and 50 were served with both ketchup and onions.

**13**

| The number of hamburgers served with onions but NOT with ketchup | The number of hamburgers sold without ketchup and without onions |

---

$$-1 < a < b < 0$$

**14**

| $ab^2$ | $ba^2$ |

---

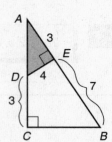

**15**

| The percent of the area of square *ABCD* that is shaded | The percent of the area of right triangle *ABC* that is shaded |

**GO ON TO THE NEXT PAGE** ▶

**Section 2**

**2**

Directions for Student-Produced Response Questions

Each of the remaining 10 questions (16–25) requires you to solve the problem and enter your answer by marking the ovals in the special grid, as shown in the examples below.

Answer: $\frac{7}{12}$ or 7/12

Answer: 2.5

Answer: 201
Either position is correct.

Write answer → in boxes

Grid in → result

←Fraction line

←Decimal point

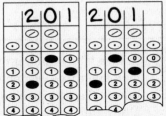

Note: You may start your answers in any column, space permitting. Columns not needed should be left blank.

• Mark no more than one oval in any column.

• Because the answer sheet will be machine-scored, **you will receive credit only if the ovals are filled in correctly.**

• Although not required, it is suggested that you write your answer in the boxes at the top of the columns to help you fill in the ovals accurately.

• Some problems may have more than one correct answer. In such cases, grid only one answer.

• No question has a negative answer.

• **Mixed numbers** such as $2\frac{1}{2}$ must be gridded as 2.5 or 5/2. (If ⬛2⬛1⬛/⬛2 is gridded, it will be interpreted as $\frac{21}{2}$, not $2\frac{1}{2}$.)

• <u>Decimal Accuracy</u>: If you obtain a decimal answer, **enter the most accurate value the grid will accommodate.** For example, if you obtain an answer such as 0.6666 . . ., you should record the result as .666 or .667. **Less accurate values such as .66 or .67 are not acceptable.**

Acceptable ways to grid $\frac{2}{3}$ = .6666 . . .

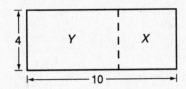

---

**16**

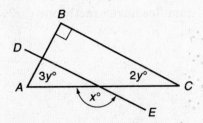

In the figure above, if $BC \parallel DE$, what is the value of $x$?

**17**

In the figure above, when square $X$ is folded back along the broken line, what will be the area of the part of rectangle $Y$ that is NOT covered by square $X$?

**GO ON TO THE NEXT PAGE** →

## Section 2

**2**

**18** The average of 14 scores is 80. If the average of four of these scores is 75, what is the average of the remaining 10 scores?

**19** In how many different ways can three men and three women be arranged in a line so that the men and the women alternate?

**20** If all of the books on a shelf with fewer than 45 books were put into piles of five books each, no books would remain. If the same books were put into piles of seven books each, two books would remain. What is the greatest number of books that could be on the shelf?

**21** If $10^k = \frac{1}{2}$, what is the value of $10^{k+3}$?

**22** The ratio of boys to girls in a room is 2 to 3. After three boys leave the room, the ratio of boys to girls becomes 1 to 2. How many girls are in the room?

**23**

If the area of $\triangle ABC$ above is 45, what is the length of $DB$?

**24** If $a$ and $b$ are positive integers, what is the SMALLEST value of $a + b$ for which $2a + 3b$ is divisible by 24?

**25** Of 200 families surveyed, 75% have at least one car and 20% of those with cars have more than two cars. If 50 families each have exactly two cars, how many families have exactly one car?

## Section 3

**3**

| Time—15 Minutes<br>10 Questions | In this section solve each problem, using any available space on the page for scratchwork. Then decide which is the best of the choices given and fill in the corresponding oval on the answer sheet. |

**Notes:**

1. The use of a calculator is permitted. All numbers used are real numbers.

2. Figures that accompany problems in this test are intended to provide information useful in solving the problems. They are drawn as accurately as possible EXCEPT when it is stated in a specific problem that the figure is not drawn to scale. All figures lie in a plane unless otherwise indicated.

Reference Information

   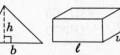    

$A = \pi r^2$  $A = \ell w$  $A = \frac{1}{2}bh$  $V = \ell wh$  $V = \pi r^2 h$  $c^2 = a^2 + b^2$  Special Right Triangles
$C = 2\pi r$

The number of degrees of arc in a circle is 360.
The measure in degrees of a straight angle is 180.
The sum of the measures in degrees of the angles of a triangle is 180.

---

**1**

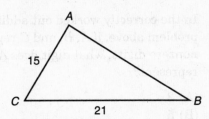

In the figure above, if the length of line segment $AB$ is $\frac{1}{4}$ of the perimeter of $\triangle ABC$, what is the length of $AB$?

(A) 6
(B) 8
(C) 9
(D) 12
(E) 18

**2** If $x^2 + 3y = x^2 - 3$, then $y =$

(A) −1
(B) 0
(C) 1
(D) $\sqrt{x}$
(E) $\frac{-\sqrt{x}}{3}$

**3** If $a = b \div c$, where $b, c \neq 0$, then, when $b$ is multiplied by 2 and $c$ is divided by 2, $a$ is

(A) unchanged
(B) decreased by 4
(C) increased by 4
(D) multiplied by 4
(E) divided by 4

**GO ON TO THE NEXT PAGE**

## Section 3                                                                    **3**

**4** The average of $a$, $b$, and $c$ is 3 times the median. If $0 < a < b < c$, what is the average of $a$ and $c$ in terms of $b$?

(A) $2b$

(B) $\frac{5}{2}b$

(C) $3b$

(D) $\frac{7}{2}b$

(E) $4b$

**5** If $x - 1$ represents an odd integer, then which of the following must represent an even integer?

   I. $x + 1$
   II. $3x$
   III. $(x - 1)^2$

(A) I only
(B) II only
(C) III only
(D) I and III only
(E) II and III only

**6**

| $x$ | $y$ |
|-----|-----|
| $-2$ | 3 |
| 0 | $-1$ |
| 3 | 8 |

For the ordered pairs of numbers $(x, y)$ in the table above, which of the following equations best describes the relationship between $x$ and $y$?

(A) $y = -x + 1$
(B) $y = x^2 - 1$
(C) $y = 2x - 1$
(D) $y = 3x - 1$
(E) $y = (x - 1)^3$

**7**

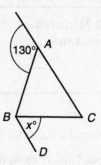

In the figure above, if $AB = AC$ and $AC \parallel BD$, what is the value of $x$?

(A) 45
(B) 50
(C) 60
(D) 65
(E) 70

**8**
$$\begin{array}{r} 3\,A\,.\,B\,5 \\ +\quad 3\,.\,A\,B \\ \hline C\,1\,.\,C\,1 \end{array}$$

In the correctly worked out addition problem above, if $A$, $B$, and $C$ represent nonzero digits, what digit does $A$ represent?

(A) 3
(B) 5
(C) 6
(D) 7
(E) 8

**GO ON TO THE NEXT PAGE**

# Section 3

**3**

**9**

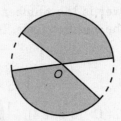

In the figure above, the sum of the lengths of the broken arcs of circle $O$ is $2\pi$. If the diameter of the circle is 18, what is the area of the shaded region?

(A) $9\pi$
(B) $18\pi$
(C) $45\pi$
(D) $63\pi$
(E) $72\pi$

**10**  George spent 25% of the money he had on lunch and 60% of the remaining money on dinner. If he then had $9.00 left, how much money did he spend on lunch?

(A) $2.50
(B) $4.00
(C) $5.00
(D) $7.50
(E) $10.00

**IF YOU FINISH BEFORE TIME IS CALLED, YOU MAY CHECK YOUR WORK ON THIS SECTION ONLY. DO NOT TURN TO ANY OTHER SECTION IN THE TEST.**

**STOP**

# ANSWER KEY FOR PRACTICE TEST 2

Note: The number inside the brackets that follows each answer is the number of the lesson in Chapter 3, 4, 5, 6, or 7 that relates to the topic or concept that the question tests. In some cases two lessons are relevant, and both numbers are given.

## SECTION 1

| | | | | |
|---|---|---|---|---|
| 1. **D** [4-1] | 6. **B** [3-5] | 11. **B** [3-3] | 16. **C** [3-5, 4-4] | 21. **C** [6-8] |
| 2. **E** [5-4] | 7. **A** [6-6] | 12. **C** [3-3] | 17. **B** [6-7] | 22. **B** [5-1] |
| 3. **A** [6-4] | 8. **A** [3-8] | 13. **C** [7-3] | 18. **A** [4-3] | 23. **B** [6-6] |
| 4. **C** [7-1] | 9. **D** [4-1] | 14. **B** [6-2, 6-6] | 19. **B** [7-4] | 24. **D** [3-8, 6-6] |
| 5. **E** [3-3] | 10. **C** [3-7] | 15. **B** [4-2] | 20. **A** [7-4] | 25. **C** [5-2] |

## SECTION 2

| | | | | |
|---|---|---|---|---|
| 1. **C** [3-6] | 6. **D** [4-5] | 11. **C** [7-3] | 16. **144** [6-1] | 21. **500** [3-2] |
| 2. **B** [3-8] | 7. **D** [6-5] | 12. **B** [7-1] | 17. **8** [6-5] | 22. **18** [5-4] |
| 3. **A** [3-2] | 8. **B** [3-5] | 13. **C** [7-2] | 18. **82** [7-1] | 23. **7** [6-5] |
| 4. **A** [4-3] | 9. **A** [4-4] | 14. **A** [3-4] | 19. **72** [7-2] | 24. **9** [3-3] |
| 5. **C** [3-2] | 10. **A** [6-5] | 15. **C** [6-5] | 20. **30** [3-3] | 25. **70** [3-8] |

## SECTION 3

| | | | | |
|---|---|---|---|---|
| 1. **D** [6-5] | 3. **D** [3-6, 4-3] | 5. **B** [3-1] | 7. **D** [6-1] | 9. **E** [6-6] |
| 2. **A** [4-1] | 4. **E** [7-1] | 6. **B** [4-3] | 8. **D** [7-4] | 10. **D** [5-2] |

---

## SELF SCORING CHART

### SECTION 1 *(25 Questions)*

| | |
|---|---|
| Number Correct | _____ (a) |
| Number Incorrect | _____ (b) |
| Number Omitted | _____ (c) |
| (a) minus $\frac{1}{4}$(b) = | **Raw Score I** _____ |

### SECTION 2 *(25 Questions)*

**Part 1 Questions 1–15**

| | |
|---|---|
| Number Correct | _____ (d) |
| Number Incorrect | _____ (e) |
| (d) minus $\frac{1}{3}$(e) = | **Raw Score II** _____ |

**Part 2 Questions 16–25**

| | |
|---|---|
| Number Correct | **Raw Score III** _____ |

### SECTION 3 *(10 Questions)*

| | |
|---|---|
| Number Correct | _____ (f) |
| Number Incorrect | _____ (g) |
| Number Omitted | _____ (h) |
| (f) minus $\frac{1}{4}$(g) = | **Raw Score IV** _____ |

**Total Score** = Raw Score I + II + III + IV _____

### Evaluation Chart

| | |
|---|---|
| 51–60 | Superior |
| 45–50 | Very Good |
| 40–44 | Good |
| 35–39 | Above Average |
| 30–34 | Average |
| 20–29 | Below Average |
| 0–19 | Below Average; needs intensive study |

# ANSWER EXPLANATIONS FOR PRACTICE TEST 2

## SECTION 1

**1. D** If $(y - 1)(1 - 3) = (5 - 13)$, then
$$(y - 1)(-2) = (-8)$$
$$y - 1 = \frac{-8}{-2} = 4$$
$$y = 4 + 1 = 5$$

**2. E** If $x$ represents the number of times per minute that the cricket will chirp when the temperature is 20° Fahrenheit, then
$$\frac{\text{temperature}}{\text{chirps}} = \frac{12}{30} = \frac{20}{x}$$
$$12x = 20(30) = 600$$
$$x = \frac{600}{12} = 50$$

**3. A** If three angles are in the ratio of $2 : 4 : 9$, their measures can be represented as $2x$, $4x$, and $9x$. Since the measures of the four angles of a quadrilateral must add up to 360°,
$$2x + 4x + 9x + 75 = 360$$
$$15x + 75 = 360$$
$$15x = 285$$
$$x = \frac{285}{15} = 19$$

Since $9x = 9(19) = 171$, the largest angle of the quadrilateral measures 171°.

**4. C** Since the average of the three quantities $y$, $2y$, and $3y$ is 12,
$$\frac{y + 2y + 3y}{3} = 12$$
$$\frac{6y}{3} = 12$$
$$2y = 12$$
$$y = \frac{12}{2} = 6$$

**5. E** The question states that, when a whole number $N$ is divided by 2, the remainder is 0. Therefore, $N$ must be even, and you can eliminate choices (A), (C), and (D). You are also told that, when $N$ is divided by 3, the remainder is 2. Test each of the remaining choices:
- (B) $30 \div 3 = 10$    remainder of 0
- (E) $74 \div 3 = 24$    remainder of 2

A possible value of $N$ is 74.

**6. B** To determine which choice is not equal to $\frac{1}{3}$, do the arithmetic for each choice until you find a result that is not $\frac{1}{3}$.
- (A) $\frac{1}{2} - \frac{1}{6} = \frac{3}{6} - \frac{1}{6} = \frac{2}{6} = \frac{1}{3}$

- (B) $\frac{1}{6} \div 2 = \frac{1}{6} \times \frac{1}{2} = \frac{1}{12}$

**7. A** The stock investment that is closest in value to \$1200 will correspond to the sector of the circle graph that is $\frac{\$1200}{\$4800}$ or $\frac{1}{4}$ of the circle. Since all the sectors of a circle must add up to 360°, look for the sector whose sides form an angle that measures approximately $\frac{1}{4} \times 360°$ or 90°. This sector is labeled stock $A$.

**8. A** If $x$ girls were present, then
$$x = 0.40(x + 15)$$
$$= 0.40x + 6$$
$$0.60x = 6$$
$$x = \frac{6}{0.60} = 10$$

Since 10 of the 12 girls were present on Tuesday, $12 - 10$ or 2 girls were absent on that day.

**9. D** Since $2^x = 16 = 2^4$, then $x = 4$. Hence, $2x + y = 16$ becomes
$$2(4) + y = 16$$
$$8 + y = 16$$
$$y = 16 - 8 = 8$$

**10. C** Since

$$\frac{3}{5} \text{ of } 500 = \frac{3}{5} \times 500 = 3 \times 100 = 300$$

there are 300 miles of city driving and $500 - 300$ or 200 miles of highway driving. In city driving, the car averages 20 miles per gallon, so the car uses $\frac{300}{20}$ or 15 gallons of gas during city driving. You are also given that the car averages 25 miles per gallon in highway driving, so the car uses $\frac{200}{25}$ or 8 gallons of gas during this part of the trip. Thus, a total of $15 + 8$ or 23 gallons of gas are used for the 500-mile trip. Since gas costs \$2 per gallon, the cost of the gas for the whole trip is $23 \times \$2$ or \$46.

**11. B** Marbles of five different colors are dropped into the jar in a fixed order: yellow, blue, red, white, and then green. Since $88 \div 5 = 17$ with a remainder of 3, the jar contains 17 of each color marble *plus* three additional marbles—one yellow, one blue, and one red marble. Hence, there are $17 + 1$ or 18 red marbles in the jar.

**12. C** Since the jar contains 18 yellow, blue, and red marbles, but only 17 white and green, one white and one green marble or two marbles must be added so the jar will contain the same number of marbles of each color.

13. **C**   Since 18 of the 88 marbles are yellow, the probability that a marble selected at random from the jar will be yellow is $\frac{18}{88} = \frac{9}{44}$.

14. **B**   In any isosceles right triangle the hypotenuse is $\sqrt{2}$ times the length of a leg. Since, in the figure, hypotenuse $AB = 6\sqrt{2}$, the radius of semicircle $AB$ is $3\sqrt{2}$. Hence:

$$\text{area semicircle } AB = \frac{1}{2}\left[\pi(3\sqrt{2})^2\right]$$
$$= \frac{1}{2}[\pi(3\sqrt{2}) \times (3\sqrt{2})]$$
$$= \frac{1}{2}\ [\pi(9 \times 2)]$$
$$= \frac{1}{2}(18\pi)$$
$$= 9\pi$$

15. **B**   If $r = s^{t+1}$ and $s \neq 0$, then

$$\frac{r}{s} = \frac{s^{t+1}}{s^1} = s^{(t+1)-1} = s^t$$

16. **C**   Determine whether each Roman numeral statement is always true when $0 < a < 1$.
- I. Since $0 < a < 1$, then $\sqrt{a} > a$. For example, if $a = 0.25$, then $\sqrt{0.25} = 0.5$ and $0.5 > 0.25$. Dividing the inequality by $a$ produces the equivalent inequality $\frac{\sqrt{a}}{a} > 1$. Statement I is always true.
- II. If $2a < 1$, then $a < \frac{1}{2}$, which is not necessarily true since you are told only that $0 < a < 1$. Statement II is not always true.
- III. Suppose $a^2 - a^3 < 0$ is true. Then $a^2(1 - a) > 0$. Since $a^2 > 0$ and $1 - a > 0$, then $a^2(1 - a) > 0$, so $a^2 - a^3 < 0$. Statement III is always true.

Only Roman numeral statements I and III are always true.

17. **B**   A rectangular container with dimensions of 3 inches by 4 inches by 10 inches can hold $3 \times 4 \times 10$ or 120 cubic inches of lemonade. If $h$ represents the number of inches in the height of the lemonade after it is poured into a cylindrical pitcher with a diameter of 8 inches, then $\pi r^2 h = 120$, where $r = \frac{8}{2} = 4$. Since the volume of the lemonade remains the same,

$$\pi(4)^2 h = 120$$
$$16\pi h = 120$$
$$h = \frac{120}{16\pi} = \frac{15}{2\pi}$$

18. **A**   Simplify each of the given fractions:

$$\frac{a+b}{a} = \frac{a}{a} + \frac{b}{a} = 1 + \frac{b}{a} = 3, \quad \text{so } \frac{b}{a} = 2$$

$$\frac{a+c}{c} = \frac{a}{c} + \frac{c}{c} = \frac{a}{c} + 1 = 2, \quad \text{so } \frac{a}{c} = 1$$

Eliminate $a$ by multiplying corresponding sides of $\frac{b}{a} = 2$ and $\frac{a}{c} = 1$ together:

$$\frac{b}{a} \times \frac{a}{c} = 2 \times 1$$
$$\frac{b}{c} = 2$$

Since $\frac{b}{c} = 2$, then $\frac{c}{b} = \frac{1}{2}$.

19. **B**   You are told that, in a certain sequence of numbers, each number after the first is 7 less than twice the preceding number, and that the third number in the sequence is 23. For each answer choice, find the third term of the sequence until you get 23 as the third term.
- (A) The second term $= 2 \times 9 - 7 = 18 - 7 = 11$, and the third term $= 2 \times 11 - 7 = 22 - 7 = 15$.
- (B) The second term $= 2 \times 11 - 7 = 22 - 7 = 15$, and the third term $= 2 \times 15 - 7 = 30 - 7 = 23$.

20. **A**   Since

$$\textcircled{x} = \frac{x^2}{2} \quad \text{and} \quad \boxed{y} = y + 3$$

then

$$k = \textcircled{6} \div \boxed{6} = \frac{6^2}{2} \div (6 + 3)$$
$$= 18 \div 9$$
$$= 2$$

Thus,

$$\textcircled{k} = \boxed{\frac{2^2}{2}} = \boxed{2} = 2 + 3 = 5$$

21. **C**   The angles at $R$ and $S$ have the same measure, so $RS = TS$. Vertex $S$ must lie on a vertical line that contains the midpoint $a$ of horizontal base $RT$, as shown in the figure below:

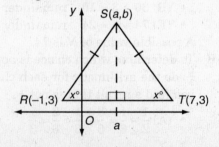

The $x$-coordinate of the midpoint of $RT$ is $\frac{-1+7}{2}$ or 3, so $a = 3$ Hence, you can rule out choices (A), (B), (D), and (E).

22. **B** If 3 times 1 less than a number $n$ is the same as twice the number increased by 14, then $3(n - 1) = 2n + 14$. Hence:
$$3n - 3 = 2n + 14$$
$$3n - 2n = 14 + 3$$
$$n = 17$$

23. **B** All points 4 inches from point $A$ lie on a circle that has point $A$ as its center and that has a radius of 4 inches. Similarly, all points 6 inches from point $B$ lie on a circle that has point $B$ as its center and that has a radius of 6 inches. Since the length of line segment $AB$ is 9 inches, the two circles intersect at two points that are 4 inches from $A$ and 6 inches from $B$, as shown in the figure below.

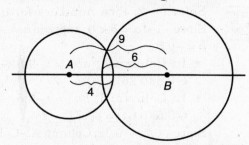

24. **D** Draw radius $OD$. Since the area of a square is equal to one-half the square of a diagonal, $\frac{1}{2}(OD)^2 = 2$ or $(OD)^2 = 2 \times 2 = 4$ so $OD = \sqrt{4} = 2$. The circumference of circle $O = 2\pi r = 2\pi(2) = 4\pi$. Since angle $AOB$ is a right angle, arc $ADB$ is $\frac{90}{360}$ or $\frac{1}{4}$ of the circumference of circle $O$. Hence, the length of arc $ADB = \frac{1}{4} \times 4\pi = \pi$.

25. **C** Since the value of $y + 2w$ exceeds $w$ by 75% of $w$,
$$y + 2w = w + 0.75w$$
$$= 1.75w$$
$$y = -0.25w$$
$$\frac{y}{w} = -0.25 = -\frac{1}{4}$$

## SECTION 2

1. **C** Since there are 60 minutes in 1 hour, there are $3 \times 60$ or 180 minutes in 3 hours and $\frac{1}{3} \times 60$ or 20 minutes in $\frac{1}{3}$ of an hour. Since there are $180 + 20$ or 200 minutes in $3\frac{1}{3}$ hours, Column A = Column B.

2. **B** In column A,
$$40\% \text{ of } \frac{x}{2} = 0.40 \times \frac{x}{2} = 0.20x$$
In column B,
$$75\% \text{ of } \frac{x}{3} = 0.75 \times \frac{x}{3} = 0.25x$$
Since $x > 0$, you can cancel $x$ in both columns. Then, since $0.25 > 0.20$, Column B > Column A.

3. **A** In Column A, multiply each number inside the parentheses by the number outside the parentheses:
$$\sqrt{3}(1 + \sqrt{3}) = \sqrt{3} + (\sqrt{3} \times \sqrt{3}) = \sqrt{3} + 3$$
Now compare $\sqrt{3} + 3$ in Column A with $\sqrt{3} + 2$ in Column B. Since both columns contain $\sqrt{3}$, cancel it. Since $3 > 2$, Column A > Column B.

4. **A** In Column A, $7w - 2w + 3w = 5w + 3w = 8w$. Since it is given that $w = \frac{3}{8}$, then $8w = 8\left(\frac{3}{8}\right) = 3$. In column B, $\frac{1}{w}$ is the reciprocal of $w$. Since the reciprocal of $\frac{3}{8}$ is $\frac{8}{3}$, then $\frac{1}{w} = \frac{8}{3}$. Since $\frac{8}{3} = 2.66\ldots$, Column A > Column B.

5. **C** In Column A, since $x^3 = 7$, then
$$x^6 = (x^3)(x^3) = (7)(7) = 49$$
In Column B,
$$x^3 + 42 = 7 + 42 = 49$$
Hence, Column A = Column B.

6. **D** Solve the quadratic equation $y^2 - 2y - 3 = 0$ by factoring:
$$y^2 - 2y - 3 = 0$$
$$(y - 3)(y + 1) = 0$$
$$y - 3 = 0 \quad \text{or} \quad y + 1 = 0$$
$$y = 3 \qquad y = -1$$
If $y = 3$, Column A > Column B. If, however, $y = -1$, Column B > Column A. Since two different answers are possible, the correct choice is D.

7. **D** In Column A, the length of a side of a square whose perimeter is 32 is $\frac{32}{4}$ or 8. In Column B, since it is given that $x^2 = 64$, then $x = 8$ or $x = -8$. If $x = 8$, the two columns are equal. If, however, $x = -8$, Column A > Column B. Since two different answers are possible, the correct answer is choice D.

8. **B** *Solution 1*: Since $a < b$, then $1 - a > 1 - b$. Hence, since the reciprocal of $1 - b$ is greater than the reciprocal of $1 - a$, Column B > Column A.
*Solution 2*: Pick numbers for $a$ and $b$. Suppose $a = 0.5$ and $b = 0.75$. Then:
• In Column A, $1 - a = 1 - 0.5 = 0.5$, so
$$\frac{1}{1 - a} = \frac{1}{0.5} = 2$$

- In Column B, $1 - b = 1 - 0.75 = 0.25$, so

$$\frac{1}{1-b} = \frac{1}{0.25} = 4$$

Since $4 > 2$, Column B > Column A.

9. **A**  Since

$$(p + r)(p - r) = p^2 - r^2 = p^2 - 0.0009$$

then $r^2 = 0.0009$, so $r = \sqrt{0.0009} = 0.03$ in Column A. In Column B, $\frac{3}{1000} = 0.003$. Since $0.03 > 0.003$, Column A > Column B.

10. **A**  Since the area of rectangle *ABCD* is 1, then

$$\left(2\frac{3}{4}\right)x = 1 \quad \text{or} \quad \left(\frac{11}{4}\right)x = 1$$

so $x = \frac{4}{11} = 0.36...$ in Column A. Since $\frac{1}{3} = 0.33...$ in Column B, and $0.36 > 0.33$, Column A > Column B.

11. **C** List all the possibilities, with H representing a head and T a tail, when three coins are tossed:

(H, H, H)  (T, T, T)
(H, H, T)  (T, T, H)
(H, T, H)  (T, H, T)
(T, H, H)  (H, T, T)

Of the eight possible outcomes, three have exactly two heads and one tail. Hence, the probability of getting two heads and one tail is $\frac{3}{8}$. Similarly, three outcomes have two tails and one head. Since the probability of getting two tails and one head is also $\frac{3}{8}$, Column A = Column B.

12. **B** Since $x + y = 3$, then

$$\frac{x+y}{2} = \frac{3}{2}$$

so the average of *x* and *y* in Column A is $\frac{3}{2}$. Adding corresponding sides of the given equations, $x + y = 3$ and $z - y = 1$, gives $x + z = 4$, so

$$\frac{x+z}{2} = \frac{4}{2} = 2$$

Since the average of *x* and *z* in Column B is 2, Column B > Column A.

13. **C** You are told that, of the 150 hamburgers sold in the student cafeteria, 100 were served with ketchup, 75 with onions, and 50 with both ketchup and onions. Represent these facts in a Venn diagram, as shown below

- 100 − 50 or 50 hamburgers were sold with ketchup but not with onions.
- 75 − 50 or 25 hamburgers (Column A) were sold with onions but not with ketchup.

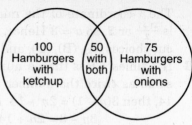

Total hamburgers sold = 150

- 50 hamburgers were sold with both onions and ketchup.
- 150 − (50 + 25 + 50) or 25 hamburgers (Column B) were sold without ketchup and without onions.

Hence, Column A = Column B.

14. **A** *Solution 1:* In Column A, $ab^2 = (ab)b$. In Column B, $ba^2 = (ba)a = (ab)a$. Since both *a* and *b* are negative, $ab > 0$, so *ab* can be canceled in both columns. Since $b > a$, Column A > Column B.

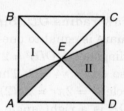 *Solution 2:* Pick numbers for *a* and *b*. Since $-1 < a < b < 0$, suppose $a = -0.8$ and $b = -0.5$.

- In Column A, $ab^2 = (-0.8)(-0.5)^2 = (-0.8)(0.25) = -0.2$.
- In Column B, $ba^2 = (-0.5)(-0.8)^2 = (-0.5)(0.64) = -0.32$

Since $-0.2 > -0.32$, Column A > Column B.

15. **C**

In Column A, regions I and II have the same area, so the sum of the areas of the two shaded regions is equivalent to the area of $\triangle AEB$, which is $\frac{1}{4}$ or 25% of the area of square *ABCD*.

In Column B, the area of right triangle *AED* is $\frac{1}{2}(3)(4)$ or 6. Since *AD* is the hypotenuse of a 3-4-5 right triangle, *AD* = 5. In right triangle *ACB*, leg *AC* = 5 + 3 or 8, and hypotenuse *AB* = 3 + 7 or 10. Since *BC* is a leg of a 6-8-10 right triangle, *BC* = 6. The area of right triangle *ABC* is $\frac{1}{2}(6)(8)$ or 24. Thus, the percent of the area of right triangle *ABC* that is shaded is $\frac{6}{24} \times 100\% = 25\%$.

Hence, Column A = Column B.

16. **144** Since the measures of the acute angles of a right triangle add up to 90,

$$3y + 2y = 90$$
$$5y = 90$$
$$y = \frac{90}{5} = 18$$

Hence, in right triangle $ABC$, $\angle C = 2y = 2(18) = 36$. An acute and an obtuse angle formed by parallel lines are supplementary. Since $BC \parallel DE$, $x + \angle C = 180$, so

$$x = 180 - 36 = 144$$

**17. 8** Since the height of rectangle $Y$ is 4, the length of each side of square $X$ is also 4. Then the base of rectangle $Y$ is $10 - 4$ or 6. Thus, when square $X$ is folded back along the broken line, $6 - 4$ or 2 units of the base of rectangle $Y$ will not be covered by square $X$. The area of the part of rectangle $Y$ that is not covered by square $X$ is $2 \times 4$ or 8.

**18. 82** If the average of 14 scores is 80, the sum of the 14 scores is $14 \times 80$ or 1120. If the average of four of these scores is 75, the sum of these four scores is 300, so the sum of the remaining 10 scores is $1120 - 300$ or 820. The average of these 10 scores is $\frac{820}{10}$ or 82.

**19. 72** In the six places on the line, the three men can be arranged in the first, third, and fifth positions in $3 \times 2 \times 1$ or 6 different ways. The three women can be arranged in the second, fourth, and sixth positions in $3 \times 2 \times 1$ or 6 different ways. Hence, the three men and the three women can be arranged so that they alternate, with a man in the first position, in $6 \times 6$ or 36 different ways. Similarly, the three men and three women can be arranged so that they alternate, with a woman in the first position, in $6 \times 6$ or 36 different ways. Hence, the total number of different ways three men and three women can be arranged in a line so that they alternate is $36 + 36$ or 72.

**20. 30** You are looking for the greatest positive integer less than 45 that, when divided by 5, gives a remainder of 0 and that, when divided by 7, gives a remainder of 2. List all multiples of 5 less than 45, and pick the largest one that, when divided by 7, gives a remainder of 2.
- The multiples of 5 less than 45 are 40, 35, 30, 25, 20, 15, 10, and 5.
- $40 \div 7$ gives a remainder of 5.

$35 \div 7$ gives a remainder of 0.

$30 \div 7$ gives a remainder of 2.

Hence, the greatest number of books that could be on the shelf is 30.

**21. 500** If $10^k = \frac{1}{2}$, then

$$10^{k+3} = 10^k \cdot 10^3 = \left(\frac{1}{2}\right) \cdot 1000 = 500$$

**22. 18** Since the ratio of boys to girls in a room is 2 to 3, let $2x$ and $3x$ equal the numbers of boys and girls, respectively, in the room. After three boys leave the room, the ratio of boys to girls becomes 1 to 2. Set up a proportion:

$$\frac{\text{boys}}{\text{girls}} = \frac{2x-3}{3x} = \frac{1}{2}$$
$$3x = 2(2x - 3)$$
$$= 4x - 6$$
$$6 = x$$

Since $3x = 3(6) = 18$, there are 18 girls in the room.

**23. 7**

$$\text{area } \triangle ABC = \frac{1}{2}(CD)(AB)$$
$$45 = \frac{1}{2}(6)(AB)$$
$$\frac{45}{3} = AB$$
$$15 = AB$$

Since $CDA$ is a 6-8-10 right triangle, leg $AD = 8$, so $DB = 15 - 8$ or 7.

**24. 9** List all positive integer values for $a$ and $b$ that make $2a + 3b = 24$. Then pick the pair of values for which $a + b$ has the smallest value.

| $a$ | $b = \dfrac{24 - 2a}{3}$ | $a + b$ |
|---|---|---|
| 1 | $\dfrac{22}{3}$ | |
| 2 | $\dfrac{20}{3}$ | |
| 3 | $\dfrac{18}{3}$ or 6 | $3 + 6 = 9$ |
| 4 | $\dfrac{16}{3}$ | |
| 5 | $\dfrac{14}{3}$ | |
| 6 | $\dfrac{12}{3}$ or 4 | $6 + 4 = 10$ |
| 7 | $\dfrac{10}{3}$ | |
| 8 | $\dfrac{8}{3}$ | |

There is no need to go further since the next value of $a$ is 9, which is the smallest value of $a + b$ obtained thus far.

25. **70** You are given that 75% of the 200 families surveyed have at least one car, so $0.75 \times 200$ or 150 families have at least one car. Since 20% of those with cars have more than two cars, $0.20 \times 150$ or 30 families have more than two cars. If 50 families each have exactly two cars, then $150 - (50 + 30)$ or 70 families have exactly one car.

## SECTION 3

1. **D** Let $x$ represent the length of $AB$. Since $AB$ is $\frac{1}{4}$ of the perimeter of $\triangle ABC$,
$$x = \frac{1}{4}(15 + 21 + x) = \frac{1}{4}(36 + x)$$
$$4x = 36 + x$$
$$3x = 36$$
$$x = \frac{36}{3} = 12$$

2. **A** If $x^2 + 3y = x^2 - 3$, then subtracting $x^2$ from each side of the equation gives
$$3y = -3 \quad \text{or} \quad y = \frac{-3}{3} = -1$$

3. **D** *Solution 1*: If $a = b \div c$, then $a = \frac{b}{c}$. When $b$ is multiplied by 2 and $c$ is divided by 2,
$$a_{\text{new}} = 2b \div \frac{c}{2} = 2b \times \frac{2}{c} = 4\frac{b}{c} = 4a$$
Hence, $a$ is multiplied by 4.
*Solution 2*: Plug in easy-to-work-with numbers for $b$ and $c$. When $b = 8$ and $c = 2$, then $a = 8 \div 2 = 4$. Multiplying $b$ by 2 gives 16, and dividing $c$ by 2 gives 1 so, $a_{\text{new}} = 16 \div 1 = 16$. Since 16 is 4 times as great as 4, $a$ is multiplied by 4.

4. **E** If $0 < a < b < c$, then the median or middle value of $a$, $b$, and $c$ is $b$. Since the average of $a$, $b$, and $c$ is 3 times the median,
$$\frac{a + b + c}{3} = 3b$$
$$a + b + c = 9b$$
$$a + c = 8b$$
Since $\frac{a+c}{2} = \frac{8b}{2} = 4b$, the average of $a$ and $c$ in terms of $b$ is $4b$.

5. **B** Determine whether each Roman numeral expression always represents an even integer when $x - 1$ represents an odd integer.
  • I. Since $x - 1$ is an odd integer, $x - 1 + 2$ or $x + 1$ is the next consecutive odd integer. Hence, expression I does not represent an even integer.
  • II. Since $x - 1$ represents an odd integer, $x - 1 + 1$ or $x$ represents the next even integer. Since any multiple of an even integer is also even, $3x$ is an even integer. Hence, expression II represents an even integer.
  • III. The square of an odd integer is always odd. For example, $3^2 = 9$. Since $(x - 1)^2$ represents the square of an odd integer, expression III does not represent an even integer.
Hence, only Roman numeral expression II is an even integer.

6. **B** To find the equation that best describes the relationship for the ordered pairs of numbers $(x, y)$ in the given table, substitute the different values for $x$ in each of the equations in the answer choices until you find the equation that yields values of $y$ that match the values for $y$ in the table. The equation in choice (B), $y = x^2 - 1$, is the only equation that works for all three pairs of numbers:
  • If $x = -2$, then $y = (-2)^2 - 1 = 4 - 1 = 3$. Thus, $(x, y) = (-2, 3)$, which matches the first ordered pair in the table.
  • If $x = 0$, then $y = 0^2 - 1 = -1$. Thus, $(x, y) = (0, -1)$, which matches the second ordered pair in the table.
  • If $x = 3$, then $y = 3^2 - 1 = 9 - 1 = 8$. Thus, $(x, y) = (3, 8)$, which matches the third ordered pair in the table.

7. **D** An exterior angle of a triangle is equal to the sum of the two nonadjacent interior angles of the triangle, so $\angle B + \angle C = 130$. Since $AB = AC$, the angles opposite these sides, angles $B$ and $C$, must be equal. Hence, $\angle B = \angle C = \frac{1}{2}(130) = 65$. You are also told that $AC \parallel BD$. Since acute angles formed by parallel lines are equal, $x = \angle C = 65$.

8. **D** In the given addition problem, consider the rightmost column, that is, the first column. In the first column, $B$ must be 6 since $5 + 6 = 11$. One is then carried over to the second column, so $1 + 6 + A = 1C$. When 1 is carried over to the third column, $1 + A + 3 = 11$, so $A$ is 7.

9. **E** Since the diameter of the circle is 18, its circumference is $18\pi$. The sum of the

lengths of the broken arcs of circle O is $2\pi$, so the sum of the lengths of the unbroken arcs that bound the shaded region is $18\pi - 2\pi$ or $16\pi$. This represents $\frac{16\pi}{18\pi}$ or $\frac{8}{9}$ of the circle. Since the radius of the circle is $\frac{18}{2}$ or 9, the area of the circle is $9^2 \times \pi$ or $81\pi$. The area of the shaded region is $\frac{8}{9}$ of the area of the circle. Since $\frac{8}{9} \times 81\pi = 8 \times 9\pi = 72\pi$, the area of the shaded region is $72\pi$.

10. **D** Suppose George started with $x$ dollars. After he spent 25% of $x$ dollars on lunch, he had $0.75x$ dollars left. Since he spent 60% of the amount of money that was left on dinner, he spent $0.60(0.75x)$ or $0.45x$ dollars on dinner. If he then had $9.00 left,

$$x - 0.25x - 0.45x = \$9$$
$$x - 0.70x = \$9$$
$$0.30x = \$9$$
$$x = \frac{\$9}{0.30} = \$30$$

Hence, George spent 25% of $30 or $7.50 on lunch.